D1807515

WITHDRAWN

Enlargement and Compaction of Particulate Solids

Butterworths Monographs in Chemical Engineering

Butterworths Monographs in Chemical Engineering is a series of occasional texts by internationally acknowledged specialists, providing authoritative treatment of topics of current significance in chemical engineering.

Series Editor

J W Mullin
Professor of Chemical Engineering, University College, London

Published titles:

Solid–Liquid Separation
Liquids and Liquid Mixtures, Third Edition

Forthcoming titles:

Mixing in the Process Industries
Fundamentals of Fluidized-Bed Chemical Processes
Diffusion in Liquids
Introduction to Electrode Materials

Butterworths Monographs in Chemical Engineering

RECKITT & COLMAN
RESEARCH & DEVELOPMENT
LIBRARY & INFORMATION
SERVICE

14 SEP 1983

STONEFERRY

Enlargement and Compaction of Particulate Solids

Editor
Nayland G. Stanley-Wood, PhD., BPharm., C.Eng., M.I. Chem. Eng., M.P.S.
Lecturer, Schools of Studies in Chemical Engineering
and Powder Technology
University of Bradford

RECKITT & COLMAN
RESEARCH & DEVELOPMENT
LIBRARY & INFORMATION
SERVICE

14 SEP 1983

STONEFERRY

Butterworths
London Boston Durban Singapore Sydney Toronto Wellington

All rights reserved. No part of this publication may be
reproduced or transmitted in any form or by any means,
including photocopying and recording without the written
permission of the copyright holder, application for which should
be addressed to the Publishers. Such written permission must also
be obtained before any part of this publication is stored in a
retrieval system of any nature

This book is sold subject to the Standard Conditions of Sale of
Net Books and may not be resold in the UK below the net price
given by the Publishers in their current price list.

First published 1983

© Butterworth & Co (Publishers) Ltd 1983

British Library Cataloguing in Publication Data

Compaction of particulate solids.—(Butterworths
  monographs in chemical engineering)
  1. Compacting   2. Bulk solids
  I. Stanley-Wood, Nayland G.   II. Series
  660.2′8429   TP156.C/

ISBN 0–408–10708–1

Filmset in Monophoto Times by Northumberland Press Ltd, Gateshead
Printed in Great Britain at the University Press, Cambridge

Preface

Sir Isaac Newton stated in his opus on 'Opticks', London 1721, that:

'The parts of all homogeneal bodies which fully touch one another stick together very strongly ... There are therefore agents in nature able to make particles of bodies stick together by very strong attractions and it is the business of experimental philosophy to find them out ...'

Since British Patent 9977 for the 'shapping of pills, lozenges and black lead by pressure dies', which is accredited to Brockenden (1843), the experimental philosophy, initiated by Newton to understand why particles can form coherent bodies, has been assiduously investigated. The pressing of particulate materials to produce a wide variety of blocks, pellets, tablets and compacts is nowadays extensively employed in many industrial processes.

The demise of skilled craftsmen, who produced tablets and compacts by empirical control methods, resulted in a need for the development of more sophisticated technological measurement and control. The awareness and the desire to be in the vanguard of such technological changes created a demand for post-experience courses specializing in the methodology of compaction and size enlargement. In 1979, 1980 and 1982 post-experience courses on the Compaction of Particulate Solids were held at the University of Bradford sponsored by the Institution of Chemical Engineers in their programme of continuing education. The policy of the Continuing Educational Sub-Committee is to encourage development of such courses to meet the continuing educational needs of the chemical engineering profession and associated professions and industries. The courses which were attended by post-experience delegates from all over Europe, were divided into the following topics: Fundamentals of Powder and Compact Characterization; Compaction and Agglomeration Techniques; and Industrial Practice and Application of Compaction and Densification. Because of the success of these courses it was decided to adopt the same organization of subject matter for this book.

Each chapter has been contributed by an industrial or an academic worker actively engaged in research and development or the industrial production of compressed or size enlarged particulate materials. The subject matter included in the book, characterization of powders and granules before and after compaction, mixing, shear testing, fluidized-bed granulation, mechanisms of size enlargement and compaction, and the instrumentation of industrial presses and industrial processes cannot, however, claim to be completely comprehensive in such a diversified and wide-ranging field as compaction. This interdisciplinary approach to the densification of materials, however, draws upon chemical engineers, physicists, powder and pharmaceutical technologists, ceramacists and metallurgists in an attempt to achieve a cross fertilization of knowledge and 'know how' between various industries.

I should like to express my thanks and gratitude to all the lecturers who participated in the post-experience courses and contributed to this book and for the time and effort taken to write their chapter. Without such endeavour this book would not have been accomplished.

Nayland Stanley-Wood

Contributors

J. J. BENBOW
Formerly, Imperial Chemical Industries plc, Agricultural Division, Billingham, Cleveland

I. K. BLOOR
The British Ceramic Research Association, Queens Road, Penkhull, Stoke-on-Trent

R. D. BRETT
The British Ceramic Research Association, Queens Road, Penkhull, Stoke-on-Trent

N. HARNBY
Schools of Studies in Chemical Engineering and Powder Technology, University of Bradford, Bradford

B. HUNTER
Imperial Chemical Industries plc, Pharmaceutical Division, Macclesfield, Cheshire

D. E. LLOYD
The British Ceramic Research Association, Queens Road, Penkhull, Stoke-on-Trent

P. J. LLOYD
Department of Chemical Engineering, Loughborough University of Technology, Loughborough

H. M. MACLEOD
United Kingdom Atomic Energy Authority, Windscale Nuclear Power Development Laboratories, Seascale, Cumbria

A. W. NIENOW
Department of Chemical Engineering, University of Birmingham, Birmingham

H. S. THACKER
Manesty Machines Limited, Speke, Liverpool

J. C. WILLIAMS
School of Studies in Powder Technology, University of Bradford, Bradford

Contents

1 **Particle characterization by size, shape and surface for individual particles—**
N. G. STANLEY-WOOD   1
  1.1   Scope   1
  1.2   Characterization of individual particles   2
  1.3   Averages   11
  1.4   Shape   18
  1.5   Application of shape factors for surface area evaluation   35
  References   39

2 **Particle characterization by size, shape and surface for contacted particles—**
N.G. STANLEY-WOOD   43
  2.1   Porosity, voidage and particle porosity   43
  2.2   Nitrogen adsorption   52
  2.3   Mercury penetration   95
  2.4   Application of nitrogen isotherms Types II and IV and mercury
        intrusion to compacted solids   99
  References   116

3 **Mixing of powders—**N. HARNBY   120
  3.1   Powder mixing   120
  3.2   The mixing process   121
  3.3   Quantitative assessment of mixture quality   126
  References   127

4 **Mechanisms of size enlargement—**P. J. LLOYD   128
  4.1   Introduction   128
  4.2   Basic mechanisms   128
  4.3   The granulation process   131
  Bibliography   135
  References   135

5 **Flow and handling of solids; the design of solid handling plants—**J. C.
WILLIAMS   136
  5.1   Introduction   136
  5.2   Types of storage hopper   136
  5.3   Measurement of the failure properties of a particulate solid   138

5.4 Design of mass flow hoppers   142
5.5 Design of a plant for mass flow   145
References   147

6 Pharmaceutical granulation and compaction—B. HUNTER   148
6.1 Introduction   148
6.2 Theoretical considerations   149
6.3 Powder preconditioning   153
6.4 Compression scale-up   155
6.5 Formulation and process optimization   158
References   159

7 Mechanisms of compaction—J. J. BENBOW   161
7.1 Introduction and scope   161
7.2 Application of pressure and frictional effects   165
7.3 Particle rearrangement   167
7.4 Deformation without rearrangement   168
7.5 Strength-producing mechanisms   171
7.6 Load removal and stress relaxation   172
7.7 Material properties   173
7.8 Powder compaction equations   174
7.9 Tabletting defects   174
7.10 Conclusions   176
References   177

8 Fluidized bed granulation—A. W. NIENOW   179
8.1 Basic fluidized bed concepts   179
8.2 Definitions and applications   180
8.3 Variations on the basic process: practical difficulties   181
8.4 Quenching   181
8.5 Mass and moisture balance   182
8.6 Heat balance   182
8.7 Particle growth mechanisms: dynamic equilibrium   183
8.8 Growth models and rates   186
8.9 Batch *versus* continuous operation   186
8.10 Pilot plant testing   187
8.11 The use of inert 'nuclei'   187
References   188

9 Compact characterization—N. G. STANLEY-WOOD   189
9.1 Strength of materials: fundamentals   189
9.2 Soil mechanics stress–strain curves for granular materials   199
9.3 Volume reduction in unidimensional consolidation   205
9.4 Compaction of powders   213
References   225

10 Instrumentation of tablet machines—H. S. THACKER   227
10.1 Introduction   227
10.2 Instrumentation of single acting machines   227
10.3 Instrumentation of rotary tablet machines   230

10.4   Force measuring systems   231
10.5   Uses of instrumentation   235
References   240

11 Compaction of ceramics—H. M. MACLEOD   241
11.1   Introduction and scope   241
11.2   Pressure transmission through powders   241
11.3   Pressure–volume relationships   253
11.4   Friction and lubrication   261
11.5   Process variables   265
References   274

12 Isostatic pressing and compacting techniques—D. E. LLOYD,
I. K. BLOOR and R. D. BRETT   277
12.1   Introduction   277
12.2   Component shapes   279
12.3   Tooling   279
12.4   Isostatic pressing of a sphere   280
12.5   Tooling for rods and discs   281
12.6   Tooling for complex shapes   282
12.7   General aspects of tool design   283
References   287

CHAPTER 1
Particle characterization by size, shape and surface for individual particles

N. G. Stanley–Wood
Schools of Studies in Chemical Engineering and Powder Technology, University of Bradford, Bradford

1.1 Scope

In the compaction of powders in a rigid mould to produce a coherent mass the fabrication process can be considered to occur in three stages.

1. The powder or assembly of particles must flow or be fed to the mould.
2. On application of a compressive pressure the particles are forced into intimate contact and areas of contact are formed between particles to produce bonds.
3. Ejection of the compacted material and the stress release in the compact to produce a coherent compact is dependent upon the resultant area of contact, fracture or deformation of the particles in addition to the shape and dimensions of the discontinuities between particles.

In all three stages outlined above information on the size, shape, surface and porosity of the material before and after compaction is required to aid characterization of compaction as well as to quantify the degree and properties of the compaction process and compacted material.

This chapter deals with the measurement of size and shape of individual particles or collections of individual particles, both spherical and non-spherical. Because the individual size and shape of particles, after compression, is altered and difficulties arise in the distinction of bonded particles in terms of size and shape, the alternative parameter of particle surface area may be used to characterize the properties of compacts or particles in contact with each other. This surface characterization can be achieved by measurement of the amount of gas adsorbed on to the unbonded solid surface by low temperature adsorption or mercury penetration techniques. Since some materials are porous the adsorption and penetration methods described can be adapted and adopted to measure the internal porosity of compressed and non-compressed solids. Applications of these surface topographical characterization methods to compacted material are discussed and appraised in Chapter 2.

1.2   Characterization of individual particles

1.2.1   Particle size

The term powder particle usually refers to an object which has a physical boundary regardless of the finite size. The upper limit of an individual particle is therefore difficult to define because a particle is a discrete portion of matter which is small in relation to the space in which it is considered. Thus planets, boulders and electrons can be encompassed within this specification. For practical purposes however an arbitrary definition of the size of powder particles has been chosen and, as stated in BS 2955, can be defined as:

'A powder shall consist of discrete particles of any material with a maximum dimension of less than 1000 micrometre (= 1 mm)'

The lower limit is commonly taken, these days, to be in the sub-micrometre to colloidal dimension (0.001 micrometre) range. The range normally encountered in compaction technology is however 1000—10 micrometres. However, with the advent of micronized comminution techniques and vapour phase condensation, sizes in the nanometre range are being encountered more and more.

Because no single method of particle measurement can be expected to be applicable over the entire size spectrum many diverse techniques and instruments are available for particle sizing and counting. The choice of method for particle characterization is therefore dependent upon the use for which collected information on size or shape is intended. Since in particle characterization the physical dimensions of regular and irregular particles are required, the methods and instruments for sizing and counting have been classified on a physical criterion.

Table 1.1

Physical criterion	Diameter measured		Size range applicability/μm
1. Image	Projected area	d_a	Optical 0.8—800
			Electron 0.002—15
2. Mechanical	Sieve diameter	d_t	Dry 40—1000
			Wet 1—40
3. Dynamic	Free-fall diameter	d_f	Gravity 1—100
	Drag diameter	d_d	Centrifugal 0.05—25
	Stokes diameter	d_{st}	
4. Attenuation	Projected area	d_a	X-Ray 0.05—100
	Stokes diameter	d_{st}	Centrifugal 0.05—50
5. Scattering	Surface volume diameter	d_{sv}	0.3—50
6. Electrical	Volume diameter	d_v	0.5—300
7. Surface	Surface diameter	d_s	Permeametry 0.1—50 (mean)
	Surface volume diameter	d_{sv}	Adsorption 0.005—50 (mean)

Table 1.2. Equivalent measurements

Key	Description	Symbol
i	A circle having the same projected area as the profile of a particle	d_a
ii	The width of a minimum square aperture through which a particle will pass	d_t
iii	A sphere having the same volume as the particle	d_v
iv	A sphere having the same surface area as the particle	d_s
v	A sphere having the same velocity of fall in a fluid if in the Stokes region	d_{st}
vi	A sphere having the same ratio of surface area to volume as the particle	d_{sv}

Table 1.1 shows seven arbitrary classifications and the diameters measured by these physical methods. In addition *Table 1.1* shows the size range of applicability for each method to aid the scientist or technologist in the selection of the appropriate and desired measurement technique.

An irregularly shaped particle has no unique dimension and its size may be expressed in terms of the diameter of a circle or of a sphere which is equivalent to the particle in some stated property.

The more common equivalents are shown in *Table 1.2*. All these equivalent diameters differ numerically except for a spherical particle. The particle size measured, except for spherical particles, depends to some extent upon the physical principles employed in the measurement process.

1.2.2 Image-forming instruments and techniques

Measurements by microscope are not solely limited to the sizing of solid particles but can also be used to size oil or liquid globules in emulsions or metal inclusions in alloys. Visual measurement of particles is recommended in many instances because it is the only absolute method of sizing. Since individual particles are measured it is at times used as the basic and standardization method for many other indirect techniques. In addition to the parameter of size, particles seen with either an optical or an electron microscope can be characterized by shape or surface topography. Direct measurement in the microscope range can be achieved either with a low power microscope for particles between 1000 and 100 micrometre or with a high power, good resolution microscope, for particles between 100 and 0.2 micrometre. The British Standard 3406[1] does however recommend that the minimum size of particles measured with an optical microscope should be limited to 0.8 μm. If there are only a few and regularly shaped particles to be measured and counted then a glass eyepiece disc engraved with a fine line divided into equal parts[2] can be used. This must be calibrated against a stage or slide micrometer[3] to give the real size of the particles under investigation. Instead of measuring the uniaxial linear dimensions of particles in terms of the number of equally spaced divisions, irregularly shaped particles can be compared with circles or discs, sometimes called globes, on an eyepiece graticule. This method classifies particles into size groups rather than measuring each particle. In this way many more particles can be sized and the labour required for measuring particles and determining particle size distributions reduced. The size of the circles or globes on the graticules are calibrated in the same way as the linear eyepiece scale by using a stage micrometer. Numerous types of ocular graticule have been produced for specialized sizing[4–6], but the graticules in general use are either the Fairs' graticule[7,8] in which each circle, and therefore group, increases in size by a root two progression or the Integrating discs I, II, III or IV graticules[9] which have 25, 100, 400 or 9000 equidistant graduations per inscribed square respectively.

Manual or semiautomatic sizing and counting aids are available. The size of particles can be measured directly from eyepiece images or by being projected on to a screen as with a projection microscope[3,10,11]. The images from a photomicrograph and a 35 mm slide can be sized and counted by using either a Zeiss–Endter Particle Size Analyser[12], or a Chatfield Particle Size Analyser[13].

With all the microscope instruments so far discussed the particle image must be seen with the human eye and the size of the particle either measured or placed into an appropriate size group which is subsequently counted. To remove the fatigue of peering down a microscope or at a projected image, instruments were produced in which the presence and size of particles were detected with a scanning light spot[14,15] or slit[16]. These instruments have now however been superseded by image analysers which use the electronic scanning beam of a television camera. One of the first particle size analysers using the scanning spot of a television screen was the Metal Research Particle Analyser[17].

It is now possible with hardware computer modules, as seen with the Quantimet 720[18], to obtain not only a statistical diameter, a count and a distribution, but also a distribution by area, projected length and chord size. Many other mathematical particle functions such as particle perimeters and shapes can now also be rapidly evaluated.

An alternative method, still using the scanning beam of a television camera, is to process and data reduce the videosignal by software computer analysis. One of the first generation software analysis image analysers, the π MC, was developed by Millipore[19] in conjunction with Bausch and Lomb. Bausch and Lomb have recently introduced the second generation software image analysers termed the Omnicon[20] which has a more powerful software programme and versatility.

In size analysis of sub-micrometre particles, electron microscopy was, until light scattering instruments became available, the only method of characterization. With an accelerating voltage of 60 kV the wavelength of radiation corresponds to 0.005 nm and gives a resolution of about 2.5 nm. As the order of accelerating voltage increases to 10^5 V then the resolution power of a transmission electron microscope (TEM) can increase to a resolution value of 0.1—0.4 nm.

Particle size distributions are not usually conducted on the positive images of the particles which appear on the electron microscope's fluorescent screen but either by analysis of sheet film exposed behind the screen to give a photoelectromicrograph or by interfacing the output from an electron microscope directly to an image analyser. This interfacing with image analysers can also be achieved with scanning electron microscopes (SEM) but a directed image storage module (DIS) is usually required to store the slow image data from SEM to deliver it at a faster rate to image analysers. In all electron microscopy work, except SEM, there is, however, considerable expertise required in specimen preparation.

1.2.3 Mechanical diameters and techniques

It is usually thought that the cheapest and probably the easiest method of particle characterization is to find the size distribution of a powder by sieve analysis. In the case of the passage of a non-spherical particle through this type of go–no go gauge there is however only a certain probability that a particle with a certain size in one orientation will arrive and pass through that aperture in that orientation. Thus although sieving has been used successfully in many circumstances, and many technological refinements have been adopted, the method of sieving and its counterpart screening is at the moment neither a precise nor an accurate technique for particle size measurement.

The factors of particle shape and surface character together with the sieve factors of variation in aperture size due to the weaving, loading rate on the sieve and whether the technique of wet or dry sieving is used, all have an effect which can over- or under-size particles. These factors introduce errors which prevent a sharp and distinct split in the size distribution of particles.

To reduce these errors Heywood[21] showed statistically that by using a standard method with different sieves the reproducibility of the sieving operation can be doubled. This work led to a British Standard[22] on the use of fine-mesh sieves and the definition, in terms of the amount of material passing a sieve in a 2 minute period, of a hand and machine sieving 'end-point'. There are however various definitions for the arbitrary end-point of sieving. The BS 12 suggests a standard time of 15 minutes for machine sieving while no time limit is specified for this method in BS 410. Since the reproducibility by using machines, instead of hand sieving, is increased by 30 per cent[21], a large number of commercial shakers which try to follow the handsieving technique mechanically are available. These mechanical shakers such as Rotap[23], Pascall Inclyno[24] and Endecott[25] usually take a stack or 'nest' of either 305 mm (12 inch) or 203—200 mm (8 inch) diameter sieves. Alternatively a nest of sieves can be vibrated electromagnetically at various degrees of amplitude as seen with the equipment from Retsch[26], Fritsch[27] and Rhewun[28].

One of the factors which affects the particle sizing of sieved materials is the variation in the aperture size that can occur in the manufacture of woven wire sieves. Numerous countries have put forward standard specifications regarding the distribution of aperture tolerances which are considered acceptable. Sieve apertures are usually obtained from the weaving of brass, rustfree steel or phosphor-bronze wire. In an effort to unify the specifications which appear in DIN 4189 (1964), DIN 4177 and DIN 4195 (Germany); NONORM 480 (Holland); GOST 3584-53 (USSR); BS 410 (Britain); AFNOR 11-501 (France); ASTM E 11-61 (USA); and CSN 15-3105 (1962) (Czechoslovakia) the International Organization for Standardization (ISO) has given consideration to the fourth-root-of-two-series and the R 20/3 derived series of Preferred Numbers[29] to produce a series of preferred aperture sizes which are available in ISO 565 (1972) and ISO 2591 (1973).

In addition to woven wire sieves there are now available precision formed plates with either circular apertures from 3.67 mm to 40 μm diameter or slit shaped apertures with dimensions of 0.93 × 4.28 mm to 0.43 × 0.03 mm. The transverse profile of these sieve apertures, instead of a straight cylindrical shape, can also be conical (VECO 30) or a double cone shape (VERO 30). Sieve sizes down to 1 μm are now available from Buckbee Mears[30], Endecotts[25], Veco[31] and Pascall[24]. One of the problems associated with sieving is the blockage of the aperture by the particles which will not pass through the go–no go gauge. This can be prevented, in some circumstances, by the addition of sieving aids but the best solution is to prevent the particles remaining in sieve apertures. This has been accomplished by passing a rotating stream of air through the powder from the underside of a sieve (Alpine Air Jet Siever[32]) or by using an oscillating column of air together with repetitive electromagnetic vibrations (Bradley, Sonic Sifter, Model LP 3 or P 60[33]).

1.2.4   Dynamic diameter by gravity sedimentation

The study of the dynamics of particles, whether they are regular or irregular, single or multiparticulate, in laminar or turbulent fluid streams, whether the motion is rectilinear as in gravity sedimentation or curvilinear as in centrifugal sedimentation, has produced an abundance of empirical, semi-empirical and theoretical equations.

The Andreasen pipette incremental method[34] which uses Stokes' equation to determine a particle diameter at known times, is one of the simplest and yet highly reproducible methods for particle size distribution and characterization, when a standard procedure is used. For complete validity of Stokes' law, however, a sphere must be falling in an infinite fluid at very low Reynold numbers. The distribution of particle sizes by the Andreasen method is obtained by extracting a quantity of the initial homogeneous settling suspension at known heights and time intervals. The weight of particles deposited at these known time intervals is determined by evaporation of the suspending fluid and subtraction of the weight of any dispersing or wetting agent initially added. A disadvantage of the Andreasen and other incremental methods is that the withdrawal of material from a settling suspension upsets the dynamic equilibrium of the system. Thus the cumulative method of analysis, where the weight of material settled out is determined without withdrawal from the system, was thought to remove this disadvantage. Both incremental and cumulative methods and the procedure of particle characterization are given in British Standard 3406 Part 2[1]. Information on the gravity sedimentation and the analysis of fine dusts is given in DIN 6611 (1971)[35] and VD1-Richtlinie Nr 2031 (1962)[36], respectively. Comparison of the methods of particle size analysis by sedimentation was undertaken by Jarrett and Heywood[37] and a critical review of all sedimentation methods has been published by the Society for Analytical Chemistry[38]. The general recommendations are that of low volume concentrations of the settling suspensions, to ensure validity of Stokes' law, and strict adherence to standard methods, to ensure reproducibility. Comparison of particle sizes determined by different sedimentation methods and at different concentrations are not justifiable because of the variable effects of interparticulate forces.

The continuous determination of the cumulative amount of powder settled out from an initial homogeneous settling suspension can be achieved with sedimentation balances. These balances have either a weighing pan immersed entirely in the settling homogeneous suspension (Shimadzu[39], Sartorius[40]) or at the bottom of a column of settling suspension (Bachmann, Sartorius, Bostock[41]). A disadvantage of the beam balances is that, as the pan collects and weighs the settled particles in incremental weight steps of 2—10 mg, a pumping action occurs which disturbs the dynamic equilibrium of the system. The introduction of continuous reading electrobalances (Cahn[42], Mettler[43]) with precise mechanical column construction and pressure-compensated sedimentation columns (Leschonski[44,45]) has led, by use of two-layer cumulative techniques, to high correlation and reproducibility with other sedimentation methods. The continuous trace from an electrobalance can, with analogue–digital transducers, be interfaced with a data logger or a small computer so that the tangents from the cumulative weight–time graph, with use of Oden[46,47] mathematical formulation, can be analysed statistically and a particle size distribution rapidly evaluated[48].

1.2.5 Dynamic diameter by centrifugal sedimentation

The pipette method of particle size analysis has been regarded as one of the most accurate and dependable sedimentation techniques for particle characterization in the range 100 to 2 μm. To determine smaller sizes, between 2 and 0.1 μm, it is therefore a logical step to substitute centrifugal force for gravitational force. Sedimentation in a centrifugal field is, however, subject to the same limits of concentration of the solid phase, particle shape and electroviscosity effects as is gravity sedimentation. There is, in addition to the above limitations, the effect of the varying centrifugal force with distance on a settling particle. As a particle settles or moves from a centrifugal axis towards a centrifuge's perimeter the force on a particle increases as the radius of rotation increases. Particles are not, therefore, settling under constant velocity conditions but are being accelerated as they settle. There is also a radial separation effect which occurs because of the fanlike motion of the centrifuging particles. Because of these two effects the size distribution is related to the concentration of the suspension by an exact linear integral equation which has no practical utility, although a practical approximation solution, the Kamack[49] equation, can be derived. The Kamack equation and a modification of the Kamack equation have been used in the design of two instruments, the Simcar Centrifuge and the Modified Pipette Analyser[50]. The former is no longer commercially available from Simon Carves Limited[51].

As the range of particle sizes decreases from 0.1 to 0.01 μm diameter, so the force essential to achieve adequate sedimentation rates must be supplied by ultracentrifugation or supercentrifuges. These are usually gas or air turbines (Sharples[52], Phywe[53]) or direct electrically driven centrifuges[54,55] producing, depending upon the distance of rotation, forces of between $200g$ and $4 \times 10^5 g$ at 60 000 revolutions per minute. At these speeds and sizes, however, the effect of molecular diffusion may influence sedimentation rates.

The Sharples centrifuge and the Hausner supercentrifuge collect fractions of different sized particles on removable liners or steps within the centrifuge bowl or cascade type rotor. These fractions can then be removed and weighed to determine the particle size distribution of sub-micrometre particles[52,56,57]. The weight of the sub-micrometre particles collected is, however, extremely small and it is therefore advantageous to use an alternative means of determining the variation in concentration other than by weight. The alternative means can be, as in the case of the Joyce–Loebl disc centrifuge[58–60] which is used in the characterization of very fine organic pigments, a colorimetric method. With the MSA Particle Size Analyser[61–63] the volumetric or sedimentation height in a standard centrifuge tube is used as a means of determining the amount of the various sized particles in a powder.

1.2.6   Attenuation diameter

The basic law for determining the attenuation of light as it passes through a suspension of particles is the Lambert–Beer law. For the Lambert–Beer law to be valid there must be perfect monochromatic radiation illuminating the absorbing material. Perfect monochromatic light is, however, impossible in practice owing to finite bandwidths centred about the required radiation wavelength and the different wavelengths arising from scattered radiations and unwanted reflections. The application of laser technology in the determination of an attenuation diameter is, however, finding increasing scope in particle characterization. Conditions for Beer's part of the Lambert–Beer law to be valid are less stringent with a broad peak of wavelength rather than with a narrow peaked wavelength. The latter is more satisfactory therefore for analytical use than the former.

Allen[64] has shown that there is no single calibration extinction curve for photosedimentation except that all curves degenerate to a common curve for particles of greater than 2 μm (absorbing) and 6 μm (transparent or non-absorbing). The rate at which the extinction coefficient decreases from two to unity is dependent upon the refractive index of the medium between the photocell detector and the suspension, the type of light source and the angle of acceptance of the photocell. With a wide angle of acceptance, white light and spherical bronze particles in the size range 3—57 μm, Allen[64] found that the extinction coefficient over this size range with non-absorbing particles was unity within experimental error. This result was subsequently used to develop a wide-angle photosedimentometer (WAP) for particle sizing and surface area distributions. To speed up the characterization of particles a Wide Angle Scanning Photosedimentometer (WASP) in which the sedimenting suspension within the sedimentation cell could be scanned during the settling period, was eventually marketed as a commercial instrument[65]. An alternative method for evaluating the sedimentation rate of particles instead of the attenuation of monochromatic light or white light is to use X-rays. The Micromeretics Sedigraph 5000 Particle Size Analyser[66,67] has been programmed to determine the size range of minerals, metal powders, pigments and abrasives from 100 to 0.2 μm quickly by scanning the sedimenting system. The instrument can therefore be regarded as a rapid response batch particle analyser. The Ladal X-ray Sedimeter[65], which uses also the principle of attenuating radiation to evaluate particle size, requires the particles to have an atomic number greater than that of carbon to be able to detect the reduction in X-ray intensity.

The X-ray beam and detector can also scan a horizontal rotor[65] spinning at speeds of 750, 1500 or 3000 rpm to obtain a size distribution of X-ray absorbing solids in the size range 1.0—0.15 μm.

1.2.7 Scattering diameter

Particle sizing by the scattering of radiation, usually light, from a single particle is one of the few methods which permit the continuous determination of size and concentration in either an air-borne or liquid-borne state. A particle passing through a light beam emits a flash of scattered light which is proportional to the volume diameter of the particle. The Mie intensity of scattered light is also a function of the scattering angle, the polarization angle, the wavelength of light and the refractive index of the system. For a given optical scattering device the scattering and polarization angle together with the wavelength of light are normally fixed. The intensity of scattered radiation is then solely dependent upon the particle diameter and the refractive index of the medium. For particle diameters up to 0.8 μm the scattering patterns, using light of wavelength 0.63 μm and a refractive index of 1.5, are without maximum and minimum values inside a mean collecting angle of 40 degrees[68] to the incident radiation. For larger sizes of airborne particles up to a diameter of 4 μm, maximum and minimum light intensities occur at values of less than, as well as greater than, 40 degrees to the incident radiation. As the value of airborne particle size decreases below 0.8 μm the absence of maximum and minimum intensities occur at greater and greater values of the mean scattering angle. The variation of light intensity with scattering angle can therefore be used to evaluate the size of narrowly classified particles in a liquid suspension cell or the size of a single particle in air with the aid of a single particle light scattering spectra atlas. The light scattering properties of particles, in liquid and gas, allow particles to be measured either by an optical or a laser system. The commercial optical instruments using this method for particle detection have a detecting sensor placed at a fixed scattering angle to the incident light beam.

1.2.8   Electrical resistance diameter

The principle of the change of electrical resistance has been extensively used to measure the size of solid particles when a dilute suspension of particles flows from one region to another region through a small aperture. On either side of the aperture is an electrode immersed in an electrolytic solution.

As the particles traverse the aperture one at a time, the electrolyte displaced changes the resistance between the electrodes which produces a voltage pulse. The pulses occurring from individual particles passing through the aperture are then electronically scaled and counted. The magnitude of the pulse is claimed to be proportional to the particle volume. The method of analysis has been adequately described by Berg[69], Kubitschek[70,71] and Allen and Marshall[72].

As a stream of particles passes through the aperture, two errors from the ideal situation of counting single particles can result. There can be a vertical interaction (secondary coincidence) where two particles in the aperture give rise collectively to a single pulse above the threshold level, which would not occur if the two particles passed through individually. The error due to the horizontal interaction of two particles passing through the aperture at the same instant of time, and thus giving rise to overlapping pulses (primary coincidence) can be minimized by using extremely dilute suspensions. The recommended sample concentration, to prevent this loss in the number of particles counted, is between 10 and 80 mg of solid per 100 cm^3. The coincidence loss can alternatively be corrected by empirical, semi-empirical[73–75] and statistical[76] equations. Coulter Electronics recommend dilute suspensions and correction for coincidence with their derived equation, whilst with the Electrozone or Celloscope instrument correction with the Wales and Wilson equation is suggested[75]. The particle size determined by the passage of a particle through an electrical resistance interrogation zone or orifice was initially assumed to be independent of position in the zone. Later it was found that the shape of the pulse generated by the passage of a particle through the centre of the orifice was different from the shape of the pulse generated by a particle passing through the orifice off-centre[77]. The technique of focusing a stream of particles along the central axis of the orifice (hydrodynamic focusing) enables a higher resolution to be attained as well as allowing smaller sizes to be determined.

1.3 Averages

A powder system is usually composed of a range of particle sizes and it is convenient at times to describe the distribution of sizes in terms of a single mean size. In the evaluation of a mean particle size from a powder containing many particle sizes it is assumed that the particles all have the same shape. Thus if the powder system is divided into small size intervals or groups of size dx an assumed diameter can be allocated to each group $x_1 \, x_2 \, \ldots\ldots$ The number of particles in each group can be designated $dN_1 \, dN_2 \, \ldots\ldots$ The average or mean size can be then obtained by summation of all the products of size and number $(\sum x \, dN)$ divided by the summation of all the number of particles $(\sum dN)$. This gives rise to various statistical averages.

1.3.1 Means

1.3.1.1 The arithmetic mean, $\bar{x}_A$

This is the sum of the diameters of the separate particles divided by the number of particles

$$\bar{x}_A \equiv x_{NL} = \frac{\sum x \, dN}{\sum dN}$$

It is applicable for normal distributions.

1.3.1.2 The geometric mean, $\bar{x}_G$

This is the Nth root of the product of the diameters of N particles

$$\bar{x}_G = (x_1 \times x_2 \times x_3 \times \cdots\cdots x_N)^{1/N}$$

$$\log \bar{x}_G = \frac{\sum dN \log x}{N}$$

It is is applicable for log-normal distributions.

1.3.1.3   The harmonic mean, $\bar{x}_H$

This is the reciprocal of the arithmetic mean of their reciprocals

$$\frac{1}{\bar{x}_H} = \frac{1}{N} \times \Sigma\left(\frac{1}{x_1} + \frac{1}{x_2} + \frac{1}{x_3} \cdots \frac{1}{x_N}\right)$$

It is applicable for surface area evaluation.

1.3.2   Particle size distribution

1.3.2.1   Histogram

The choice of class widths in the graphical representation of distributions is important, because the basic requirement is that the resolution, which is defined as the class interval divided by the mean class size, must remain fairly constant. With narrowly classified powder the arithmetic distribution is justified but with most powders the geometric progression is more acceptable.

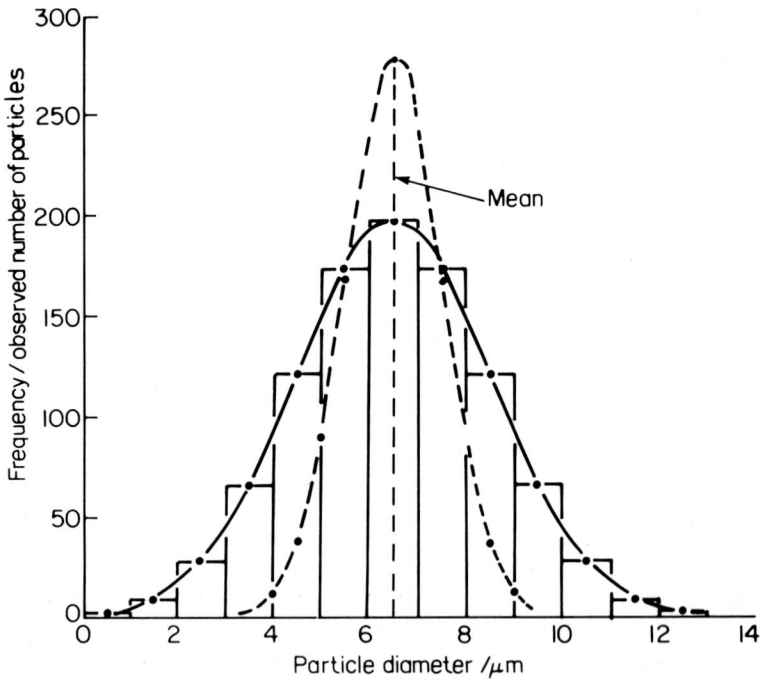

Figure 1.1.   *Histogram of a normal probability size distribution:* ———— *wide distribution;* – – – *narrow distribution. The data are as below:*

Class interval/μm	0—1	1—2	2—3	3—4	4—5	5—6	6—7
Observed number	3	9	28	66	121	174	198
Class interval/μm	7—8	8—9	9—10	10—11	11—12	12—13	
Observed number	174	121	66	28	9	3	

Although an average value represents a group of individual values it does not, by itself, completely define a particle size distribution. A term is therefore required to describe the scatter of the sizes or observations about a mean value—that is the standard or root-mean square deviation of values about the mean (*see Figure 1.1*).

1.3.2.2 Normal and log-normal distributions

Real particle size distributions seldom give symmetrical curves. Instead the size frequency curve may be asymmetrical or skewed (*see Figure 1.2*).

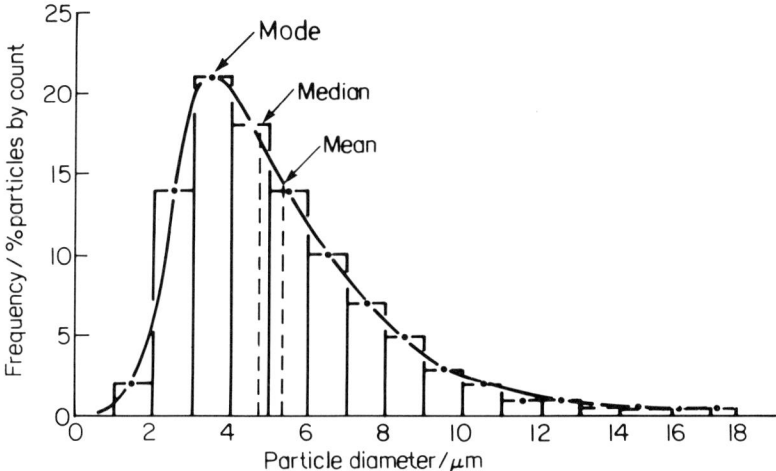

Figure 1.2. Histogram of a skewed particle size distribution. The data are as below:

Class interval/μm	0—1	1—2	2—3	3—4	4—5	5—6	6—7
Observed number	0	8	56	84	72	56	40
Frequency f/per cent	0	2	14	21	18	14	10
Class interval/μm	7—8	8—9	9—10	10—11	11—12	12—13	13—14
Observed number	28	20	12	8	4	4	2
Frequency f/per cent	7	5	3	2	1	1	0.5
Class interval	14—15	15—16	16—17				
Observed number	2	2	2				
Frequency f/per cent	0.5	0.5	0.5				

In a symmetrical or gaussian distribution the mode (the most commonly occuring value, i.e., the value at which the frequency density is a maximum), the median (the value that divides the area under the curve into equal parts) and the mean (the moment of the sum of all values of the distribution is zero) are equal numerically to each other. In a skewed distribution these three terms are not equal. The median is usually the more useful average.

Most asymmetrical particle size frequency curves can be converted into symmetrical curves which resemble a normal probability curve when the logarithm of size is substituted for the particle dimension. There is no

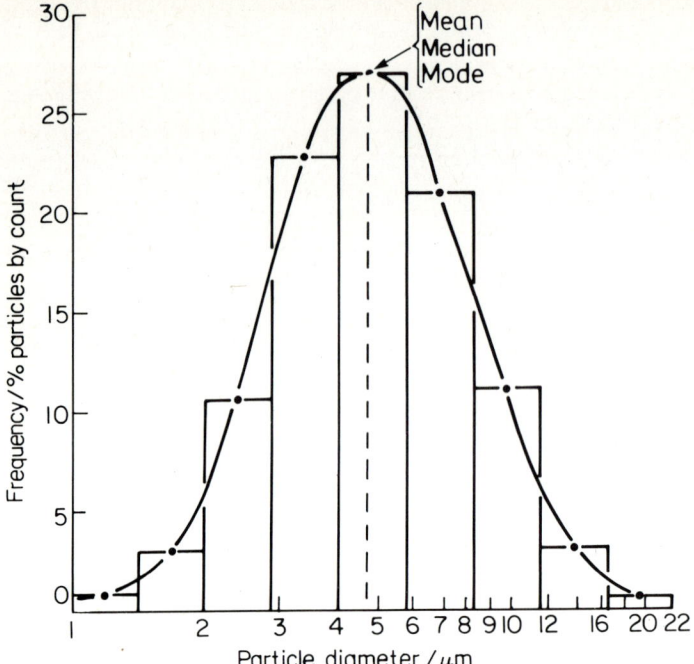

Figure 1.3.   Histogram of a log-normal size distribution. The data are as below:

Class interval/μm	1—1.41	1.41—2	2—2.83	2.83—4
f/per cent	0.6	3.0	10.6	22.5
Class interval/μm	4—5.66	5.66—8	8—11.2	11.3—16
f/per cent	27.0	21.0	11.2	3.1
Class interval/μm	16—22.6			
f/per cent	0.7			

fundamental reason for particle data approximating to a logarithmic normal frequency distribution, but most dust produced by comminution, industrial and natural powders fall into this category (*see Figure 1.3*).

In using a log size class interval, instead of arithmetical size class interval, equal prominence is given to data in all parts of the size range, i.e., the class interval from 1 to 2 μm has the same space as the class interval 10—20 μm on a logarithmic scale, in contrast to the arithmetic scale which would require 0—10 and 10—20 to give equal space.

The cumulative plot is a useful type of graph from which the mean value together with the standard deviation of the distribution can be obtained (*see Figure 1.4*).

The geometric standard deviation (σ_g) is indicated by the ratio:

$$\frac{50 \text{ per cent size}}{15.87 \text{ per cent size}} \quad \text{or} \quad \frac{84.13 \text{ per cent size}}{50 \text{ per cent size}}$$

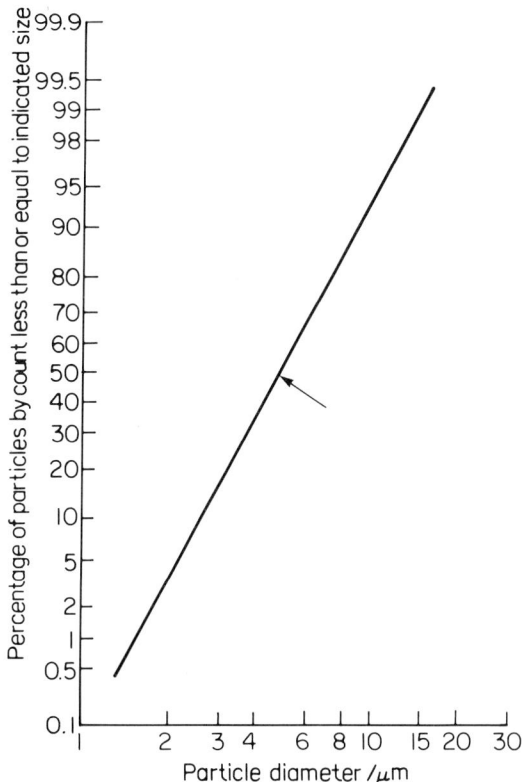

Figure 1.4. Cumulative particle size distribution plotted on logarithmic probability graph paper. The arrow indicates the median $M_g = 4.7\mu m$ with a geometric standard deviation $\sigma_g = 84$ per cent size/50 per cent size = 1.6

1.3.2.3 Normal and log-normal probability

The cumulative distribution curve is best represented by a straight line. This can be achieved either by plotting the cumulative percentage frequency on normal or log-normal probability graph paper (*see Figure 1.4*).

A useful property of the log-normal graph is that a number distribution can be converted to a surface, volume or weight distribution by means of Kapteyn's transformation equations[78] (*see Section 1.5*).

1.3.3 Rosin–Rammler, Alyavdin and Weibull equations

The Rosin–Rammler[79], Alyavdin[80] and Weibull[81] equations all have the algebraic form

$$Y = 1 - \exp\left(-\frac{X}{k}\right)^n$$

which linearizes to[82].

$$\log \ln \left(\frac{1}{(1 - Y)} \right) = n \log X - n \log k$$

The statistical basis of this equation has been given by Fisher and Tippett[83]. The variable Y in the Rosin–Rammler and Alyavdin equations is the fraction of particles less than a specific size in comminution studies while in the Weibull equation the variable Y is generally taken as the probability of failure. The variable X is the particle size in the Rosin–Rammler equation, time in the Alyavdin equation and some independent variable—time, stress or a linear function of either—in the Weibull equation. The terms k and n are distribution parameters. The value of n can be obtained from the slope of the linearized equation and k is equal to X at a point where the straight line crosses a horizontal line through $Y = 63.21$ per cent. The use of the above equation avoids the congestion of points caused by logarithmic scales, especially above values of $Y = 50$ per cent, and also semi-logarithmic graphs.

The crowding of points can be seen in *Table 1.3* where the distance between scale lengths ΔL from the Rosin–Rammler, Alyavdin or Weibull equations are uniform (10 units) while with either logarithmic or log–log reciprocal scales the distances either decrease or are non-uniform, thus distorting the graphic interpretation of the distribution.

The degree of resolution between points is a better criterion for the spacing of experimental points than the physical distance between points on graph paper. *Table 1.4* shows the degree of resolution, obtained by the division of the class size interval by the mean class size, for an arithmetic and geometric size progression. With the arithmetic progression, because the mean class size increases with the same class interval, the degree of resolution decreases with increase in size. In a geometric progression the degree of resolution remains constant.

The Rosin–Rammler distribution was initially used for the sieve analysis of coal and has now been applied to other particulate systems. The equation can be expressed as:

$$R = 100 \exp \left(-bd^n \right)$$

where R is the percentage by weight greater than sieve diameter size d, n and b are constants, b being a measure of the range of particle size present and n a character of the substance analysed.

$$\log \left(\frac{R}{100} \right) = -bd^n \log e$$

$$\log \left(-\log \frac{R}{100} \right) = +n \log d + \log b + \log \log e$$

which yields a straight line when $\log [\log (100/R)]$ is plotted against $\log d$.

Table 1.3. Distance between points, ΔL, for various logarithmic scales as a method to indicate the non-uniformity of spacings between points

Scalar points at positions	1	10	20	30	40	50	60	70	80	90	99.9
ΔL for Rosin–Rammler distribution / Alyavdin distribution / Weibull distribution		9	10	10	10	10	10	10	10	10	9.9
Logarithmic scale ΔL	0	8.5	11.1	12.6	13.6	14.5	15.1	15.7	16.2	16.6	17.0
		8.5	2.6	1.5	1.0	0.9	0.6	0.6	0.5	0.4	0.4
log–log reciprocal scale ΔL	0	8.6	11.4	13.1	14.4	15.5	16.6	17.5	18.6	19.9	23.9
		8.6	2.8	2.7	1.3	1.1	1.1	0.9	1.1	1.3	4.0

Table 1.4. Degree of resolution for arithmetic and geometric progressions

Arithmetic progression

Size	1	2	3	4	5	6	7	8	9
$\dfrac{\text{Class interval}}{\text{Mean class size}}$	—	$\dfrac{1}{1.5}$	$\dfrac{1}{2.5}$	$\dfrac{1}{3.5}$	$\dfrac{1}{4.5}$	$\dfrac{1}{5.5}$	$\dfrac{1}{6.5}$	$\dfrac{1}{7.5}$	$\dfrac{1}{8.5}$
Degree of resolution	—	0.66	0.40	0.28	0.22	0.18	0.15	0.13	0.12

Geometric progression

Size	1	$\sqrt{2}$	2	$2\sqrt{2}$	4	$4\sqrt{2}$	8	$8\sqrt{2}$	16
$\dfrac{\text{Class interval}}{\text{Mean class size}}$	—	$\dfrac{\sqrt{2}-1}{(\sqrt{2}\times 1)^{1/2}}$	$\dfrac{2-\sqrt{2}}{(2\times\sqrt{2})^{1/2}}$	—	—	—	$\dfrac{8-4\sqrt{2}}{(8\times 4\sqrt{2})^{1/2}}$	$\dfrac{8\sqrt{2}-8}{(8\sqrt{2}\times 8)^{1/2}}$	$\dfrac{16-8\sqrt{2}}{(16\times 8\sqrt{2})^{1/2}}$
Degree of resolution	—	0.347	0.348	—	—	—	0.348	0.348	0.348

The peak of the distribution curve, when $n = 1$, occurs at 36.8 per cent which is obtained from $100/e$ and denotes the mode of the distribution.

1.4 Shape

Particle shape is a fundamental characteristic of a particle, in a manner similar to that of the size of a particle, and can be assessed from individual particles by the measurement of the length or diameter of the particle. In BS 2955 (1958)[1] various types of particle shape have been described in words and subsequently defined, but these verbal descriptions are inadequate for incorporation in calculations for ascertaining the effect of particle shape in various particle systems and on particle properties. An attempt was made in BS 512 (1966)[1] to designate arbitrary ratios of particle breadth to length, B/L, and particle thickness to breadth, T/B, to various three-dimensional particle shapes such as discs, rods, blades and equidimensional particles so that numerical values could be assigned to shape rather than literal descriptions.

Herdan[84] has stated that 'shape factors' have a three-fold function.

1. They are factors of proportionality between the particle sizes determined by different methods of particle characterization. Thus the particle size determined by microscopical techniques can be compared with that determined by sedimentation.
2. They are conversion factors for expressing the results of different particle measurement techniques as an 'equivalent sphere' parameter.
3. They transform the second and third power of the measured particle diameter into the particle surface and particle volume respectively.

Thus the volume or surface of a particle can be written as:

> Volume of particle ≡ Volume coefficient (volume shape factor) multiplied by the *cube* of some characteristic dimension.

> Surface of particle ≡ Surface coefficient (surface shape factor) multiplied by the *square* of some characteristic dimension.

Algebraically the above definitions can be written as:

$$V = \alpha_v x^3$$

and

$$S = \alpha_s x^2$$

where x is the size characteristic of the method of measurement. From a microscope count method the size characteristic in a one-dimensional quantity can be either a Feret's, d_F, Martin's, d_M or projected area, d_a, diameter.

Since the volume and surface for a given particle are fixed, the numerical values of coefficients of shape α_v and α_s are dependent on the chosen size characteristic and subscripts to α_s and α_v must be amplified to denote the method of particle size measurement.

Hence for a projected area diameter, d_a; the volume shape factor becomes α_{va} and the surface shape factor α_{sa}.

$$V = \alpha_{va} \times d_a^3$$

and

$$S = \alpha_{sa} \times d_a^2$$

The two dimensional quantity (area) can be measured from the microscopical image of the particle and the three dimensional quantity (volume) can be measured from either the sedimentation or electrical resistance technique. The surface of an irregularly shaped particle can be represented as $\alpha_{sa} d_a^2$ when the characteristic dimension is the projected area diameter d_a and by definition is equivalent to πd_s^2 when the dimension is that of a sphere of the same surface. Likewise the volume of an irregular particle is $\alpha_{va} d_a^3$ and equivalent, for a sphere, to $\pi d_v^3/6$.

The surface and volume shape factors, α_s and α_v, are numerically equal to 3.14 and 0.52 respectively when the particle is spherical because of the relationship of surface and volume of sphere being πd^2 and $\pi d^3/6$ respectively. It has been found empirically[85] that for irregular and non-uniformly shaped particles the ratio v/x^3, where v is the average particle volume and x the mean particle size, remains sensibly constant for various grades of particles. It has also been found empirically that the ratio s/x^2, where s is the average particle surface and x the mean particle size, remains sensibly constant for different sizes of the same material. It is thus possible at times to speak of an average particle shape although variations of shape with size for some materials have been found[86]. A surface–volume shape coefficient can be obtained by using the above two ratios.

$$\alpha_{sv, x} = \frac{\alpha_{sx}}{\alpha_{vx}}$$

where x is the arbitrary symbol which depends on the method used to characterize the size of particles.

Although Rose[87] concluded in 1961 that:

'Perhaps one of the most abortive searches yet made
is for a means of defining the shape of a particle'

many workers are currently appraising past work on shape factors. Beddow, Vetter and Sissan[88] have reviewed the various methods of particle shape analysis and proposed a generalized scheme in which past and current methods for shape analysis could be grouped into four classes:

Class I   Specific characteristic of individual particles
Class II   Bulk properties of powders
Class III   Mathematical techniques for shape generation
Class IV   Literal definitions of shape

Davies[89] has stated, however, that since particle shape information is needed to describe the behaviour of particles in many industries in which particles are mixed, reacted, stored or transported and although the main parameters of particle size, density, shape and surface interact in a complex manner, for purposes of clarity particle shape can be represented by either:

1.   The general particle form, i.e. aspect ratio, proportionality constants

or

2.   The surface topography, i.e. roughness, angularity or roundness or a combination of the above two categories.

Past and current techniques to evaluate shape factors can also be categorized either by direct or primary methodology for the measurement of two- or three-dimensional parameters of individual particles or the methodology of indirect or secondary measurement for a collection of particles.

1.4.1   Direct measurement of particle shape

Heywood[90] derived an expression for the shape of a particle in terms of the general geometrical form from the ratio of length, L, breadth, B, and thickness, T (*Figure 1.5*).

To separate the dimensional proportions of the particle from the geometric shape Heywood used the ratios of elongation, n (ratio L/B) and flatness, m (ratio B/T) together with a term defined as the equidimensional

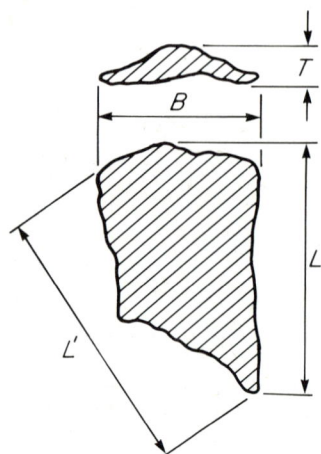

Figure 1.5.   Meaning of T, B, L and L' in the Heywood method

shape factor, α_0, which represented part of the α_v shape factor due to the geometric shape only.

$$\alpha_0 = \alpha_v m \sqrt{n} \tag{1.1}$$

where α_v is the volume shape factor.

Also from a variety of known geometrical shapes and for irregularly shaped particles Heywood[91,92] used the expression.

$$\alpha_s = 1.57 + C\left[\frac{\alpha_0}{m}\right]^{4/3}\left(\frac{n+1}{n}\right) \tag{1.2}$$

or, combined with equation (1.1)

$$\alpha_s = 1.57 + C\alpha_v^{4/3}\left(\frac{n+1}{n^{1/3}}\right)$$

where α_0 and C can be determined experimentally (*Tables 1.5 and 1.6*). Values of L, B and T can now easily be obtained with an automatic scanning microscope attachment.

This approach for a collection of powders has been used by Church[93], and Dwyer *et al.*[94] and Mandelbrot[95] to obtain relationships between particle parameters and linear diameters.

Davies[89] by measurement of the three dimensions of 500 particles with a microlathe and microscope used the triangular representation of Sneed and Folk[96] to produce a contour map of shaped sorted sands (*Figures 1.6 and 1.7*). Such triangular plots—used extensively by workers in the field of sedimentology—give useful information on the form distribution of particles within a powder but could not, however, distinguish between cubes or spheres, or between cuboids or ellipsoids.

Table 1.5. Shape coefficients for some geometrical and irregular shapes

Shape group		α_0	C	$\alpha_0 C^{4/3}$
Geometric form:	Tetrahedral	0.328	4.36	0.986
	Cubical	0.696	2.55	1.571
	Spherical	0.524	1.86	0.785
Approximate form:	Angular (a) Tetrahedral	0.38	3.3	0.91
	(b) Prismoidal	0.47	3.0	1.10
	Subangular	0.51	2.6	1.06
	Rounded	0.54	2.1	0.92

Table 1.6. Typical values for the shape coefficients of particles of various shapes

Material	α_0	α_s	ψ
Rounded particles, water-worn sand	0.32—0.41	2.7—3.4	0.817
Angular particles, pulverized minerals	0.20—0.28	2.5—3.2	0.655
Flaky particles, talc, gypsum	0.12—0.16	2.0—2.8	0.543
Thin flakes, mica, graphite	0.01—0.03	1.6—1.7	0.216

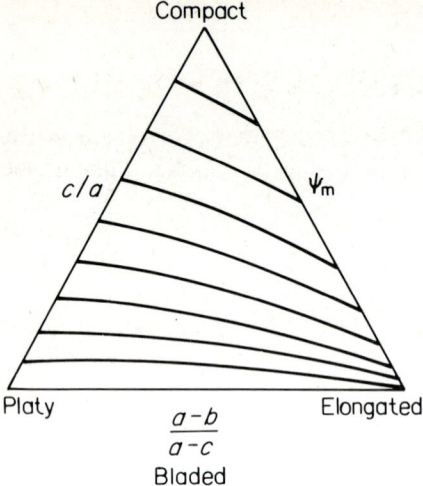

Figure 1.6.   *Triangular plot by Sneed and Folk*[96]. *a is the long diameter, b is the intermediate diameter and c is the short diameter.* ψ_m *is the maximum projection sphericity* $= 3\sqrt{(c^2/ab)}$

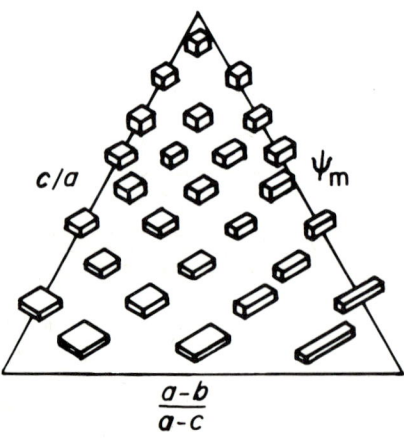

Figure 1.7.   *Shape relationships by Sneed and Folk*[96]

To distinguish between cubes and spheres requires an alternative technique which involves the comparison of an irregularly shaped particle with that of a more symmetrical body and subsequent measurement of the degree of asymmetry of the irregular shaped particle from that of the symmetrical body.

Wadell[97–99] defined the shape of a particle as either having

1. Sphericity ϕ

or

2. Roundness P

Sphericity was defined as the ratio of the surface area of a sphere of the same volume as the irregular particle to the actual surface area of the irregular particle

$$\phi = \frac{\pi d_v^2}{\pi d_s^2} \qquad (1.3)$$

which as a working formula is

$$\phi = \frac{d_c}{D_c} = \frac{\text{Diameter of a circle equal in area to the projected area of a particle in a stable plane}}{\text{Diameter of the smaller circle circumscribing the projection of the particle}}$$

Treasure[100] showed that the surface coefficient α_s can be related to the Heywood equidimensional shape factor α_0 by Wadells' sphericity

$$\phi = 4.38 \, \frac{\alpha_0^{2/3}}{\alpha_s} \qquad (1.4)$$

Roundness was obtained by measurement of the radius of curvature (r) of

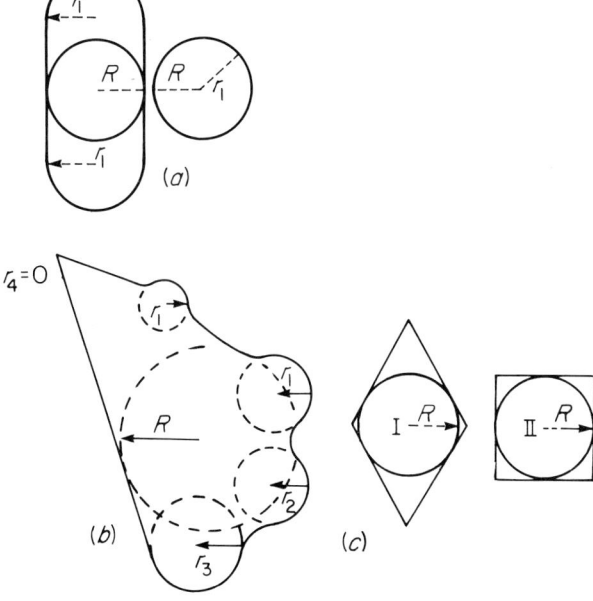

Figure 1.8. Concept of roundness by Wadell[97-99]

each projection in the perimeter of an irregular particle which was then related to the radius of the maximum inscribed circle (R). Thus

$$P = \frac{\sum \dfrac{r}{R}}{N} \tag{1.5}$$

where N is the number of projections measured (*Figure 1.8(b)*).

As the corners of a particle are worn down r tends to R and P tends to unity (*Figure 1.8*). Wadell's roundness factor can be used to illustrate the need to distinguish between the form of the particle and the roundness or surface perimeter, roughness, of a particle. *Figure 1.8(a)* shows two particles of identical Wadell roundness, $(P = 1)$, but of different form while *Figure 1.8(c)* shows two particles of roundness $P = 0$ with different form and angularity.

Davies[89] has proposed a simple space representation of form and roundness by the incorporation of the triangular plots of Sneed and Folk (*Figure 1.9*) and Wadell's roundness ratio. The roundness ratio is, at the moment, favoured instead of Wadell's sphericity because of the insensitivity of the sphericity ratio which requires large changes in relative particle pro-

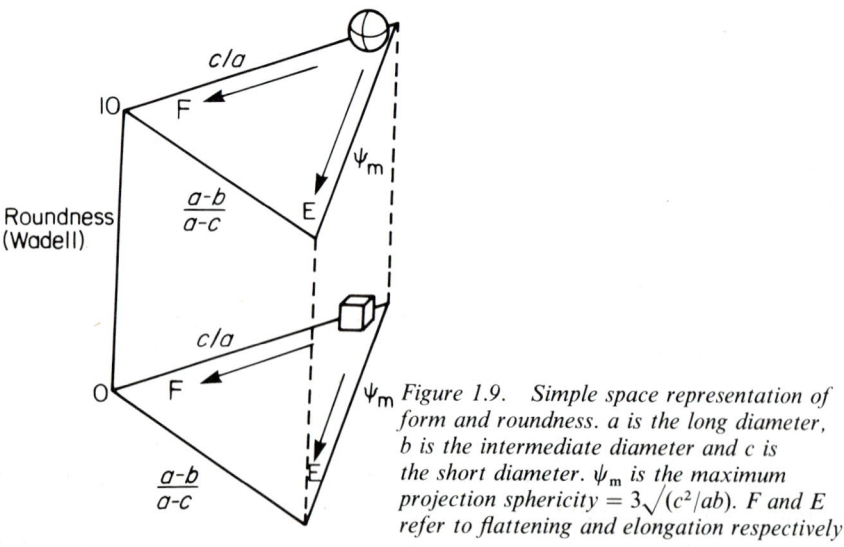

Figure 1.9. Simple space representation of form and roundness. a is the long diameter, b is the intermediate diameter and c is the short diameter. ψ_m is the maximum projection sphericity = $3\sqrt{(c^2/ab)}$. F and E refer to flattening and elongation respectively

Table 1.7.   Shape coefficients for sections of a cube

Relative dimensions of cube	α_0	α_s	ϕ
Thickness = breadth	0.696	4.71	0.806
Thickness = breadth/2	0.348	3.14	0.761
Thickness = breadth/10	0.070	1.88	0.434
Thickness = breadth/100	0.007	1.60	0.110
Thickness = 0	0	1.57	0

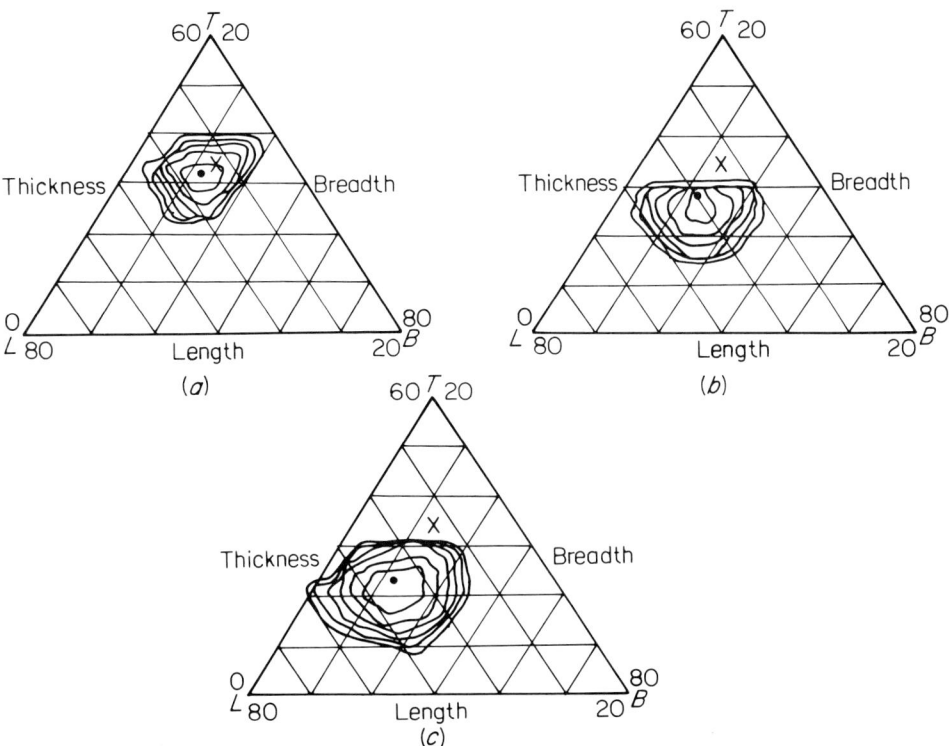

Figure 1.10.   Triaxial contour diagrams of shape-sorted sands. Diagram (a) *is for spherical sand,* (b) *for an intermediate sand and* (c) *for an irregular sand. L = length, B = breadth, T = thickness and × represents a uniaxial point*

portions of length and breadth to thickness to produce significant changes in ϕ (*Table 1.7*)(*Figures 1.10 and 1.11*).

Combination of Wadell's roundness shape factors with the Sneed and Folk[96] triangular representation overcomes the criticism of Medalia[101] with regard to the inadequacy of the triangular particle profile to distinguish between prolate and oblate ellipsoids and that the Heywood shape factors of elongation and flatness ratios are more satisfactory than triangular particle profiles. The Davies method of feature-space representation cannot however account for the inherent roughness of the surface of most real materials. Any shape factor calculated becomes dependent on the fineness or surface area of particles measured because the evaluation of the surface area or perimeter of an irregular particle varies with the measurement technique. The sphericity value of a porous spherical particle is too low and may therefore be indistinguishable from that of a smooth but irregular and highly convoluted particle. Medalia, who represented a particle as an ellipsoid of equivalent radii of gyration according to the dynamic of a rigid body—the dynamic method—regarded this dynamic method as an improvement over the methods based on surface area of particles meas-

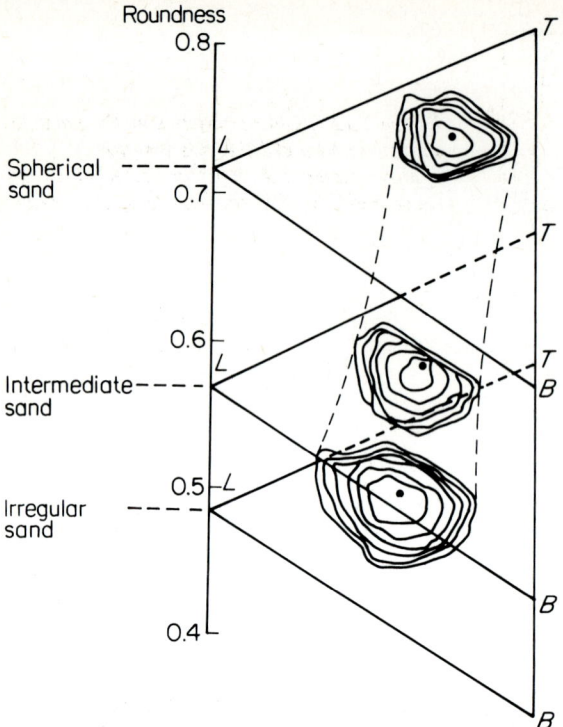

*Figure 1.11.   Form-roundness feature-space
representation of shape-sorted sands*

ured by permeametry or nitrogen adsorption. The use of the ellipsoidal
form for calculation of the surface area of a particle necessitates however
the use of elliptic integrals which are not readily evaluated. Aschenbrenner[102] used the regular geometric form of a 14 face-sided solid to
overcome the difficulties of surface area calculation by elliptic integrals.
The concept of a surface shape factor, from the area and perimeter of an
irregular particle, was used by Hausner[103] who encompassed the complicated shape of the particle image with a rectangle (*Figure 1.12*). The characteristics measured were the side lengths of the rectangle, *a* and *b*, the
projected area, *A*, and the circumference of this area, *C*.

Since the cross-sectional area and circumference of a spherical particle is
$A = d^2\pi/4$ and $C = d\pi$ respectively the relationship $C^2 = 4\pi A$ exists. For a
spherical particle, which gives the minimum surface for a given volume, a
shape factor can be defined as:

$$Z = \frac{C^2}{12.6A}$$

(1.6)

which has a value of unity for a spherical shaped particle. *Table 1.8* shows
for various shaped particles (*Figure 1.13*) that *Z* increases considerably

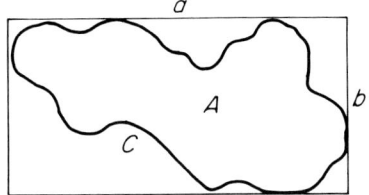

Figure 1.12. Characteristics of the particle shape: a and b are the side lengths of the rectangle; A is the projected particle area; and C the circumference of this area

I

II

III

IV

V

VI

VII

VIII

IX

X

XI

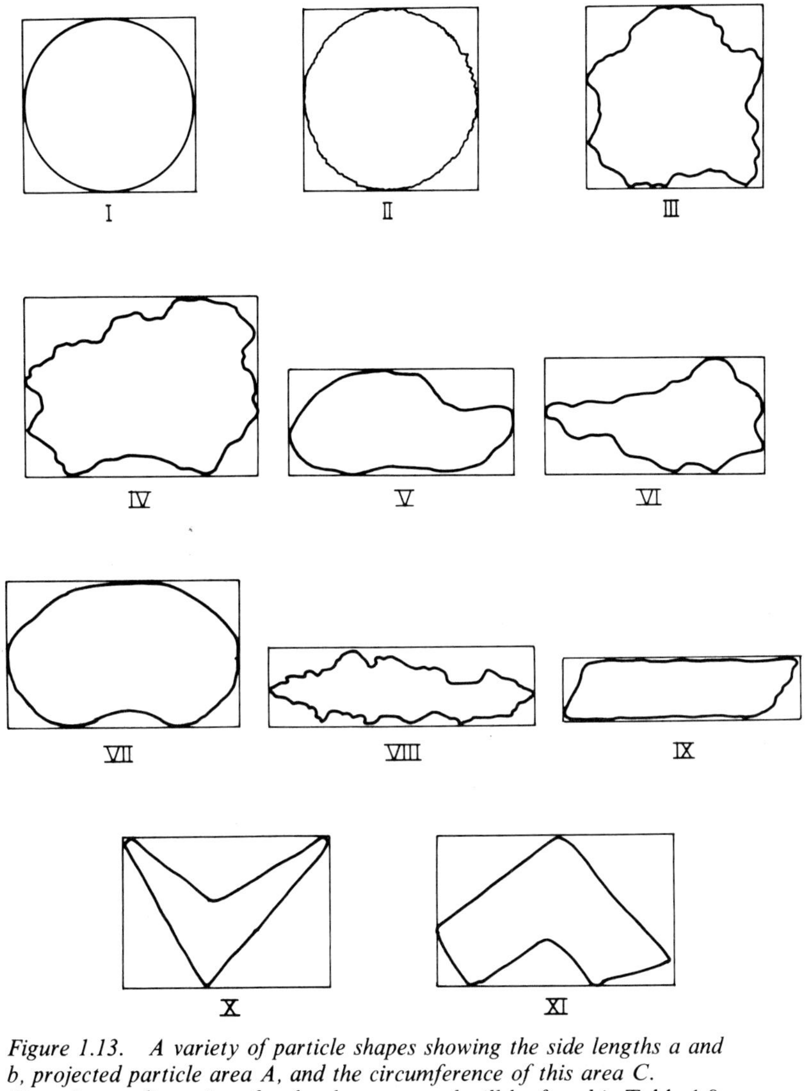

Figure 1.13. A variety of particle shapes showing the side lengths a and b, projected particle area A, and the circumference of this area C. The relevant dimensions for the shapes quoted will be found in Table 1.8

Table 1.8. Relative dimensions and characteristic results of powder particles of various shapes

Shape	Relative dimensions				Particle characteristics		
	a	b	A	C	x	y	z
I	6	6	28.2	18.8	1	0.78	1
II	6	6	28.2	26.5	1	0.78	1.98
III	6	6	25.4	25.5	1	0.71	2.04
IV	8	6	36.2	29.4	1.34	0.76	1.89
V	7.6	3.4	18.7	20.5	2.23	0.72	1.77
VI	7.5	3.8	15.5	21.6	1.97	0.54	2.2
VII	8	5	29.4	22.9	1.6	0.74	1.4
VIII	9	2.5	11.8	33.5	3.6	0.52	7.5
IX	8	2	13.2	17.8	4.0	0.83	1.89
X	7	5	10.7	19.5	1.4	0.31	2.8
XI	8	5	20.2	21.7	1.6	0.5	1.84

when the surface is corrugated, the values being between 1 and 10, with dendritic shaped particles having the highest values.

Hausner also defined an elongation factor:

$$x = a/b \tag{1.7}$$

which was found, however, not to be informative enough to characterize shape because particles used in powder metallurgy varied between $x = 5$ and 10. Hausner's bulkiness factor defined as:

$$y = \frac{A}{a + b} \tag{1.8}$$

can be used in relation to the packing density of material although values above 0.8 are seldom seen and with values of $y < 0.6$ indicate specimens with large indentations and deep notches (*Table 1.8* and *Figure 1.13*).

The relationship between perimeter and the linear diameter of Feret and Martin, before and especially after the advent of the image analyser, have been used to determine shape factors[93,94].

Because automatic image analysers can simply and rapidly measure particle area, with and without porosity, perimeter, Feret and Martin diameter, longest chord length and number of particles, numerical values of the shape and form of particles can now be obtained from ratios of these various image parameters. Dwyer[94] using the π-MC system determined the shape population of spherical resin particles and sugar maple pollen from an aspect ratio T, defined as $\pi L^2/4A$ where L was the longest dimension and A the projected microscope image area. The aspect ratio was related to average area of the particles measured, $\bar{A}$, the average projected area diameter $\bar{P}$ and the variance or standard deviation, σ, of the area–size distribution of all sample particles by:

$$\sigma^2 = \frac{4\bar{A}}{\pi} - \frac{\pi^2(\bar{P})^2}{4TE^2} \tag{1.9}$$

which could be used to ascertain the change in particle form rather than roundness and angularity, of fibrous shapes from other forms. Nystrom and Stanley-Wood[104,105] described a ring gap sizer, intially used for characterization of fragile explosive materials, which could measure the thickness of particles, as opposed to an intermediate diameter between length and thickness in a sieve classification. The geometric mean particle size obtained from the ring gap sizer was within 7 per cent of the mean thickness of particles measured by microscopy. Comparison between the ring gap sizer diameter and the dry sieve diameter could be used to establish shape factors of free flowing materials.

With the advent and development of automated microscope systems linked into computers, the description of irregular particle shapes and characterization of profiles by complex mathematical functions has now become available. Schwartz and Shane[106] were the first to describe a two-dimensional profile of an irregularly shaped particle by a geometric signature waveform which could undergo Fourier analysis. The signature waveform was generated by locating, at a central position within the particle profile, a vector which touched the outline profile of the particle. The magnitude of the vector at various angles when moved around the irregular profile could either produce a graph of magnitude against angle or be treated as a continuous periodic harmonic waveform. The complex particle signature waveform can then be broken down into a list of simple harmonic waves by Fourier analysis[107–109]. The list of the Fourier harmonic series within the complex signature waveform gives a dimensionless shape description of the particle. The central location of the vector can be, as in the work of Schwartz and Shane, the centre of the smallest circle inscribing the particle, or can be at the centre of gravity of the particle[110,111]. Kaye[112] reviewed the work on Fourier shape analysis and also introduced the work of Mandelbrot[113] on fractals as an alternative mathematical method to obtain an index of ruggedness or texture of particles. The length of a convoluted curve can be estimated from the product of the number of known fixed sized dimensional steps required to traverse a convoluted curve. As the step dimension decreases the closer the zig-zag stepped curve will be to the convoluted curve until eventually a constant value of the length of the convoluted curve is reached. Mandelbrot has shown, however, that for a fractal curve a plot of the logarithm of the estimated length of a curve against the logarithm of the step size gave a linear relationship which extrapolated to an infinite length. The slope of the log–log plot gives however a quantity $(1—D)$ which is defined as the fractal dimension of the rugged boundary curve. The Mandelbrot fractional dimension—a fractal— is solely a number between 1 and 2 to describe the structure or shape of an irregular profile.

Flook[114] applied fractal dimensions to characterize the texture and structure, using a Quantimet fitted with a 2D Amender module, of geometrically constructed model particles and profiles of carbon black aggregates. He found that for carbon black aggregates the plot of logarithmic

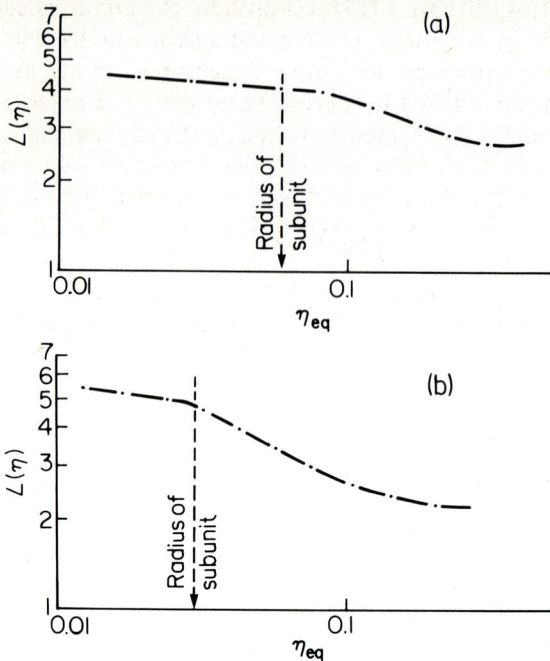

Figure 1.14.   Fractal dimension for (a) *carbon black aggregate and* (b) *stimulated floc. $L(\eta)$ is the estimated length of the perimeter and η_{eq} is the equivalent step size*

perimeter length against logarithmic step size showed departure from linearity at large and small step sizes. At small step sizes the fractal dimension—from the slope of the log–log plot—was dominated by the structure of the sub-unit while at large step sizes the fractal dimension was due to the coarse structure of the agglomerate. Flook concluded that it was now a practical proposition to make fractal dimension measurements to characterize the textured and structured profiles of a wide variety of materials.

1.4.2   Indirect measurement of particle shape

One of the inherent restrictions in the characterization of particle size is that measurement of a particle dimension is dependent upon the measurement technique. An irregular particle characterized by sieve analysis, sedimentation or by an electrical resistance technique will not have identical diameters. The difference between measured sizes can, however, be advantageous. The factor of proportionality between the particle sizes determined by different analysis techniques can be regarded as a shape factor.

BS 3406 Parts 2, 3 and 4 (1963) state that a 75 μm particle by sieve analysis may have a mean projected diameter of 105 μm or a mean Stokes

diameter of 70.5 μm. In BS 4359 Part 3 (1970), Appendix B, various calculated and experimental values of surface, volume and specific-surface coefficients (shape factors) are tabulated for given dimensional forms and
irregular particle shapes as determined by sieving, permeability, sedimentation, electrical resistance and light extinction methods of measurement of
particle size and surface area. In all cases the evaluated numerical values of
3.14, 0.52 and 6.0 for α_{sv}, α_{vv} and α_{svv} respectively for spherically defined
forms are exceeded for non-spherical or irregular-shaped particles over a
range of sizes (2000—0.5 μm) and materials.

From Heywood's[91,92] flatness ratio, $m = B/T$, and elongation ratio,
$n = L/B$, a relationship between shape and sieve characteristics of a powder
can be established:

$$d_a = 0.98 \left[\frac{2nm^2}{m^2 + 1} \right] d_t \qquad (1.10)$$

Table 1.9 shows the ratios of the projected area diameter, d_a, to the sieve
diameter d_t for various values of m and n.

Table 1.9.   Values of d_a/d_t for various values of the flatness
ratio m and the elongation ratio n

n	$m = 1.0$	1.2	1.5	2.0	2.5	3.0
1.0	1.00	1.06	1.14	1.23	1.28	1.32
1.2	1.09	1.17	1.25	1.35	1.41	1.44
1.5	1.22	1.30	1.39	1.51	1.57	1.61
2.0	1.41	1.51	1.61	1.74	1.87	1.87
2.5	1.57	1.68	1.80	1.95	2.03	2.07
3.0	1.73	1.84	1.97	2.13	2.23	2.28

The surface area of a powder can be calculated for a non-uniformly sized
particle distribution from shape factors or conversely, if the particle size
together with the surface area of a powder is known, shape factors can be
calculated.

Allen[86] has shown that the shape factors—determined from the ratio of
surface area obtained by photosedimentation, S_w to sieve size, d_t, for graded
sand over the particle size range 75—2000 μm, varied with sieve size and
gave shape factors, α_{svt} less than 6.0 (the accepted value for spherical particles)
at low particle size, when the density, ρ_s, is indepent of size:

$$\alpha_{svt} = S_w \rho_s d_t \qquad (1.11)$$

where S_w was obtained from the maximum optical density D_0 through a
length L of various concentrations (C) of suspended material:

$$S_w = \frac{9.2}{L} \left(\frac{D_0}{C} \right).$$

Figure 1.15 shows the shape factor α_{svt} divided by 6 which becomes unity
for a spherical particle form. For molochite the shape factor α_{svt} was not

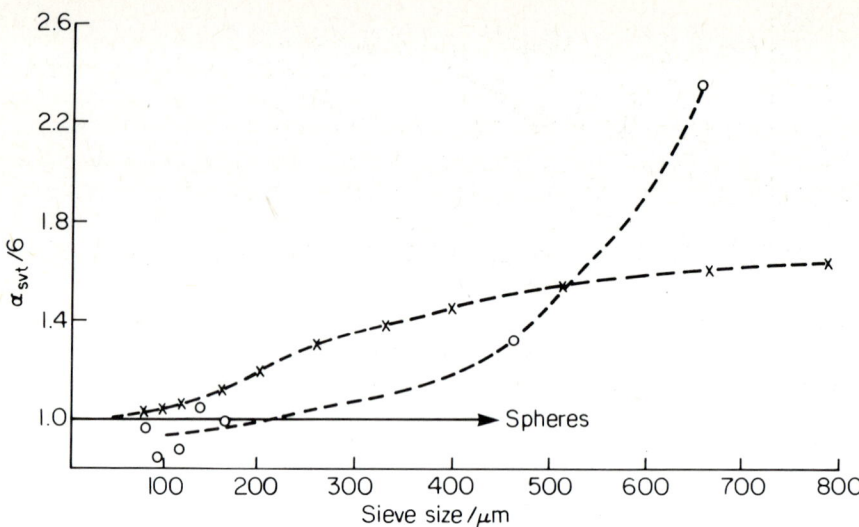

Figure 1.15.   Relationship between surface-volume shape coefficient as determined by wide-angle photosedimentation and by sieve size for × molochite and ○ sand

less than 6 but did show a variation with size; the shape factor increases with increase in particle size.

Stein and Corn[115], in the analysis of respirable coal mine dust from three locations, determined the aerodynamical diameter with a Bahco centrifugal classifier and also a settled projected diameter, d_{sp}, by optical microscope; a randomly projected area d_{rp} from measurement of light passing through a suspension and the total surface area by krypton gas adsorption at liquid nitrogen temperature from which a d_{BET} was calculated. The ratios d_{sp}/d_{rp} and d_{BET}/d_{rp} (equivalent to shape factors) were determined but only the d_{BET}/d_{rp} shape factor showed any variation with aerodynamical size and location (*Figures 1.16 and 1.17*). The sensitivity of the surface area shape factor was thus regarded as the only parameter which could be used to find the influence of the physiological action of dust after inhalation in man.

Kotrappa[116] determined the shape factors from quartz particles within the size range 0.2—2.0 μm from measurements of the aerodynamic projected area d_p, the Stokes diameter, d_{st}, surface and volume diameters and the BET low-temperature adsorption of nitrogen. The effective shape factors were however expressed as regression equations (*Table 1.10*) rather than proportionality constants. The volume and surface shape factors and the ratio of d_p/d_{st} remained fairly constant over the size range examined, similar to other works (Hamilton[117], Cartwright[118], Timbrell[119]). The dynamic and projected area resistance shape factors (K and K_p respectively) tended to decrease with decreasing particle size indicating that smaller sizes of particles behave like spheres. Pilpel[120] has shown that the shape of small particles obtained by comminution is dependent on the method of size

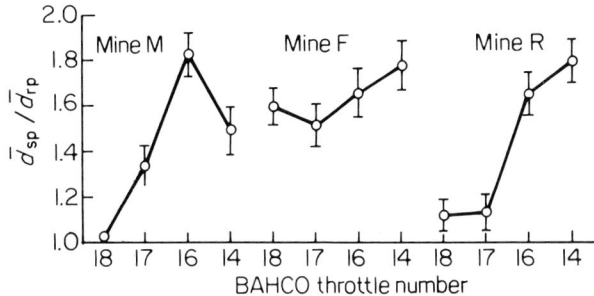

Figure 1.16. Ratios of average settled oriented area particle diameter $(\bar{d}_{sp})$ *and average random projected area particle diameter* $(\bar{d}_{rp})$ *for size fractions of coal mine dust from coal mines* M, F *and* R

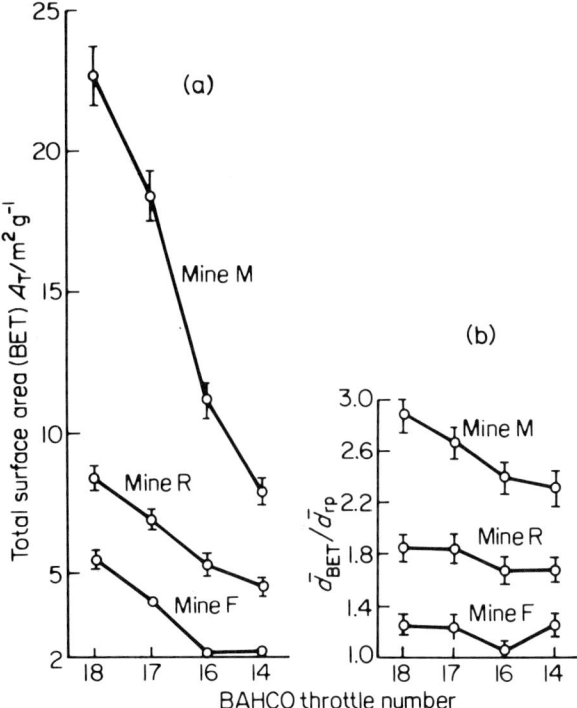

Figure 1.17. Size fractions of coal mine dust from coal mines M, F *and* R. *Curves* (a) *show the total surface areas* A_T *of coal mine dust, and curves* (b) *show ratios of average surface area particle diameter from* BET *measurements* $(\bar{d}_{BET})$ *to average random projected area particle diameter* $(\bar{d}_{rp})$

Table 1.10. Effective shape factors applicable over a size range 0.2—2.0 μm (d_p)

Shape factors	Regression equation	Error factor*
Volume shape factor (a_v)	$0.342d_p^{0.07}$	1.139
Surface shape factor (a_s)*†	5.2 ± 0.52	—
Stokes diameter (d_{st})	$0.65d_p^{0.95}$	1.039
Diameter ratio (d_p/d_{st})	$1.53d_p^{0.05}$	1.027
Dynamic shape factor (K)	$1.82d_p^{0.17}$	1.063
Projected area resistance shape factor (K_p)	$1.42d_p^{0.16}$	1.08

* Multiply the estimate (column 2) by the error factor (column 3) and its reciprocal to determine the range of the standard error of the estimate.
† Error in surface shape factor is the error associated with the measurement of specific surface areas.

Table 1.11. Effect of micronization on the shape of pigment particles

Pigment	Measured surface area/m² g⁻¹	Calculated surface area/m² g⁻¹	Shape factor
Prussian blue	61.3	2.3	0.04
Prussian blue, micronized	58.9	12.7	0.23
Red iron oxide	7.6	3.4	0.44
Red iron oxide, micronized	7.3	4.9	0.68
Burnt umber	138	1.2	0.01
Burnt umber, micronized	94	5.0	0.05
Yellow iron oxide	17.6	1.4	0.08
Yellow iron oxide, micronized	17.7	2.5	0.14

reduction. Micronized pigments are often less irregular and more spherical in shape than particles which have been repeatedly fractured and sheared. *Table* 1.11 shows the shape factors for a number of materials either micronized or fractured. The shape factor—the ratio of BET measured surface to calculated surface area from electron microscopy—when closer to unity indicates a particle with a more spherical form.

Use was made of Heywood's shape factors by Pilpel and Walton[121] in the effect of size and shape on the flow and failure properties of penicillin powders which had an angular prismoidal form. The particle surface coefficient, K_{sa} was the ratio of S to d_a^2 where values of S (the mean particle surface area) were obtained from air permeability and the number of particles per gram obtained from a Coulter counter analysis. The mean particle projected diameter was obtained by photographing 200 particles, cutting out and weighing the photographic image. The particle volume coefficient K_{va} was the ratio of V (calculated from particle density and number of particles per gram) to d_a^3. The shape factor α_{sva} was then taken as K_{sa}/K_{va}. The analysis yielded relationships between the particle size as expressed by V/S and particle shape as expressed by α_{sva} with cohesion, tensile strength, internal friction and a failure factor. Shotton and Obiorah[122], also using Heywood's shape factors, found that compacts formed from dendritic crystals of sodium chloride, which had a large shape factor, were stronger than compacts formed from cubic crystals of sodium chloride which had a smaller shape factor. Care must be taken in the use of Coulter counting technique in ascertaining the number of particles because of the non-uniformity of the electrical field and the effect of particle shape on the conductance of the orifice containing the particle[123]. These effects can be adsorbed in practice, however, by calibration with monosized spheres.

1.5 Application of shape factors for surface area evaluation

1.5.1 Image diameter

When a microscopy number–length count is shown to have a log-normal distribution, the surface and weight distributions are also log-normal with the same standard deviation, σ_g. Conversion from a number–length, NL, distribution to either a number–surface, NS, or number–volume, NV, distribution is possible by using the geometric mean size x_{gN} and the following equations:

$$\ln x_{NL} = \ln x_{gN} + 0.5 \ln^2 \sigma_g \tag{1.12}$$

$$\ln x_{NS} = \ln x_{gN} + 1.0 \ln^2 \sigma_g \tag{1.13}$$

$$\ln x_{NV} = \ln x_{gN} + 1.5 \ln^2 \sigma_g \tag{1.14}$$

Since a surface volume mean diameter can be written as:

$$x_{SV} = \frac{\Sigma d_V}{\Sigma d_S} \equiv \frac{\Sigma x_i^3}{\Sigma x_i^2} \equiv \left[\frac{\Sigma x_i^3}{\Sigma n_i} \bigg/ \frac{\Sigma x_i^2}{\Sigma n_i} \right] = \frac{x_{NV}^3}{x_{NS}^2} \tag{1.15}$$

where Σd_V, Σd_S and Σn_i are the cumulative volume, surface and number of particles in the assembly of particles, x_i is the size of the particle of class i and x_{NV}^3 and x_{NS}^2 are the characteristic dimension of particles in terms of number volume and number surface respectively. From equations 1.12–1.15, since:

$$x_{SV} = x_{NV}^3 / x_{NS}^2$$

$$\ln x_{SV} = 3 \ln x_{NV} - 2 \ln x_{NS}$$

$$\ln x_{SV} = (3 \ln x_{gN} + 3 \times 1.5 \ln^2 \sigma_g) - (2 \ln x_{gN} + 2 \ln^2 \sigma_g)$$

$$= \ln x_{gN} + 2.5 \ln^2 \sigma_g \tag{1.16}$$

The log-normal distribution number count can thus be converted to a log-normal surface distribution and the specific surface of the particulate system can be determined from

$$S_V = \frac{\phi}{x_{sv}} \qquad (1.17)$$

where ϕ is the shape factor which is taken as 6 for spherical particles.

1.5.2 Mechanical diameter

The surface area per unit volume can be defined as:

$$\frac{S}{V} = \frac{\sum N \pi d_s^2}{\sum N \frac{\pi}{6} d_v^3} \qquad (1.18)$$

for surface and volume diameters. Thus the specific surface, S_V, can be calculated for irregular and spherical particles to be

$$S_V = \frac{\alpha_s}{\alpha_v} \cdot \frac{1}{d_t} \equiv \frac{6d_s^2}{d_v^3}$$

$$= \frac{\alpha_{sv,t}}{d_t} \equiv \frac{6}{d_{sv}} \qquad (1.19)$$

The value 6 is only applicable to spherical particles and d_{sv} is the surface volume diameter and d_t the mechanical sieve diameter.

A relationship between surface volume mean diameter d_{sv} and the weight harmonic mean diameter $1/(d_H)$ exists which is important when the surface area of particles is required.

The weight harmonic mean can be written:

$$\frac{1}{d_H} = \frac{1}{G} \frac{\sum g_i}{d_1} \qquad (1.20)$$

where G is the total weight of sample comprising various sized particles, g_i is the weight of particles of size d_1; $\sum g_i$ is the summation of the weights of all particles.

For particles of size d_1

$$\frac{\sum g_i}{d_1} = \frac{\alpha_v \rho \sum d_1^3 \, dN_1}{d_1}$$

and

$$G = \alpha_v \rho \sum d_1^3 \, dN_1$$

$$\frac{1}{G} \frac{\sum g_i}{d_1} = \frac{\sum d_1^2 \, dN_1}{\sum d_1^3 \, dN_1} = \frac{1}{d_{sv}}$$

But

$$\frac{\sum g_i}{G} = \text{fraction by weight of the sample of particles of size } d_1$$
$$\text{(from sieve analysis)}$$

Thus

$$\frac{1}{d_{sv}} = \frac{1}{d_H} = \sum \frac{Z_i}{d_i} \qquad (1.21)$$

where $Z_i = g_i/G$ (*see Table 1.12*).

1.5.2.1 Numerical example of calculation of specific surface

This example calculates the specific surface from a measured distribution of a weight distribution obtained by sieve analysis.

In calculating specific surface it is convenient to use size classes in a geometric progression. The arithmetic mean of the class limits is used as the particle diameter in each class.

Table 1.12. Surface area calculation from a sieve analysis

Sieve size range/mm	Mean diameter of class/mm	Wt in class/g	Fractional wt in class	Proportional surface area in class/m^2 m^{-3}
d_t	d_1	g_i	$g_i/G = Z_i$	Z_i/d_1
2.36—1.70	2.00	3.2	0.08	39
1.70—1.18	1.44	6.8	0.17	118
1.18—0.85	1.02	10.0	0.25	245
0.85—0.60	0.725	10.0	0.25	344
0.60—0.43	0.512	6.8	0.17	332
0.43—0.30	0.362	2.0	0.05	138
0.30—0.21	0.256	0.8	0.02	78
0.21—0.15	0.181	0.4	0.01	55
			Totals	
		40.0	1.00	1349
		$\sum g = G$		$S_v = \sum Z_i/d_1$

From the data in *Table 1.12*

$$S_v = \alpha_{sv,t} \sum (Z_i/d_1)$$

If shape factor $\alpha_{sv,t}$ is 8.0 and the density of particles 2.3×10^3 kg m^{-3}, then

$$\frac{S_v}{\rho} = S_w = 8.0 \times \frac{1349}{2.3 \times 10^3}$$

$$= 4.69 \text{ m}^2 \text{ kg}^{-1}$$

Hence the total proportional surface area in each class size (sieve size range) is

$$S_v = \alpha_{sv,t} \sum (g_i/d_1)/G$$

where $\alpha_{sv,t}$ is the surface volume shape factor determined by sieve analysis.

1.5.3   Dynamic diameter

The shape of irregular particles can be expressed in geometrical or dynamical terms. The dynamical size is usually found by measurement of the rate of fall u and classification of a powder into fractions by sedimentation. The mean size of each fraction is expressed by a Stokes diameter:

$$d_{st} = (18\mu u/(\rho_s - \rho_p)g)^{1/2} \tag{1.22}$$

A dynamical expression of shape is obtained by comparison of the Stokes diameter with the diameter d_v of a sphere which has the same volume as the mean of the particles[124]. The dynamic shape factor is then defined at:

$$\alpha = d_v^2/d_{st}^2 \tag{1.23}$$

This is equal to 1 for spheres and increases up to 5 or more for elongated or flat particles with a concentration of less than 2 per cent v/v. The equation ceases to be valid when particles are less than 2 μm owing to Brownian motion. The dynamical behaviour of a spheroid is related to its geometry by viscous flow hydrodynamics; thus it is possible to correlate the dynamics and geometry of irregularly shaped particles by finding a spheroid which will replace the mean sized irregular particle.

The evaluation of the specific surface of a powder from a size analysis determined by Stokes diameter distribution by weight is similar to that of the surface area from a sieve analysis. The proportional surface area in each size class of the Stokes diameter is calculated and the cumulative area obtained from the distribution and the geometrical surface volume–Stokes shape factor.

1.5.4   Attenuation diameter

When a parallel beam of light passes through a settling suspension the decrease in intensity is related to the cross-sectional area of the particles in its path by the Lambert-Beer law:

$$I = I_0 \exp\left[-S_p cL\right] \tag{1.24}$$

where I_0 and I are the intensities of the light in the absence of obscuring particles and that transmitted by the presence of particles in suspension respectively. S_p is the projected area of particles, c is the concentration and L the length of suspension in the direction of the light beam. Replacement of light intensities by optical density, D, gives:

$$\log \frac{I}{I_0} = \frac{S_p cL}{2.303} = D \tag{1.25}$$

On the assumption that the total surface area of individual particles S_w is four times their random projected area S_p, equation (1.25) can be rearranged to give the weight specific surface area, S_w, from a single measure-

Table 1.13. Shape factors from optical density and sieving*

Sieve size/μm	Mean size/μm	Optical density	Concentration $\overline{\mathrm{g\ cm^{-3}} \times 10^{-2}}$	Surface area/cm² g⁻¹	Surface volume shape factor
d_t	$\bar{d}_t$	D_{max}	c	S_w†	$\alpha_{sv.t}$
75—89	82	0.44	0.304	266	5.59
89—105	97	0.41	0.390	192	4.85
105—125	115	0.39	0.400	180	5.34
125—150	137	0.466	0.476	178	6.25
150—180	165	0.37	0.478	142	6.00
420—500	460	0.40	0.652	113	8.00
650—210	650	0.42	0.922	84	14.10

* The density ρ_s of sand is 2.56 g cm⁻³ and the length of the light beam L is 5.0 cm.
† From equation 1.26.

ment of maximum optical density D_{max} when the suspension is fully dispersed.

$$S_w = \frac{4 \times 2.3}{L} \left(\frac{D_{max}}{c} \right) \tag{1.26}$$

From samples graded by sieving, having an arithmetic average sieve diameter $\bar{d}_t$, the surface volume shape factor may be determined from:

$$\alpha_{sv,t} = S_w \rho_s \bar{d}_t \tag{1.27}$$

Table 1.13 gives the results for shape factors of sand determined by maximum optical density and sieve analysis from Allen[86].

References

1. British Standards Institution, British Standard House, 2 Park St., London
2. Polaron, 4 Shakespeare Road, Finchley, London
3. Carl Zeiss D-7082 Oberkochen/Wiertt
4. PATTERSON, H. S., and CAWOOD, W., *Trans. Faraday Soc.* **32**, 1084 (1936)
5. MAY, K. R., *J. scient. Instrum.* **22**, 187 (1965)
6. HAMILTON, R. J., HOLDSWORTH, J. F. and WALTON, A., *Br. J. appl. Phys.* Suppl. 3, s101 (1954)
7. Graticules Ltd., Botany Trading Estate, Tonbridge, Kent, England
8. FAIRS, G. L., *J. R. microsc. Soc.*, **71**, 209 (1951)
9. Zeiss Information Sheet No. 60 (1966) and No. 70 (1968)
10. Vickers Instruments Limited, Haxby Road, York, Yorkshire, England
11. Wild–Heerbrugg A.G. CH-9435 Heerbrugg, Switzerland
12. ENDTER F. and GEBAUER H., *Optik, Stuttg.* **13**, 87 (1956)
13. CHATFIELD, E. J., *J. scient. Instrum.* **44**, 615 (1967)
14. Mullards Equipment Ltd., Manor Rd, Crawley, Sussex, England
15. WALTON W. H., *Br. J. appl. Phys.* Suppl. 3, s121 (1954)
16. Casella Automatic Particle Counter, Cooke Troughton and Simm Ltd
17. Imanco (Image Analysing Computer Ltd) Metals Research Ltd., Melbourne, Royston Cambridge, England

18. Metal Research Ltd, Melbourne, Royston, Cambridge, England
19. Millipore (UK) Ltd, Millipore House, Abbey Road, London NW10 75P
20. Bausch and Lomb GmbH, Munchner Strasse 72, D-8043, Unterfohring
21. HEYWOOD, H., *Trans. Instn. Min. Metall.* **55**, 373 (1945/46) *see* ref. 84
22. British Standard 410, *see* ref. 1.
23. Rotap, Garder Laboratory, Bethesda, Maryland, USA
24. Pascall Ltd., Gatwick Road, Crawley, Sussex, England
25. Endecotts Ltd., Lombard Road, London SW19, England
26. Retsch KG, Neuer Market 25, D-5657 Haan (Rheinland)
27. A. Fritsch and Co, Hamptstrasse 542, D-6580, Idar-Oberstein 1
28. Rhewun Maschinenfabrik GmbH, Postfach 120440, D-5630, Remscheid-Luttringhausen
29. MULLIN, J. W., *Chem. and Ind.* 1435 (1971)
30. Buckbee Mears 245 E 61st St, St Paul 1, Minnesota, USA
31. Veco, Zeefplatenfabrik n.v. Eerbeek, Holland
32. Alpine AG, Gogginger Landstrasse 66, Postfach 629, D-89, Augsburg-2
33. SUHM, H. D., *Powder Technol.* **2**(6) 356 (1969)
34. ANDREASEN, A. H. M., *Kolloidchem. Beih.* **27**(6), 5384 (1928)
35. DIN 6611 (April 1971) 'Sedimenationsanalyse in Schwerfeld', Grunlagen
36. V.D.1, Richtlinie Nr. 2031, Ausgabe 1962
37. JAPRETT, B. A. and HEYWOOD, H., *Br. J. appl. Phys.* **5** Suppl. 3, s21 (1954)
38. Royal Society of Chemistry, Analytical Division, Burlington House, London W1V 0BN
39. Shimadzu Sasakusho Ltd, Kanda, Hitoshirocho, Chiyodra, Tokyo, Japan
40. Sartorius-Werke, A.G. Gottingen, Holland
41. Gallenkamp Ltd, Portrack Lane, Stockton on Tees, Co. Durham, UK
42. Ventron Corp. Cahn Instruments, 16207S, Carmenita Road, Cerritos, California
43. Mettler Instruments AG, CH-8606 Grerfensee-Zurich, Switzerland
44. PRETORIUS, S. T. and MANDERSLOOT, W. G. B., *Powder Technol.* **1**, 23 (1967)
45. LECHONSKI, K. and ALEX, W., *'Proceedings of Second International Conference on Particle Size Analysis'*, Society for Analytical Chemistry; London (1970)
46. ODEN, S., *Kolloidzeitschrift*, **18**, 33 (1916)
47. ODEN, S., *Kolloidzeitschrift*, **26**, 100 (1919)
48. BURGERS, J. M., *Proc. K. ned. Akad. Wet.* **44**, 1045 (1941)
49. KAMACK, H. H., *Analyt. Chem.* **23**, 844 (1951)
50. ALLEN, T., *'Particle Size Measurement'*, 2nd ed., Chapman and Hall, London (1975)
51. Simon Engineering Ltd., Cheadle Heath, Stockport, Cheshire SK3 0RT
52. Sharples Centrifuges Ltd., Camberley, Surrey, England
53. Phywe AG, Gottingen BRD, Holland
54. Escher Wyss AG, Zurich, Switzerland
55. HELDRETH, J. D. and PATTERSON, D., *J. Soc. Dyers Colour.* **80**, 474 (1964)
56. BRADLEY, D., *Chem. Process Engng* **43**, 591 (1962)
57. BROWNING, H. A. and GAST, T., *Chemie-Ingr-Tech.* **43**, 523 (1971)
58. ATHERTONE, E., COOPER, A. C. and FOX, M. R., *J. Soc. Dyers Colour.* **26**, 62 (1964)
59. Joyce Loebl and Co. Ltd., 2 Noble Corner, Great West Road, Heston, Middlesex, UK
60. BERESFORD, J., *J. Oil Colour Chem. Ass.* **50**, 594 (1967)
61. WHITBY, K. T., *Am. Soc. Test. Mater. Sp. Pub.* 234, 117 (1958)
62. WHITBY, K. T., *Heat Pip. Air Condit.* **27**, 231 (1955)
63. Mine Safety Appliances Co., Pittsburgh, Pennsylvania 15208, USA; Queenslie Industrial Estate, Glasgow E3, Scotland
64. ALLEN, T., *Powder Technol.* **2**, 134, 141 (1968)
65. Ladal (Scientific Equipment) Ltd., Warley Edge Farm, Warley, Halifax, Yorkshire
66. Micromeritics Instrument Corporation, 800 Goschen Spring Road, Norcross, Georgia 30071, USA
67. ORR, C., HENDRIX, W. and SMITHWICK, J., *'Proceedings of Second International Conference on Particle Size Analysis'*, Society for Analytical Chemistry, London (1970)
68. OLAF, J. and ROBOCK, R., *Staub, Reinhalt. Luft.* **21**, 495 (1901)
69. BERG, R. H., *Am. Soc. Test. Mater. Sp. Pub.* 234 and 245 (1965)
70. KUBITSCHEK, H. E., *Nature* **182**, 234 (1958)
71. KUBITSCHEK, H. E., *Research, Lond.* **13**, 128 (1960)
72. ALLEN, T. and MARSHALL, K., *'The Electrical Sensing Zone Method'*, University of Bradford Press, Bradford (1972)
73. PRINCEN, L. H. and KWOLEK, W. F., *Rev. scient. Instrum.* **36**, 646 (1965)

74. PISANI, J. F. and THOMSON, G. H., *J. Phys (Fr) E* **4**, 359 (1974)
75. WALES, M. and WILSON, J. N., *Rev. scient. Instrum.* **32**, 1132 (1961)
76. EDMUNDSON, I. C., *Proc. Soc. anal. Chem.* **5**, 240 (1968)
77. SPIELMAN, L. and GOREN, S. I., *J. Colloid Interface Sci.* **26**, 175 (1968)
78. KAPTEYN, J. C., *'Skew Frequency Curves in Biology and Statistics'*, Groningen (1903); *see* ref. 84
79. ROSIN, P. and RAMMLER, E., *J. Inst. Fuel* **29**, 109 (1933/34)
80. ALYAVDIN, V. V., *Proceedings of Topics on Breakage in the Cement Industry*, Gyproce-ment, USSR (*see* ref 84)
81. WEIBULL, W., *J. appl. Mech.* **18**, 293 (1951)
82. HARRIS, C. C., *Chem. Technol.* **1**, 446 (1971)
83. FISHER, R. A. and TIPPETT, L. H. C., *Proc. Camb. phil. Soc. math. phys. Sci.* **24**(2), 180 (1928)
84. HERDAN, G., *'Small Particle Statistics'*, 2nd Ed., Butterworths Scientific Publications, London (1960)
85. British Standard 4359, Part 3 (1970)
86. ALLEN, T., *'Proceedings of Second International Conference on Particle Size Analysis'*, Society for Analytical Chemistry, London (1970)
87. ROSE, H. E., *'Powders in Industry'*, p. 130, Society for Chemical Industry, London (1961)
88. BEDDOW, J. K., VETTER, A. F. and SISSAN, K., *Powder Metall. Int.* **8**(2), 69 (1976)
89. DAVIES, R., *Powder Technol.* **12**, 111 (1975)
90. HEYWOOD, H., *'Symposium on Particle Size Analysis'*, Suppl. 25, 14, Institution of Chemical Engineers, London (1947)
91. HEYWOOD, H., *J. imp. Coll. chem. Engng Soc.* **8**, 25 (1954)
92. HEYWOOD, H., *J. Pharm. Pharmac.*, Suppl. 15, 56т (1963)
93. CHURCH, T., *Powder Technol.* **2**(1) 27 (1968)
94. DWYER, J. L., MANALON, D. A. and MERBON, R. R. A., *'Proceedings of Second International Conference on Particle Size Analysis'*, Society for Analytical Chemistry, London (1970)
95. MANDELBROT, B., *Science, N.Y.* **156**, 636 (1957)
96. SNEED, E. D. and FOLK, R. L., *J. Geol.* **64**, 114 (1958)
97. WADELL, H., *J. Geol.* **43**, 250 (1935)
98. WADELL, H., *J. Geol.* **41**, 310 (1933)
99. WADELL, H., *J. Geol.* **40**, 443 (1932)
100. TREASURE, G., *'Storage and Recovery of Particulate Solids'*, Ed. Richards, sect 5.2, Institution of Chemical Engineers, London (1966)
101. MEDALIA, A. I., *Powder Techol.* **4**(3), 117 (1971)
102. ASCHENBRENNER, B. C., *J. sedim. Petrol.* **26**, 15 (1956)
103. HAUSNER, H. H., *'Proceedings of Conference on Particle Size Analysis'*, Society for Analytical Chemistry, London (1966)
104. NYSTROM, C. and STANLEY-WOOD, N. G., *Acta Pharm. Suecia* **13**, 277 (1976)
105. NYSTROM, C. and STANLEY-WOOD, N. G., *Acta Pharm. Suecia* **14**, 181 (1977)
106. SCHWARTZ, H. P. and SHANE, K. C., *Sedimentology* **13**, 273 (1969)
107. NAYLOR, A. G. and WRIGHT, C. D., *'Proceedings of Third International Conference on Particle Size Analysis'*, Ed. Groves, Heyden, London (1977)
108. MELOY, T. P., *Powder Technol.* **17**, 27 (1977)
109. BEDDOW, J. K., VETTER, A. F. and SISSON, K., *Powder Metall. Int.* **8**, 69, 107 (1978)
110. ERLICH, B. and WEINBERG, B., *J. sediment. Petrol.* **40**, 205 (1970)
111. BEDDOW, J. K., PHILIP, G. C. and VETTER, A. F., *Powder Technol.* **18**, 19 (1977)
112. KAYE, B. H., *Powder Technol.* **21**, 1 (1978)
113. MANDELBROT, B. P., *'Fractals, Form, Chance and Dimensions'*, Freeman, San Francisco (1977)
114. FLOOK, A. G., *Powder Technol.* **21**, 295 (1978)
115. STEIN, F. and CORN, M., *Powder Technol.* **13**(1), 133 (1975)
116. KOTRAPPA, P., *Aerosol Sci.* **2**, 353 (1971)
117. HAMILTON, R. J., *Br. J. appl. Phys.* **3**, 590 (1954)
118. CARTWRIGHT, J., *Ann. occup. Hyg.* **5**, 163 (1962)
119. TIMBRELL, V., *Br. J. appl. Phys.* **3**, 586 (1954)
120. PILPEL, N., *Paint Mf.*, July 23 (1969)
121. WALTON, C. A. and PILPEL, N., *British Pharmaceutical Conference*, Keele, paper 3 (1972); *J. Pharm. Pharmac.*, **24**, 10P–16P (1972)

122. SHOTTON, E. and OBIORAH, B. A., *British Pharmaceutical Conference*, London, paper 22 (1973); *J. Pharm. Pharmac.*, **25**, 37P–43P (1973)
123. ROSS, H. N. and HALBURT, H. M., *Ind. Eng. Chem. Fundam.* **9**, 658 (1970)
124. DAVIES, C. N., *Nature*, **201**, 905 (1964)

CHAPTER 2

Particle characterization by size, shape and surface for contacted particles

N. G. Stanley-Wood
Schools of Studies in Chemical Engineering and Powder Technology, University of Bradford, Bradford

2.1 Porosity, voidage and particle porosity

For the characterization of a collection of particles in the form of a compact it is not sufficient solely to know the particle size, distribution and shape of individual particles. Knowledge of the structural form of the initial packing formation of spherical or irregular shapes is also advantageous before detailed analysis of the topography of uncompressed and compressed particles by nitrogen adsorption and mercury intrusion techniques is undertaken.

The space within a compact—tablet, pellet or briquette—is not completely occupied with solid material. Between the solid particles there are voids filled with gas or liquid and there may even be pores within the individual particles. The controlled packing of particles is of two types, namely, packing in which sticking of the particles occurs at the particle surface contacts, termed coherent packing; or packing in which sticking does not occur—noncoherent. Fusion can occur in a coherent compact by the application of heat to give a sintered product. The size of the space within a compact or a packed arrangement is a function of the particle size, different packing arrangements being achieved by particle orientation and the degree of compaction.

The distribution of void space within compacts or pores within particles plays an important role in the characterization of compacts, but it is the cumulative effect of the pores or voids which has a greater effect on the strength, compressibility and permeability of the product.

A number of properties of a powder, such as bulk density, heat conduction and adhesion, are also related to the closeness of packing.

2.1.1 Relationships between voidage, porosity and density

The density of a specimen is commonly defined as the mass divided by its volume. When a specimen has no internal pores then no difficulty arises in density determination, but when porosity is present in particles various subdivisions of volume can be seen (*Figure 2.1*), and therefore different densities determined.

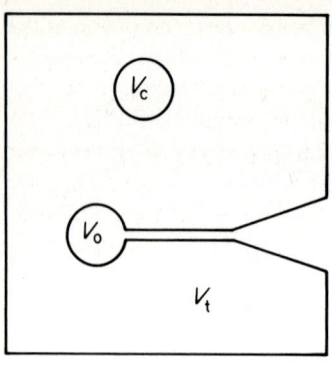

Figure 2.1.   To show the volume of closed pores (V_c), open pores (V_o) and their relation to V_t, the volume of solid and true volume

2.1.1.1   True density

The density of the particle or specimen measured by a method which will measure the internal pores of the particulate material gives

$$\rho_t = \frac{\text{Mass}}{V_t}$$

Two methods exist for the determination of true density. The first is an X-ray method[1,2] which is the measurement of the volume of the unit cell of the crystal. The density is then derived from

$$\rho_t = \frac{nM_w}{NV}$$

where

n = number of molecules in unit cell

M_w = molecular weight

N = Avogadro's number

V = unit cell volume

Secondly, single crystal densities can be measured by utilizing Archimedes principle[3,4]. During this determination, care must be taken to minimize the errors arising from surface tension effects, adsorption of liquid used (water, xylene) and bubble formation.

2.1.1.2   Particle density

The particle density is determined by measurement of the volume of a fluid which does not enter the closed internal pores of the material:

$$\rho_p = \frac{\text{Mass}}{V_p}$$

where V_p is the particle volume and equivalent to $V_t + V_c$ (*Figure 2.1*). It is essential to remove air from the specimen before admittance of the probe fluid. If the fluid is liquid the specimen can be boiled with the probe liquid as specified in BS 1902 (1952). The use of gases as a probe fluid make volume measurement, at times, more difficult owing to the compressibility and low density of gases. Particle densities can be obtained by the evacuation of air or other gases from the specimen pores in an appropriate apparatus. Helium can then be introduced and allowed to penetrate small pores and the displacement volume measured. Helium is the most common gas used because of its small molecular size and assumed lack of adsorption on solid surfaces at room temperature.

2.1.1.3   Bulk density (apparent density)

The bulk density is defined as the mass of particles per unit pore volume when formed under stated, freely poured, conditions:

$$\rho_b = \frac{\text{Mass}}{V_b}$$

where V_b is the bulk volume which is equivalent to $V_t + V_c + V_o$. This can be achieved by direct volume measurement, open pore filling with liquids or low melting point solids, or a mercury displacement method[5,6]. At one atmosphere pressure mercury will penetrate circular capillaries of approximately 15 μm diameter.

2.1.1.4   Tap density

The tap density of a powder is obtained when a stated amount of powder in a container of stated dimensions is vibrated or tapped under definitive conditions.

In the space occupied by the packing of particles, the column is not completely filled with solid matter. The degree to which the space is filled with solid matter depends upon the orientation and arrangement of particles and the existence or non-existence of the shape and size of pores within particles. Much confusion exists with the terms voidage and porosity, and although the interparticular space can be regarded as a voidage, with the intraparticular space being regarded as a porosity, particle porosity (intraparticle space) when present has an effect upon the overall space (voidage) in a compact or packing of particles. Voidage can be defined as either:

1. The ratio of open space volume to the combined open space and solid material volume, or
2. The ratio of the interparticle space volume to the mechanical volume of a coherent mass.

Likewise, powder or compact porosity can be defined as voidage or as the ratio of the volume of voids plus the volume of open pores to the total volume occupied by the powder or compact mass. Thus

$$\text{Voidage} = e = \frac{\text{volume of voids}}{\text{volume of bed or compact}} = \frac{V_v}{V_v + V_s}$$

where V_s and V_v are void and solid volume respectively, and

$$\text{Powder porosity} = \frac{\text{total volume} - \text{solid volume}}{\text{total volume}}$$

$$= 1 - \frac{\text{solid volume}}{\text{total volume}}$$

$$= 1 - \frac{V_t}{V_b}$$

where $V_b = \text{Mass}/\rho_b$ and $V_t = \text{Mass}/\rho_t$, ρ_t and ρ_b being the true density of solid and bulk density of bed respectively.

Then the porosity of a compact or a powder can be equivalent to

$$e = 1 - \frac{\text{bulk density}}{\text{true density}} = 1 - \frac{\rho_b}{\rho_t}$$

The interparticle volume for a non-porous solid is the difference between the reciprocals of bulk density and the true density, the particle density for a non-porous solid being equivalent to the true density. For a porous material there is, in addition to the interparticle volume, an internal pore volume. This internal pore volume is the difference between the reciprocals of true density and the particle density because the particle density does not include internal pores.

Particle porosity can be defined as the ratio of the space volume within an individual or collection of individual uncompressed particles to the over-all displacement volume of the individual or collection of individual uncompressed particles.

For compacts of porous materials the sum of voids and internal volume is the difference between the reciprocals of bulk density and true density.

2.1.2   Porosity as a function of particle arrangement

When the non-coherent arrangement of monosized non-porous particles is considered, then the degree of porosity is independent of particle size and porosity is solely a function of arrangement.

The volume of the unit cell to be considered and the pore volume per unit cell is however dependent on particle size. This is obvious because a property like a ratio is dimensionless while a property like an area or volume which includes dimensions will vary with size.

The four commonest regular packings for spheres are as follows (*see Table 2.1*).

Table 2.1. Characteristics of some regular packing

Type of packing	Points of contact or Co-ordination number	Volume of unit cell*	Porosity	Pore volume per cm³ of spheres cm³/cm³	Pore volume per unit cell*	Radius of sphere inscribed in the cavities†	Radius of circle inscribed in the throats connecting cavities†
	n		e				
Hexagonal close packed (rhombohedral)	12	$0.71d^3$	0.260	0.350	$0.18d^3$	$0.225R$-octahedral $0.414R$-tetrahedral	$0.155R$
Body centred tetragonal	10	$0.75d^3$	0.302	0.432	$0.23d^3$	$0.291R$	$0.265R$ $0.155R$
Primitive hexagonal (orthorhombic)	8	$0.87d^3$	0.395	0.654	$0.34d^3$	$0.527R$	$0.414R$ $0.155R$
Primitive cubic	6	d^3	0.476	0.910	$0.48d^3$	$0.732R$	$0.414R$
Tetrahedral	4	—	0.660	1.94	—	$1.00R$	$0.732R$

* d = diameter of particle
† R = radius of particle

2.1.2.1   Regular plane packing

2.1.2.1.1   Cubic packing

Over a layer of spheres another layer of the same sized spheres is placed in such a way that the centre of the corresponding spheres in the two layers are mounted vertically above each other. Thus each sphere has six points of contact.

2.1.2.1.2   Orthorhombic

If the second layer is spread so that the spheres of one layer fits into the 'grooves' of the other, then the sphere centres form a lattice in which there are eight points of contact.

2.1.2.2   Hexagonal plane packing

2.1.2.2.1   Tetragonal—spheroidal

Two layers of hexagonal plane packing with the centres of the spheres vertically above one another. This has ten contact points.

2.1.2.2.2   Rhombohedral

The second plane of hexagonal packed spheres fits into the grooves of the lower plane. Thus this has 12 points of contact.

Each type of packing is associated with a characteristic geometry of the interstices and for each type of packing there is a characteristic porosity which is given by the ratio of the volume of the interstices to the volume of the complete unit cell. The porosity of 25.95 per cent is a minimum only for non-coherent packing of monosized non-porous spheres.

When a powder consists of particles having a distribution of sizes the degree of porosity may be reduced to 15 per cent or less. Most contained powders have a porosity between 30 and 50 per cent, the values for rhombohedral and cubic packing.

2.1.3   Voidage and pore size as a function of particle size

In nature and with industrial powders which contain several particle sizes, the simple concept of monosized spherical voidage can be altered to imagine that the interstices or spaces within a packing of large particles of a given size may be occupied by smaller particles, thus diminishing overall voidage (*see also* Section 11.3.1).

2.1.3.1   Selected spherical particles

Furnas[7], Fraser[8], and White and Walton[9] have examined the volume reduction which occurs when selected quantities of spherical particles of different sizes and amounts are assembled together in a known volume.

Insertion of secondary, tertiary, quaternary and quinary sized spheres into the interstices of an initial regular assemblage of primary sized spheres decreases the voidage from 26 to 15 per cent. No practicable method of distributing spherical particles sized smaller than $0.116d$ into a compact has however been suggested (*Table 2.2*).

Table 2.2.   Hypothetical packing of spherical particles of different selected sizes

Size of sphere	Primary	Secondary	Tertiary	Quaternary	Quinary
Diameter*	d	$0.414d$	$0.225d$	$0.177d$	$0.116d$
Relative no.	1	1	2	8	8
Voidage/per cent	26.0	20.7	19.0	15.8	14.9
Cumulative no. of particles reqd.	12	24	48	144	242
Cumulative per cent oversized	4.95	9.9	19.8	59.5	100
Cumulative per cent undersized	95.1	98.1	80.2	40.5	0

* d is the diameter of the primary spherical particle in an assembly.

2.1.3.2   Experimental packing of spherical particles

McGeary[10] considered a practical method of achieving minimum voidage with the application of either force or vibration. This was achieved by placing large particles into a container and vibrating them to a minimum voidage. He subsequently added a second component of smaller size and vibrated the system to another voidage. In this way a denser pack was obtained than by the vibration of all sizes together (*Table 2.3*). The experimental voidage was a remarkably dense compact which, however, could be readily removed from the container. The packing fraction, P_f, which is equivalent to $(1 - e)$ was calculated by McGeary to be:

$$P_f = \frac{e_n V_n/(1 - e_n)}{e_n V_n/(1 - e_n) + V_1 + V_2 \ldots V_n} \tag{2.1}$$

where e is the volume of voids associated with V, the volume of particulate material. Subscript n refers to the finest component. Ayer and Soppet[11] extended this work and found that for spherical particles the packing efficiency, P_e, in a container could be expressed as:

$$P_e = 0.635 - 0.216 \exp(-0.313D/d_1) \tag{2.2}$$

where D and d_1 are the diameters of container and particle respectively. The figure 0.635 is the limiting value of the packing fraction for a single component system.

The void fraction (V_f) which remains is therefore unity minus the packing fraction

$$V_f = 1 - P_e$$

$$= 1 - (0.635 - 0.216 \exp(-0.313D/d_1)) \tag{2.3}$$

$$= 0.365 + 0.216 \exp(-0.313D/d_1) \tag{2.4}$$

Ayer and Soppet[11] found from 60 data points that the packing of a second component within the previous voids can be expressed as:

$$P_e = 0.635 - 0.737 \exp(-0.201d_1/d_2) \tag{2.5}$$

where d_1/d_2 is the ratio of the largest to smallest particle size in the system.

The total packing fraction then becomes, for a three component system, from equations 2.2, 2.3 and 2.5:

$$V_f = 0.635 - \frac{0.079}{\exp(0.313D/d_1)} + 0.365 + \frac{0.216}{\exp(0.313D/d_1)}$$

$$+ 0.635 - \frac{0.737}{\exp(0.201d_1/d_2)} \tag{2.6}$$

which on simplification gives

$$V_f = 0.866 - \frac{0.079}{\exp(0.313D/d_1)} - \frac{0.269}{\exp(0.201d_1/d_2)}$$

$$- \frac{0.159}{\exp(0.313D/d_1) \times \exp(0.201d_1/d_2)} \tag{2.7}$$

For irregularly shaped particles the packing efficiency was shown to be a function of particle shape (Section 1.4) and absolute size (Section 1.2).

Table 2.3. Packing fraction of vibrated particles after McGeary and Ayer and Soppet*

Ratio of diameter sizes in particle assembly		d_1 316	d_2 38	d_3 7	d_4 1
Voidage/per cent, for single sized particle assembly		39.5	37.5	37.5	42.5
Composition of bed of various sized particles to give experimental packing fraction	d_1	1.00	0.726	0.647	0.607
	d_2		0.274	0.244	0.230
	d_3			0.109	0.102
	d_4				0.061
Packing fraction P_f McGeary (exptl.)		0.580	0.800	0.898	0.951
		0.605	0.859	0.942	0.975
Ayer and Soppet (exptl.)		0.635	0.866	0.951	—

* See references 10 and 11

For ball-milled grit the equation for packing in a container was deduced as:

$$P_e = 0.484 - 0.192 \exp (-0.098d_1/d_2) \tag{2.8}$$

Since the matrix formed by consolidated coherent particles can be regarded as being a combination of porous and void structures, information can be obtained from a study of the adsorption of gases on to the surface or into the porous space of compacts by use of the concepts of monolayer, multilayer and capillary condensation.

Ramsay and Avery[12] compressed extremely small particles (4 nm) of spherical silica and zirconia, prepared by vapour phase condensation, with a narrow size distribution and determined the effect of compaction both on the specific surface area generated and the production of voids formed by the interstices of these non-porous materials when compressed over a pressure range 3—100 ton/in² (46.2—1540 MN m⁻²). It was found that the adsorption isotherms measured (*Figure 2.2*) were in accord with the experiments of Shull[13] and Pierce[14]. The uncompacted materials produced a Type II, BDDT classification[15] isotherm (Sections 2.2 and 2.3) while on application of compaction pressure to these spherical non-porous particles the isotherm measured changed from Type II to a Type IV isotherm and event-

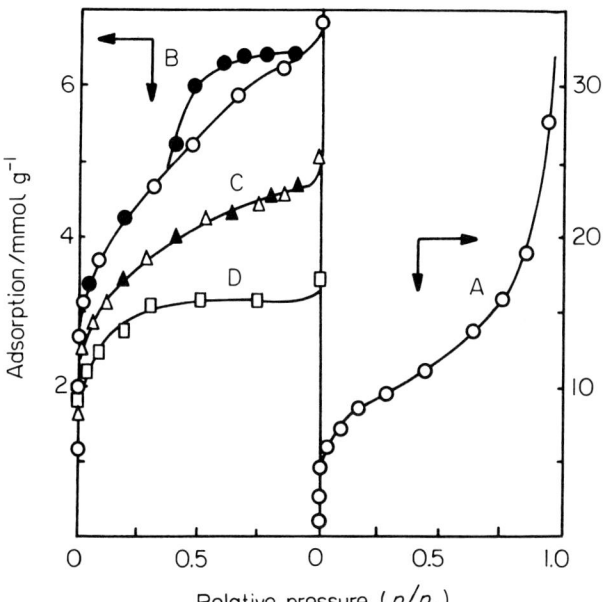

Figure 2.2. Adsorption isotherms of nitrogen at 77 K on silica powder and its compacts
Curve A, uncompressed. Curve B, compressed at 0.77 GN m⁻². Curve C, compressed at 1.16 GN m⁻². Curve D, compressed at 1.54 GN m⁻². Filled symbols denote desorption

ually at higher compaction pressure to a Type I BDDT classification isotherm. Type II isotherm is typical of nitrogen isotherms found with non-porous materials while Type IV occurs with mesoporous and Type I is obtained with microporous materials (Sections 2.2.4 and 2.2.5).

2.2   Nitrogen adsorption

Although the compaction of powders has been studied for many years[16–20] the emphasis of research has been mainly on either the mechanisms of compaction[20–25] or the bonding of solid particles rather than the changes within particles or the effects on particle topography that occur on compression of an assembly of particles.

Evaluation of surface area, porosity and voidage from the physical adsorption isotherms of gases adsorbed into and on to the matrix formed by an assembly of compressed or noncompressed particles has provided some insight into the structural form of compacted materials.

Brunauer, Deming, Deming and Teller[15] originally proposed that the adsorption isotherms measured on a wide variety of solids could be classified into five types of isotherm. *Figure 2.3* shows the BDDT isotherm classification. Although it is convenient to classify isotherms into these well-known types, there are cases where there is difficulty in assigning some isotherms into any specific class. In addition to the classification of various types of isotherm produced from the adsorption of nitrogen on to and into

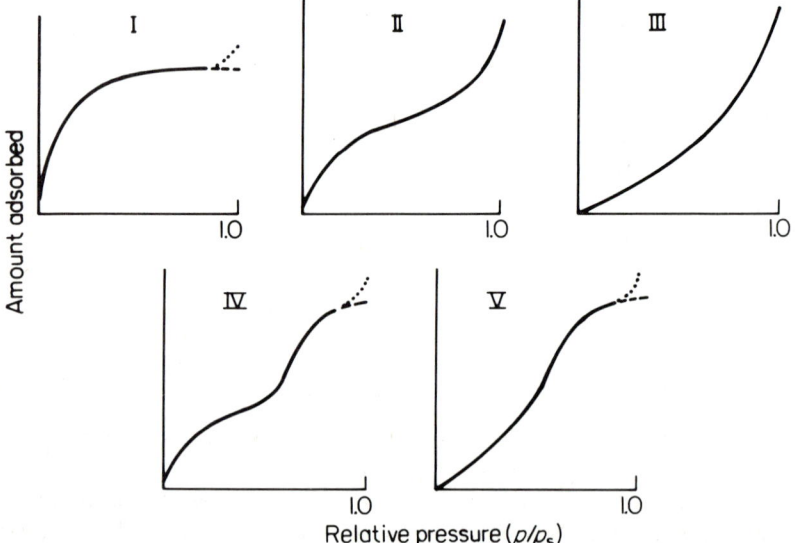

Figure 2.3.   The five types of adsorption isotherm in the classification of Brunauer, Deming, Deming and Teller (also called the Brunauer–Emmett–Teller, BET, classification)

various structural and topographical material, the nitrogen adsorption isotherm can be used to measure the surface area and porosity of a solid or compact. Numerous techniques have been used in isotherm analysis[26-29], the most common being the Brunauer, Emmett and Teller[30] (BET) equation for surface area evaluation, and the computational methods of Barrett, Joyner and Halenda (BJH)[26], Cranston and Inkley (CI)[29], for cylindrical pores or Innes[27], and Pierce[14], for parallel plates (Section 2.2.4.3).

In general it is possible, by using simplified models and equations, to express the experimental isotherm in terms of volume adsorbed and relative pressures and subsequently to calculate the monolayer coverage and surface area of a solid from Type I, II and IV isotherms. Owing to the concavity of the isotherm with the relative pressure axis, Types III and V pose difficulty in monlayer coverage evaluation and therefore surface area calculations and measurement. Types IV and I can also be used to calculate pore size distributions in the mesoporous and microporous size range.

Evaluation and measurement of very small pores (micropores) in solids can be achieved from the nitrogen adsorption isotherm by application of the Lippens and de Boer V_a–t model[31-34] or the Brunauer and Mikhail (MP) method[35]. The physical adsorption in micropores may proceed by a process of volume filling in contrast to the method of surface coverage which occurs in mesopores and on plane solid surfaces. The equation of Dubinin, Kaganer and Radushkevich (DKR)[36-38] can then be used to determine the uptake of nitrogen into micropores and to determine a monolayer adsorption coefficient which can be used to calculate specific surface areas.

The general classification of pore sizes originates from Dubinin[36,39-41] and is now incorporated into an IUPAC standard (*Table 2.4*). Isotherms of Types III and V, like surface area calculations, do not lend themselves to the calculation of pore size distributions.

Table 2.4. Classification of pore sizes

Type	*Pore radius*/nm
Micropores	0.5—1.6
Mesopores (transitional pores)	1.6—100
Macropores	> 100

2.2.1 Sample pre-treatment prior to characterization

Before the surface area of a solid can be characterized by adsorption techniques, the solid has to be pre-treated to clean the solid material of physically adsorbed gases adhered to the surface. The action of pre-treatment may, however, change the nature of the adsorbent. Urwin[42] showed that outgassing (degassing) of samples of titanium dioxide at various temperatures and times caused variation in the surface area measured. Increase in the degassing temperature over the range 23 °C—150 °C caused an in-

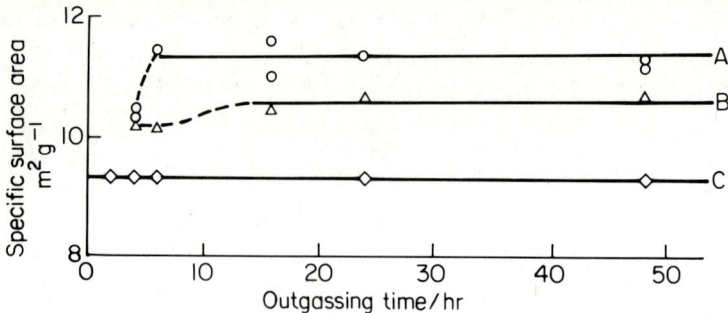

Figure 2.4. *Changes in the surface area of a titanium dioxide sample with outgassing conditions*

A, 150 °C—200 °C
B, 100 °C
C, 23 °C

crease in the surface area values which was independent of the length of time of degassing. Between the temperatures of 150 °C and 200 °C the temperature of pre-treatment became time-dependent (*Figure 2.4*).

Rouquerol et al[43] showed that for a mesoporous solid (Gasil I) heated to temperatures of 150, 400 and 650 °C, the amount of gas subsequently adsorbed increased, which indicated that outgassing extended the surface available to the adsorbates argon and nitrogen. At 900°C the amount of adsorbed gas was less, indicating that the solid had undergone some degree of sintering or thermal decomposition in the mesoporous samples. Neither the α_s plot of Sing, nor the BJH pore size analysis of the desorption branch of the isotherms, showed, however, any significant changes in the pore sizes of the mesoporous solid. With microporous material (Gasil II), the α_s plots showed microporosity and the available microporosity volume changed with outgassing temperature. The external surface area, determined from the upper part of the α_s plot, was constant up to temperatures of 600 °C. Ball and Norwood[44], who studied the surface areas and porosity changes of calcium sulphate dihydrate heated between 300 and 635 K over a period of time between 10 min and 220 h, found that at 352 K the surface area of calcium sulphate dihydrate reached a constant value after 100 hours. Heating at 420 K gave a maximum surface area after 240 hours. All *t*-plots of the isotherms showed a steady change in the statistical thickness of the adsorbed layer. Although there was an increase in surface area this was not accompanied by changes in porosity. The increase in surface area without porosity changes was attributed to the variation in the surface energy of the adsorbent, because the C_{BET} values calculated from the isotherms increased. Johansson[45] determined the surface area of compacted and uncompacted magnesium trisilicate and found, like Urwin, that although the available surface area increased with temperature, the increase was not dependent on time in the range 18—40 hours. The time necessary to degas powders compressed over the range 17.5—250 MN m^{-2} at a temperature of 296 K was, however, longer the higher the compaction pressure (17.5 MN m^{-2} after 7 h, 70 MN m^{-2} after 14.5 h, and 280 MN m^{-2} after 18 h).

2.2.2 Physical adsorption equations

The physical adsorption of gases on to the surface of a solid can be used to determine the surface area of that solid providing the area of the adsorbate is known and the volume of a unimolecular layer (monolayer) can be measured. Measurement of the monolayer can either be by volume or by weight. British Standard 4359: Part I provides a reference method for the determination of the surface area of a degassed solid by measuring the quantity of physically adsorbed nitrogen which completely covers the surface of a solid with a single layer of nitrogen molecules.

2.2.2.1 *Langmuir equation*

The first theoretical equation which described the relationship between the amount of gas adsorbed and the equilibrium pressure of the gas at constant temperature, was advanced by Langmuir[46,47]. Langmuir's original equation 2.9 may be used for both physical and chemical adsorption.

$$V_a = \frac{V_m bP}{1 + bP} \qquad (2.9)$$

From equation 2.9

$$\frac{1 + bP}{V_m} = \frac{bP}{V}$$

where V_a and V_m are the adsorbed and monolayer volumes respectively, P the experimental pressure and b an adsorption coefficient. Langmuir's equation in linear form becomes:

$$\frac{P}{V_a} = \frac{1}{V_m b} + \frac{P}{V_m}$$

or in terms of relative pressure P/P_s where P_s is the saturated vapour pressure of the adsorbate:

$$\frac{P}{P_s V_a} = \frac{1}{V_m b} + \frac{P}{P_s V_m} \qquad (2.10)$$

From equation 2.9 it can be seen that

1. At low pressures or small adsorption the bP term is small compared with 1, thus $V_a \propto P$, and Henry's law is valid (linear relationship).
2. At high pressure high adsorption occurs and 1 is small compared with the bP term and $V_a = V_m$. Therefore adsorption approaches saturation.

2.2.2.2 The BET equation

Brunauer, Emmett and Teller[30] showed that the adsorption of gases and vapours (i.e., gases below their critical temperature) formed multimolecular layers and that the adsorption isotherms could be classified into five types to cover the majority of isotherms observed to date. The BET theory can thus be regarded as a generalization of the original Langmuir theory.

The BET equation is

$$V_a = \frac{V_m C_{BET} P}{(P_s - P)[1 + (C_{BET} - 1)P/P_s]}$$ (2.11)

which in linear form becomes:

$$\frac{P}{P_s - P}\frac{1}{V_a} = \frac{1}{V_m C_{BET}} + \left(\frac{C_{BET} - 1}{V_m C_{BET}}\right)\frac{P}{P_s}$$ (2.12)

or

$$\frac{x}{(1 - x)V_a} = \frac{1}{V_m C_{BET}} + \frac{(C_{BET} - 1)x}{V_m C_{BET}}$$ (2.13)

where $x = P/P_s$ and C_{BET} is the BET coefficient.

The linear form of the BET equation assumes that C_{BET}, the adsorption coefficient, is constant. There are no isotherms, however, in which C_{BET} remains constant for all values of P from $P = 0$ to $P = P_s$. In the majority of isotherms C_{BET} is constant for P/P_s from 0.05 to 0.35 or for $\theta = 0.5$—1.5, where θ is the fraction of surface covered by adsorbed nitrogen molecules.

The adsorption coefficient, C_{BET}, is similar to 'b' in the Langmuir equation since the BET equation is based on similar assumptions. The adsorption coefficient can be written thus:

$$C_{BET} = \exp\left(\frac{E_1 - E_L}{RT}\right)$$ (2.14)

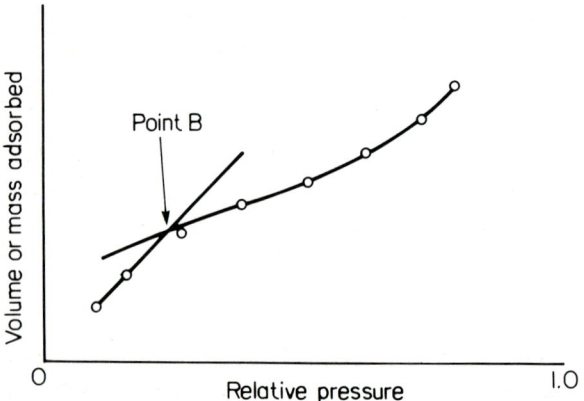

Figure 2.5. *Explanation of 'point B'*

where E_1 is the energy of adsorption of the vapour into a bare surface and E_L is the energy of condensation of vapour.

The aim of the BET equation is to find the so-called point B[48], or 'knee bend'[42] which arises from the completion of the monolayer formation of adsorbate molecules on to the adsorbent (*Figure 2.5*). This monolayer is normally situated between P/P_s 0.05 and 0.15, which is within the limits of the BET equation. At relative pressures above the limit of $P/P_s = 0.35$ the BET equation gives values which are greater than experimental data.

There are two forms of the BET equation:

1. The two parameter equation. This is the most popular equation applicable for adsorption on to a free solid surface.
2. The three-parameter equation. This is used when adsorption is restricted to parallel-plate pores in a solid and not on a free surface.

In either form the BET equation in the linear form can only be used if the C_{BET} coefficient remains constant in the region $\theta = 0.5$—1.5. For C_{BET} to remain constant the free energy of adsorption must also remain constant. As the number of adsorbed layers increases from 0.5 to 1.5, the heat of adsorption will decrease for each successive layer. The entropy of the adsorbed molecules must then increase to balance the effect of the decrease in the heats of adsorption[49]. Because of the non-specific nature of nitrogen adsorption the above balance of energies and the consistency of C_{BET} over the limited relative pressure range occurs for practically all adsorbents. If the monolayer capacity is expressed as milligrams of adsorbate adsorbed per gram of solid, x_m, then the surface area of a solid S in m^2 per gram can be expressed as

$$S = \frac{x_m N A_m}{M_w} \qquad (2.15)$$

where N is Avogadro's number (6.023×10^{23} mol^{-1}), A_m is the cross sectional area of an adsorbed molecule (in nanometre^2) and M_w the molecular weight of adsorbate. If the uptake of the adsorbate is expressed as millimoles of adsorbate per gram of solid n_m^s, equation 2.15 becomes

$$S = n_m^s N A_m \qquad (2.16)$$

Equation 2.16 is used in the new British Standard (BS 4359) and supersedes the original equation used in which the surface area was evaluated from the monolayer volume at STP of adsorbate per gram of solid (V_m)

$$S = \frac{V_m N A_m}{M} \qquad (2.17)$$

where M is the molar volume (22.4214 m^3 kmol^{-1}). The validity of the surface area calculated from nitrogen to that determined independently or geometrically can be observed by comparison of the calculated nitrogen adsorption surface area to known geometric areas of carefully prepared metal samples or glass filaments.

Alternatively electron microscopy can be used to measure the particle size of non-porous solids and calculation of the geometric surface area. Knowledge of the shape factor and density of each particle fraction examined and measured is required and a large number of particles must be counted to ensure a statistical mean diameter. British Standard 3406 recommends that 625 particles for a number count be measured while for a weight count this number can easily be exceeded. Comparison of the surface area determined by nitrogen adsorption with that calculated from the size of particles measured by electron microscopy shows, for carbon blacks, a variation between nitrogen surface area in the region of 25 per cent while with glass microspheres the variation is of the order of 17 per cent. Suprynowicz, Gorgal and Wojak[50] measured the surface area of glass capillaries by dynamic thermal desorption of nitrogen at 77 K and used the BET adsorption equation. They found that for capillary lengths of 0.82—21.9 m the ratio of measured area A_{measured} to geometric area $A_{\text{geom.}}$ was between 1.17 and 1.005, the ratio decreasing with increase in capillary length.

2.2.2.3   Huttig's equation

Huttig[51], who developed an identical equation to Baley[52], assumed that evaporation of molecules from the ith layer is completely unimpeded by the presence of molecules adsorbed in the $(i + 1)$th layer. Huttig's equation extends the range P/P_s from 0.35 to 0.7. Huttig's equation in terms of volume of adsorbate adsorbed per gram of adsorbent (V_a) is:

$$\frac{V_a}{V_m} = \frac{\dfrac{P}{P_s}\left(1 + \dfrac{P}{P_s}\right)}{1 + C_{\text{BET}}\left(\dfrac{P}{P_s}\right)} \tag{2.18}$$

which in the linear form is

$$\frac{P\left(1 + \dfrac{P}{P_s}\right)}{V_a} = \frac{P_s}{V_m C_{\text{BET}}} + \frac{P}{V_m} \tag{2.19}$$

The BET equation predicts too large an adsorption volume at high relative pressure while the Huttig equation predicts a too small adsorption volume. Lopez-Gonzales and Deitz proposed an equation which simply averages the two equations[53].

2.2.2.4   Harkin and Jura's relative method (HJr)

Harkin and Jura[54,55] found that the surface area of a solid could be found from a graph of log (P/P_s) against $1/V_a^2$, as shown in *Figure 2.6*. This is an

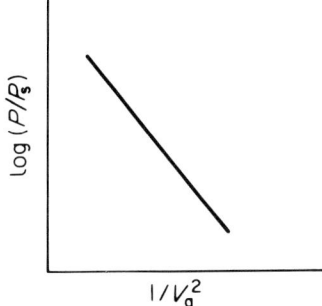

Figure 2.6. Harkin and Jura's method

empirical method based on the fact that condensed monolayers on water exhibit a linear pressure–volume–area relationship ($P = b - aA_m$). In this method the surface area is calculated without a value of V_m.

2.2.2.5 Statistical equations

Barrer, Mackenzie and Macleod[56,57] used the general equation of Dole[58] which was initially derived from the statistical method of Fowler and Guggenheim[59] for the sorption of vapours on solids. The equations recommended for nitrogen adsorption were, in the terminology of Barrer *et al.*, where P/P_s is written as x:

Equation A

$$\frac{1}{V_m C_{BET}} + \frac{xC_{BET}^x}{V_m} = \frac{xC_{BET}^x(1 + x)}{V_a} \tag{2.20}$$

Equation D

$$\frac{1}{V_m} C_{BET}^{2x} + \frac{1}{V_m} \frac{(C_{BET} - 1)}{C_{BET}} = \frac{2x\, e^{2x}}{V_a} \tag{2.21}$$

Using Dole's equations, Anderson[60] considered that the heat of adsorption in the second to ninth layers was different from the heat of liquefaction by a constant amount. Thus the BET equation is modified to:

$$\frac{x}{V_a(1 - kx)} = \frac{1}{V_m C_{BET}k} + \frac{(C_{BET} - 1)x}{V_m C_{BET}} \tag{2.22}$$

where the value of k was found to be between 0.6 and 0.7.

2.2.3 Type II isotherm, BDDT classification

Gregg and Sing[61] have stated that for non-porous solids (those solids without internal surface) there are two different types of isotherm, Types I and II. The most common equation used for surface area calculation from the Type II isotherm is the BET equation which is usually written in terms of volume at STP adsorbed (equation 2.13).

The BET equation can however be expressed in terms of the weight of adsorbate adsorbed on the surface of the solid, x_a, to give a monolayer capacity x_m in milligrams adsorbate/gram of solid or in terms of the number of moles adsorbed, n^s:

$$\frac{P}{n^s(P_s - P)} = \frac{1}{n^s_m C_{BET}} + \left(\frac{C_{BET} - 1}{n^s_m C_{BET}}\right)\frac{P}{P_s} \tag{2.23}$$

where n^s is the number of moles adsorbed, n^s_m is the number of moles in a monolayer and C_{BET} is the adsorption coefficient. When the appropriate functions:

$$\frac{P}{n^s(P_s - P)} \qquad \frac{P}{V_a(P_s - P)} \qquad \frac{P}{x_a(P_s - P)}$$

are plotted against P/P_s, straight lines should result with slope s, equivalent to

$$\frac{C_{BET} - 1}{n^s_m C_{BET}} \qquad \frac{C_{BET} - 1}{V_m C_{BET}} \qquad \frac{C_{BET} - 1}{x_m C_{BET}}$$

respectively, and intercept i, equivalent to

$$\frac{1}{n^s_m C_{BET}} \qquad \frac{1}{V_m C_{BET}} \qquad \frac{1}{x_m C_{BET}}$$

respectively. Solution of these equations gives C_{BET}, n^s_m, V_m and x_m

$$n^s_m \equiv V_m \equiv x_m \equiv 1/(s + i)$$

and

$$C_{BET} = (s/i) + 1$$

The BET equation has proved remarkably successful in the calculation of specific surface area from Type II isotherms but not however with Type III. The linearity of the BET function is usually between relative pressures of 0.05—0.35 as shown in the original work of Brunauer and Emmett[48] with the adsorption of N_2, CO_2, CO, Ar, and O_2 on charcoal. The linearity, however, is not always extended as high as 0.35 relative pressure. MacIver and Emmett[62] showed that for nitrogen on sodium chloride the linearity was between 0.01 and 0.1 relative pressure, while work by Isirikyan and Kiselev[63] on graphitized carbon black showed that the linear region was as low as 0.003—0.09 relative pressure, for N_2 at $-195\,°C$. The BET equation (equations 2.11 and 2.12) cannot, however, reproduce the nitrogen isotherm over a wide range of relative pressure. Graphical validation of this was obtained by Harris and Sing[64] who, for N_2 on non-porous silica, experimentally obtained curve I (*Figure 2.7*), but produced by calculation from the BET equation, curve II. The similarity between Type II and Type III isotherms can be shown by plotting the BET equation with various values of C_{BET}.

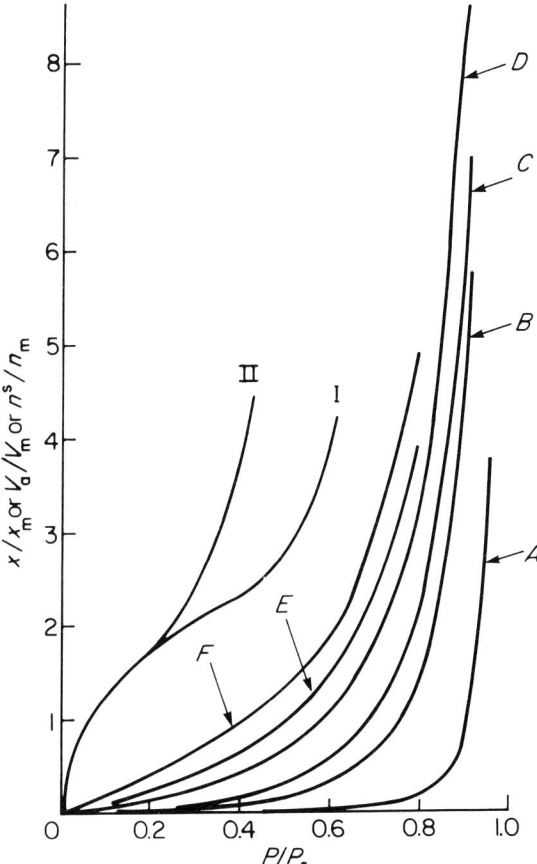

Figure 2.7. Isotherms calculated according to the
BET *equation for values of* C_{BET} *between 0 and 2.*
Values of C_{BET} *are:* A, 0.01; B, 0.1; C, 0.2; D, 0.5;
E, 1.0; F, 2.0

2.2.3.1 The C_{BET} parameter

This parameter determines the shape of the adsorption isotherm and also
the magnitude of the adsorption at a given relative pressure. The number of
molecules evaporating from a given site per second cannot be evaluated
from the properties of adsorbent and adsorbate, thus a simpler expression
for C_{BET} is normally used (equation 2.24).

$$C_{BET} = \exp\left[(E_1 - E_L)/RT\right] \tag{2.24}$$

where $E_1 - E_L$ is regarded as the net heat of adsorption, the heat of adsorption in the first layer being E_1 and that at liquefaction E_L. The Type II
isotherm exists when the C_{BET} value is greater than 2, while the Type III
isotherm results when C_{BET} is less than 2. The sharpness of the knee bend

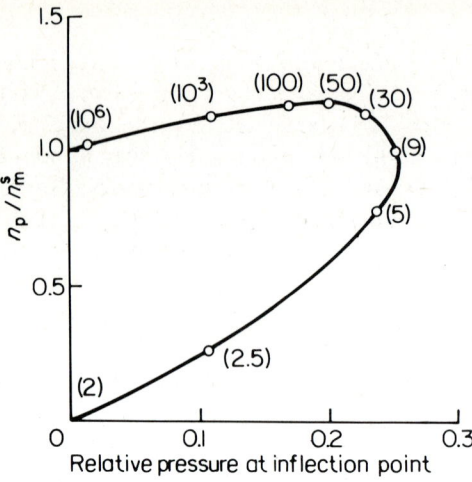

Figure 2.8.   The BET *equation. The numerals in parentheses on the curves are values of* C_{BET}

becomes more pronounced as the value of C_{BET} increases. For the BET equation to be valid and applicable to Type II isotherm, the experimental adsorption isotherm must have a point of inflection. This point of inflection can only be obtained when the C_{BET} value is ·greater than 2 and can be regarded approximately as the completion of the build up of molecules on the solid surface to form a monolayer. The point of inflection is not how- ever coincidental to the position where the volume adsorbed at the point of inflection divided by the monolayer capacity volume (calculated from the BET equation) is unity. The variation in the point of inflection occurs because of the variation in the values of C_{BET}. *Figure 2.8* shows on the y axis the ratio of n_p/n_m^s (the ratio of the moles adsorbed at the point of inflection to the calculated moles in the monolayer from the BET equation) against, on the x axis, the relative pressure at which inflection occurs. Coincidence between n_p and n_m^s is thus seen to occur at C_{BET} values of ~ 9 and over 10^6, the point of inflection at values of $C_{BET} \sim 9$ occurring at a relative pressure of about 0.25. The degree of variation or error between n_p and n_m^s decreases as C_{BET} values increase. Thus, although the point of inflection is similar to the monolayer capacity calculated by the BET equa- tion at C_{BET} values of 9 and 10^6, between C_{BET} values 9 and 10^6 an error of 20 per cent may occur. For values of C_{BET} between 9 and 2 the difference between n_p and n_m^s becomes greater and greater until, when C_{BET} is 2, the monolayer completion (monolayer capacity) is coincidental with the origin of *Figure 2.8* and the BET equation is invalid from this Type III isotherm. For C_{BET} values between 1 and 2 the point of inflection occurs at negative relative pressure values and gives from the BET equation a negative inter- cept which has no physical meaning. Equation 2.24 can be rearranged to give equation 2.25:

$$\ln C = (E_1 - E_L)/RT \qquad (2.25)$$

At values of C_{BET} equal to unity $E_1 = E_L$ since $\ln C$ is zero. At C_{BET} values of 2 and adsorption temperature 77 K, the value of $E_1 - E_L$ becomes 0.105 kcal mol^{-1} (0.441 kJ mol^{-1}) when R is taken as 1.98 cal mol^{-1} K^{-1} (8.31 J mol^{-1} K^{-1}). The sensitivity of the monolayer to C_{BET} can be seen by rearrangement of the BET equation (equation 2.12) in terms of moles adsorbed. Since the slope of equation 2.12, s, is equivalent to $(C_{BET} - 1)/(n_m^s C_{BET})$, then $1/n_m^s = s/[(C_{BET} - 1)/C_{BET}]$ and values of n_m^s are grossly affected by values of C_{BET} near unity.

2.2.3.2 *The molecular area of the nitrogen molecule, A_m*

Once point B (*Figure 2.5*), which is usually taken as the volume (or weight) of the adsorbate just covering the surface solid by means of a unimolecular layer, is found, the surface area of the solid can be estimated if the area assigned to each adsorbate molecule is known. Because the first unimolecular layer is completed at a relative pressure of approximately 0.1 and multilayers begin to form at relative pressures of 0.35, these molecular adsorbed assemblies can be compared with liquid or solid condensed states. It is logical, therefore, to assign the same molecular area to an adsorbed molecule as to that which the molecule would have in either the solid or liquid state. In normal practice the BET surface area determination uses nitrogen as the adsorbate. Since measurements are determined at the temperature of liquid nitrogen ($-196°C$) the molecular area may be taken either as 0.162 nm^2 (liquid nitrogen state) or 0.138 nm^2 (solid nitrogen state).

Emmett[65] stated that there was no way of knowing whether the correct value for the area covered by a nitrogen molecule was either 0.162 or 0.138 nm^2. Anderson[66] stated that evidence was accumulating to show 0.162 nm^2 was the better value although this molecular area for nitrogen adsorbate may vary from solid to solid. Livingston[67] suggested the value of 0.154 nm^2, this being calculated from the two-dimensional van der Waals

Table 2.5. Dimensions of adsorbates adsorbed on to solid surfaces

Adsorbate	Molecular area/nm^2	
	McC *and* H*	D *and* T†
N$_2$	0.162 ($-195°C$)	0.164 (77.4 K)
		0.166 (90.2 K)
Ar	0.138 ($-195°C$)	0.134 (77 K)
	0.138 ($-183°C$)	0.139 (90.2 K)
Kr	0.202 ($-195°C$)	0.202 (77 K)
		0.202 (90.2 K)
Benzene	0.430 (20°C)	
n-Butane	0.444 (0°C)	
Water	0.125 (25°C)	

* McClennan and Harnsberger, ref. 68
† Deitz and Turner, ref. 69

constant. This value was, however, dependent on the orientation of the nitrogen molecule. McClennan and Harnsberger[68] compared and evaluated by four different methods the area of a range of adsorbates. They recommended, using the four following methods, the values quoted in *Table 2.5*. Deitz and Turner[69] also determined the cross sectional area of adsorbed molecules by use of krypton, nitrogen and argon at 77 K and 90 K on a pristine glass fibre; their values are also in *Table 2.5*.

2.2.3.2.1   Method (a)

This uses the two-dimensional van der Waals constant, *b*, derived from Hill's[70] formula:

$$b = 6.354 \left(\frac{T_c}{P_c}\right)^{2/3}$$

Values of critical pressure and temperature, P_c and T_c, are obtained from International Critical Tables.

2.2.3.2.2   Method (b)

The nitrogen molecule is assumed to be spherical and packed on the adsorbent in a close hexagonal form:

$$A_m = 1.091 \left(\frac{M_w}{N\rho}\right)^{2/3}$$

where *M* is the molecular weight (relative molecular mass), *N* is Avogadro's number, and ρ the density.

2.2.3.2.3   Method (c)

A molecular model is made of each adsorbate and the area found from a shadow graph using a light source 6 ft distance away. This gave an underestimate of the molecular area.

2.2.3.2.4   Method (d)

From the shadow, an area is graphed filling a drawn polygon. This gave an overestimate of the molecular area.

2.2.4   Type IV, BDDT classification isotherm

With non-porous material multilayers of adsorbed molecules can build up on the plane surface of the solid and with the majority of solids allow condensation of the vapour to occur at the saturated vapour pressure P_s of the gas. When a solid is porous, however, the adsorbed layers, although capable of building up into multilayers, reach a stage when capillary condensation takes place below the saturated vapour pressure P_s of the gas within the porous matrix. This mode of nitrogen adsorption gives rise to a Type IV isotherm which can be regarded as a modification of a Type II isotherm. In a Type IV isotherm, the adsorption isotherm curve, instead of approaching the relative pressure unity value asymptotically, approaches this value usually horizontally (*Figure 2.3*). As the adsorption curve approaches P/P_s equal to unity it is assumed that the pores in a porous solid are filled by liquid adsorbate. The quantity of liquid adsorbate present in the pores should therefore be independent of the adsorbate and this is usually defined as the Gurwitsch volume. This independency has been verified from the evaluation of the volume of liquid ($cm^3 g^{-1}$) of various adsorbates taken up by different porous solids, such as iron(III) oxide and silica gels, which gave agreement of the Gurwitsch volume with various adsorbates within 5—6 per cent. In addition, a Type IV isotherm can produce a hysteresis loop, the adsorption volume–pressure increasing curve being below the desorption volume–pressure reducing curve. Mathematical analysis or interrogation of either the adsorption or desorption isotherm curves can provide a measurement of the internal structure of the solid. It is generally accepted that the pores within a solid making up the internal surface are, according to the Dubinin classification (Section 2.2) mesoporous with a size range between radii of 1.0 and 100 nm. The size of the pores, in which capillary condensation occurs, can be calculated from the Kelvin equation which is the basis of the computational pore size analysis of Barrett, Joyner and Halenda (BJH) or Cranston and Inkley (CI).

2.2.4.1 Kelvin equation

The adsorption curve of the nitrogen isotherm produced at low temperature cannot only be used to determine the surface area of a solid but with the desorption curve can be used to determine pore size distributions in porous solids.

With many adsorbents, a hysteresis loop occurs between the adsorption and desorption curve after monolayer completion. This hysteresis loop is associated with capillary condensation of the adsorbate within the pores and voids of an adsorbent. As the relative pressure is reduced from saturation to a lower pressure, the adsorbate does not evaporate as readily from a capillary pore, owing to the existence of a hemispherical meniscus in cylindrical pores, as it does from a flat surface. The adsorption curve and the filling of cylindrical pores can be presented by the Cohan equation[71] (equation 2.26). Cohan supplemented the theories of Patrick[72] and Zsigmondy[73] to explain the mechanisms of the hysteresis sorption and the vapour pressure of concave surfaces.

The vapour pressure of an annular film inside a cylindrical capillary of radius r_p due to the change in free energy accompanying the isothermal transfer of a small volume of liquid to a capillary becomes:

$$\ln \frac{P_f}{P_s} = -\frac{\gamma V}{RTr} \tag{2.26}$$

where P_f and P_s are the vapour pressure of the film and bulk liquid respectively, γ, V, R and T are surface tension, molar volume, the universal gas constant and absolute temperature. r is the radius of the annular film which is less than r_p. The pressure at which evaporation or desorption will occur in open-ended capillaries can, however, be calculated from the classical Kelvin equation (equation 2.27; *see also* equation 2.28):

$$\ln \frac{P_K}{P_s} = -\frac{2\gamma V}{RTr_K} \tag{2.27}$$

where P_K is the vapour pressure similar to P_f and r_K the Kelvin radius of the adsorbed film.

The difference in the two equations is due to the presence of a complete hemispherical liquid meniscus in the desorption process, the radii of curvatures for the desorption meniscus being $r_1 = r_2 = r_K$. In the adsorption process since the hemispherical meniscus is not complete one radius is along the capillary length which is thus infinity and the value of 2 disappears.

Most of the work done on the computation of pore sizes with porous material is based on the pore model of straight cylindrical non-intersecting capillary tubes. The Kelvin equation can then be used to determine the radius of a capillary which will be filled by liquid at a specific pressure when surrounded by a vapour of known saturated vapour pressure.

The classical Kelvin equation for sorption can be expressed as:

$$\ln \frac{P}{P_s} = -\frac{2V_L \gamma \cos \phi}{RTr_K} \tag{2.28}$$

where

P and P_s are experimental and saturated vapour pressure of adsorbate respectively
V_L is the liquid molar volume of adsorbate
γ is the surface tension of liquid
T is the Kelvin (absolute) temperature
R is the universal gas constant
r_K is the Kelvin radius
ϕ is the contact angle

In most solid–liquid interfaces with nitrogen it is assumed that the liquid wets the walls of the pores, thus $\phi = 0$ and $\cos \phi$ is unity.

From the Kelvin equation, taking the values for nitrogen of

$\gamma = 8.4 \times 10^{-3} \text{ N m}^{-1}$
$V_L = 34.65 \text{ cm}^3 \text{ mole}^{-1}$

$T = 77$ K
$R = 8.31$ J mole^{-1} K^{-1}
$\phi = 0$ degrees

a cylindrical pore with open ends of size 3 nm (3.0×10^{-9}) will be filled at a relative pressure of 0.86:

$$\frac{34.6 \times 10^{-6} \times 8.4 \times 10^{-3}}{77 \times 8.31 \times 3 \times 10^{-9}} = \frac{\ln P}{P_s} = -0.151$$

Therefore

$$\frac{P}{P_s} = 0.86$$

For cylindrical pores of size 0.3 and 30 nm the relative pressure at filling is 0.220 and 0.985 respectively. On desorption however the pore will empty when the pressure is given by:

$$\ln \frac{P}{P_s} = -\frac{2\gamma V_L}{R Tr_K}$$

Thus, for a cylindrical pore of size 3.0×10^{-9} m with a hemispherical meniscus:

$$\frac{2 \times 34.6 \times 10^{-6} \times 8.4 \times 10^{-3}}{77 \times 8.31 \times 3 \times 10^{-9}} = \ln \frac{P}{P_s} = -0.303$$

Therefore

$$\frac{P}{P_s} = 0.74$$

The relative pressure of emptying becomes 0.74 which gives rise to the hysteresis loop. For parallel plates or open slit shaped capillaries, a meniscus cannot be formed during adsorption but only after saturation. During the desorption process a meniscus is present, hence the desorption curve is delayed to produce hysteresis. The shape of the hysteresis loop has been used by de Boer to classify various shapes of pore (*see* Section 2.2.4.8).

2.2.4.2   *Thickness of adsorbed layers*

Before capillary condensation occurs in any pore there is a build up of adsorbed molecules on the walls of the pores. The thickness of this adsorbed layer, t, must be taken into account in the evaluation of the pore size distribution of porous material by any of the computation methods using the capillary condensation equations. The relationship between the actual pore size r_p, the Kelvin pore size r_K and the multilayer adsorbed thickness t for cylindrical pores is:

$$r_p = r_K + t \tag{2.29}$$

and for pores in solids consisting of parallel sided plate Γ distance apart:

$$\Gamma = r_K + 2t \tag{2.30}$$

Since the validity of the BET equation is restricted to relative pressure values of 0.35 or less the calculation of the thickness of the adsorbed layers at relative pressure greater than 0.35 should not be attempted with porous material. The thickness, t, can however be estimated from the monolayer capacity value n_m^s by assuming that a porous material behaves as a non-porous material with regard to the multilayer thickness build up.

Thus:

$$t = \frac{n^s \sigma}{n_m^s} \quad \text{or} \quad \frac{x_a \sigma}{x_m} \quad \text{or} \quad \frac{V_a \sigma}{V_m} \tag{2.31}$$

where n_m^s is the monolayer capacity obtained from the BET equation over the relative pressure range 0.05—0.35 and n^s is the adsorption value from porous materials at pressures up and beyond 0.35 relative pressure; σ is the average thickness of a single layer of molecules. The average thickness, σ, of a single layer of adsorbed molecules is not identical to the molecular dimensions of molecules because of the packing arrangements of molecules on the surface of a solid. The average thickness of a single layer is usually taken as 0.354 nm while the collision diameter of N_2 from the Lennard–Jones potential is, on an average solid, 0.3064 nm and on aluminium 0.317 nm[74].

Shull[13] used the value of $\sigma = 0.43$ nm based on the assumption that the adsorbed nitrogen molecules were arranged, in a given layer, vertically above the lower layer molecules. Packing arrangements, either hexagonal or cubic array, in which the molecules in an upper layer sit in the depressions between molecules in the lower layer, give an average thickness less than the diameter of a single molecule.

Wheeler[75,76] stated that the adsorption on the walls of fine pores is probably greater than that on a flat surface at low pressure and used the Halsey equation[77]:

$$t = \sigma \left[\frac{5}{\ln \dfrac{P_s}{P}} \right]^{1/3} \tag{2.32}$$

to evaluate the thickness of the adsorbed layer on porous solid where σ was taken as 0.43 nm.

Empirical values of the thickness of nitrogen adsorbed on non-porous solids (*t*-curves) have been experimentally found by Harris and Sing[64] for silica and alumina; Lippens, Linsen and de Boer[31] for aluminium oxides and hydroxides; Cranston and Inkley[29] for a variety of non-porous solids ranging from glass spheres to silver and tungsten metals; and Pierce[14] for anatase, potassium chloride, silica, quartz and aluminium foil. The original relationship of Pierce[14] between the number of adsorbed molecules and relative pressure was obtained with the use of non-porous solids, or carbon blacks of small particle size. These materials were later shown to produce

Table 2.6. Number of statistical layers, n, and thickness of adsorbed nitrogen layer, t, from various workers

$\dfrac{P}{P_s}$	Source									
	Shull		Pierce I		Pierce II		Lippens		Cranston and Inkley	
	n	t/nm	n	t/nm	n	t/nm	n	t/nm	n	t/nm
0.2	1.15	0.495	1.22	0.44	1.25	0.450	1.23	0.435	1.20	0.480
0.25	1.22				1.32		1.31		1.27	
0.3	1.30	0.559	1.38	0.50	1.39	0.500	1.42	0.503	1.34	0.536
0.35	1.37				1.47		1.51		1.42	
0.4	1.46	0.628	1.56	0.56	1.54	0.554	1.61	0.569	1.52	0.608
0.45	1.55				1.62		1.72		1.61	
0.50	1.64	0.705	1.78	0.64	1.70	0.612	1.83	0.648	1.70	0.680
0.55	1.74				1.80		1.95		1.79	
0.60	1.86	0.799	2.08	0.75	1.90	0.684	2.08	0.736	1.90	0.760
0.65	2.02				2.02		2.23		2.00	
0.70	2.18	0.937	2.50	0.90	2.17	0.781	2.42	0.857	2.13	0.852
0.75	2.39				2.35		2.66		2.30	
0.80	2.64	1.135	3.50	1.30	2.58	0.929	2.98	1.054	2.52	1.008
0.825	2.81				2.76		3.22		2.66	
0.85	2.99	1.286	4.40	1.60	2.92	1.051	3.46	1.225	2.80	1.120
0.875	3.24				3.12		3.82		2.95	
0.90	3.50	1.510	5.6	2.0	3.33	1.199	4.22	1.494	3.14	1.256
0.95			8.0	3.2						
0.975			13.0	4.7						

adsorption isotherms with capillary condensation between the interparticle spaces at high relative pressure and this relationship therefore (Pierce I, *Table 2.6*) between adsorbed layer thickness and relative pressure cannot be allied to the work on non-porous material.

Lippens and de Boer[78] observed that with aluminas and aluminium hydroxides and oxides the multimolecular layers of adsorbed nitrogen molecules could be formed freely on all parts of a non-porous surface.

Lippens, Linsen and de Boer by assuming, unlike Shull[13], that molecules in successive layers of a multilayer adsorption assembly are not situated on top of the nitrogen molecule of the previous layer, derived an equation for the statistical thickness of an adsorbed layer of adsorbate molecules. Taking the density of the adsorbed layer as that of a capillary condensed liquid, which for nitrogen molecules is that of the density of liquid nitrogen, they calculated the thickness of an absorbed layer from equation 2.33:

$$t = \frac{10^{-3}\,X}{S}\ \text{nm}$$

$$= \frac{10^{-3}\,M_w\,V_{sp}\,V_a}{22414S}\ \text{nm} \tag{2.33}$$

Table 2.7.   Variation of adsorbed layer thickness, t, with relative pressure P/P_s from various workers on porous solids

$\dfrac{P}{P_s}$	*Thickness of adsorbed layer, t/nm*			
	Halsey* *(eqn. 2.32)*	*Mingle and Smith†* *(eqn. 2.48)*	*Butt‡* *(eqn. 2.52)*	*Girgis* *(eqn. 2.53)*
0.2	0.63	0.49	0.34	0.51
0.3	0.69	0.57	0.39	0.58
0.4	0.76	0.67	0.45	0.66
0.5	0.83	0.78	0.52	0.76
0.6	0.92	0.92	0.59	0.90
0.7	1.04	1.12	0.71	1.09
0.8	1.21	1.44	0.88	1.38
0.9	1.56	2.19	1.27	1.81

* $\sigma = 0.43$ nm
† $\bar{r} = 3.68$ nm, $C_{BET} = 130$

$$‡ \left(\frac{P}{P_s}\right)_m = \frac{1 - \sqrt{C_{BET}}}{1 - C_{BET}}$$

where

> t is the statistical thickness of the adsorbed layer
> X is the adsorbed volume in cm^3 of *liquid adsorbate* per gram of adsorbent
> S is the specific surface area in m^2 g^{-1} of adsorbent (usually S_{BET})
> M_w is the molecular weight of adsorbate
> V_{sp} is the specific volume of adsorbate in cm^3 g^{-1}
> V_a is the adsorbed volume of the adsorbate in cm^3 gas at STP/g of adsorbent

which then gives the statistical thickness of nitrogen layers in terms of volume adsorbed and surface area:

$$t = \frac{1.547 V_a}{S_{BET}} \text{ nm} \tag{2.34}$$

When the area occupied by a nitrogen molecule is taken as 0.1627 nm^2 and the BET surface area is expressed in terms of monolayer capacity, equation 2.35 is obtained.

$$S_{BET} = 4.37 V_m \text{ m}^2 \text{ g}^{-1} \tag{2.35}$$

The statistical thickness can then be expressed as:

$$t = \frac{0.354 V_a}{V_m} \text{ nm} \tag{2.36}$$

The value of 0.354 nm is comparable with the value of 0.43 nm which was the diameter or thickness of the nitrogen molecular layer found by Shull[13] and also Barrett, Joyner and Halenda[26].

Lippens and de Boer[31], using the values of t calculated from equations 2.34, 2.35 and 2.36, found for various samples of aluminium hydroxides and oxides a relationship between the thickness of the adsorbed nitrogen molecules and relative pressure. At values of relative pressure less than 0.60 a common curve was found for the thickness of the adsorbed layer on the various samples.

Comparison of the Lippens and de Boer t-curve with the work of Barrett, Joyner and Halenda, even when corrected by the factor 0.354/0.43 due to the difference in nitrogen diameters of the two sets of workers, places de Boer's t-curve slightly above the curve obtained by Barrett *et al.* Cranston and Inkley[29] published a regression curve, t against relative pressure, for many substances, which was higher than the other two curves especially at low relative pressures. The range of t values measured experimentally by Lippens and de Boer on alumina was from $P/P_s = 0.08$ where $t = 0.3551$ nm to $P/P_s = 0.76$ where $t = 0.965$ nm. The Lippens and de Boer t-curve was verified by using, in addition to alumina, the solids MgO, SiO_2, TiO_2, and $BaSO_4$. The thickness of the layer of nitrogen on a free surface may not, however, be dependent upon the type of solid but upon the adsorption constant C_{BET}, since the t-curve method is based on the BET method and requires that between the relative pressures of 0.08 and 0.35 all curves coincide[79]. The experimental curve of de Boer has an adsorption coefficient between the values $300 > C_{BET} > 100$. The equation which best describes the experimental t-curve is Anderson's equation:

$$V_a = \frac{V_m C_{BET} k \left(\dfrac{P}{P_s} \right)}{\left[1 - k \left(\dfrac{P}{P_s} \right) \right] \left[1 + (C_{BET} - 1) k \left(\dfrac{P}{P_s} \right) \right]} \tag{2.37}$$

where $C_{BET} = 53$ and $k = 0.76$.

Lecloux[80] stated that the t-curve of Lippens, Linsen and de Boer should not be used if the value of the C_{BET} coefficient was less than 100 because the original curve was found with a solid—gas system which gave C_{BET} values in the region $100 < C_{BET} < 300$. It is recommended that solids be grouped by C_{BET} coefficient values and that the appropriate t-curve for pore size analysis be selected on a C_{BET} value basis rather than on the chemical nature of the solid. The deviation of other t-curves from the Lippens, Linsen and de Boer t-curve, due to C_{BET} value variation, occurs at low relative pressures $(P/P_s < 0.6)$. At relative pressure greater than 0.6, however, the other t-curves tend to merge into the Lippens, Linsen and de Boer t-curve.

Table 2.6 shows the variation of the thickness of the adsorbed nitrogen layer as a function of relative pressure together with n, the number of statistical layers. Pierce concluded after summarizing the n values of Shull, Lippens, Cranston and Inkley and his own that the volume adsorbed *versus* number of statistical layers (V_a/n) plots are preferable to the equivalent V_a/t plots suggested by de Boer because the t values are based on n values but involve assumptions regarding the thickness of an adsorbed layer. The initial n and t values of Pierce (Pierce I, *Table 2.6*) were obtained for a solid which was known to produce interparticle nitrogen gas condensation and thus showed a large deviation at high relative pressures. The equation used by Pierce subsequently[14] to represent the number of layers on a variety of solids ranging from carbon, metal oxides and ionic crystals can be expressed as

$$n^{2.75} = \left(\frac{V}{V_m}\right)^{2.75} = \frac{1.30}{\log(P/P_s)} \tag{2.38}$$

which was obtained from a composite curve from the published work of Pierce and others. The values used to convert n values to t values were 0.43 nm for Shull, 0.40 nm for Cranston and Inkley, 0.36 nm for Pierce, and 0.354 nm for Lippens. Only values starting at a relative pressure 0.2 are shown in *Table 2.6*, since at lower pressures there is greater deviation. The differences between n or t values from various workers are not important when used to compute wall film thicknesses for subsequent pore size analysis since at high relative pressures the wall film thickness is much smaller than the total pore radius. In other applications, however, such as micro-pore analysis, differences in the various sets of n values may lead to a wide disagreement in results. Although the multilayer thickness of adsorbed layers has been correlated with relative pressure by many workers for non-porous material, it can be argued that the structural characteristics of pores in porous materials affect the magnitude of multilayer adsorption inside capillaries. The V_a/t or V_a/n curves obtained from a flat non-porous solid tend to be unsuitable for porous material as there is no certainty that the mechanism of multilayer build-up inside pores, where the surrounding walls provide overlapping surface forces, is the same as for non-porous material.

Mingle and Smith[81], using the recommendation of Wheeler that a Halsey type equation (equation 2.32) can predict the thickness of adsorbed layers and the data of Johnson[82], derived an expression for the thickness of adsorbed layers on porous alumina catalysts (*see* equation 2.48). The range of surface area of powders used was 55.3—588 m^2 g^{-1}, a mean pore size range of 1.79—9.6 nm and C_{BET} values of 64.4—230. Butt[83] who used the same solids as Lippens and de Boer, also derived an expression correlating the multilayer thickness with both the C_{BET} value and the mean pore radius (equations 2.49, 2.50 and 2.52).

Alternatively, Broekhoff and de Boer[84] derived an expression, similar to the Kiselev and Karnaukhov expression[85], which corrected the porous adsorbed layer from that of the ideal t-value. Girgis[86], using porous silica gels, obtained a correlation between adsorbed thickness and relative pressure completely independent of structural characteristics (equation 2.53). The surface area of silica gels ranged from 178 to 814 m^2 g^{-1} with pore radii in the range 1.26—7.44 nm and C_{BET} values from 74 to 290. *Table 2.7* shows the variation of the adsorbed layer as a function of relative pressure obtained from the relationships of Halsey, Mingle and Smith, Butt and Girgis (equations 2.32, 2.48, 2.52, 2.53, respectively).

2.2.4.3 *Computational methods and models for pore size distributions*

From either the Cohan equation or the Kelvin equation, the size of a capillary pore, r_c, can be calculated from values of the relative pressure which, together with knowledge of the adsorbed quantity in terms of gaseous volume V_1, liquid condensed volume V_c, weight x_1 or moles n_i^s at any given relative pressure can result in a relationship between pore size and adsorbent uptake in porous solids.

Ignoring, for the present, the effect of the multilayer adsorption film, the volume of liquid adsorbent condensed by capillary condensation V_c can be calculated from the gaseous adsorption volume V_1 by:

$$V_c = \frac{M_w V_1}{\rho M}$$

where M_w is the molecular weight of adsorbate, ρ is the density of the liquefied gas at its saturated vapour pressure and M is the molar gas volume at STP.

A plot of the values of V_c against the radii of capillary pores can, by differentiation of the curve, give a pore size distribution (dV_c/dr_c) which follows the initial work by Foster[87].

2.2.4.4 *Pore size distribution by Kelvin or Cohan equations*

As desorption progresses more and more pores lose capillary condensed liquid and the total capillary wall area covered by liquid layers does not remain constant.

In reality, the process of desorption, when the relative pressure falls from values of P_1/P_s to P_2/P_s, occurs in two parts.

1. Evaporation of the capillary condensed liquid adsorbate occurs from the 'inner core' of capillaries of size r_c between values of r_{c2} and r_{c1} respectively.
2. The average thickness of the adsorbate multilayer decreases from t_1 to t_2 from pores which have previously lost the 'inner core' capillary condensed adsorbate. The thickness t_1 and t_2 can be estimated from an adsorption isotherm obtained from a reference non-porous adsorbent or if the adsorbate is nitrogen from an appropriate t-curve or equation.

Wheeler[75,76] summarized this situation together with the contribution of the change in the average thickness of adsorbate by:

$$V_T - V_p = \int_{r_p}^{\infty} \pi(r_p - t)^2 L(r_p) dr_p \tag{2.39}$$

where V_T is the total pore volume, and V_p is the liquid volume adsorbed at a specific relative pressure at which all pores of less than radius r_p are filled.

The evaluation of the 'inner core' capillary size is dependent however on the model chosen for the structure of a porous solid.

For open cylindrical pores, the Cohan equation is usually applied, while for cylindrical pores closed at one end, the classical Kelvin equation is used. The choice of model thus affects the calculated value of the 'inner core' radius, as the radius from equation 2.26 is half the radius r_K from the Kelvin equation (equation 2.27).

In an attempt to produce a structural analysis of porous solids, regardless of shape or assumed model, Brunauer[35] proposed the use of the Kiselev equation[85], which determined the surface area of pores within porous solids:

$$S_p = -\frac{1}{\gamma} RT \ln \frac{P}{P_s} \int_{n_H}^{n_s} dn \tag{2.40}$$

where S_p is the surface area of the 'inner cores', γ is the surface tension of the adsorbate as a liquid, and n_H and n_s are the number of moles adsorbed at the beginning of the hysteresis loop and number of moles at saturation pressure P_s.

Graphical integration of equation 2.40 gives the surface area of the *cores* of the pores and not the pores so the method is only applicable when the thickness of the adsorbed film on the walls is negligible. Although of limited value for the determination of total surface area, because the core surface is the measured parameter, the Kiselev equation (equation 2.40) has been adopted by Brunauer to describe a pore size analysis method independent of the shape of the pores. Division of the volume of the core by the surface of the core gives a volume–surface radius r_{vs} which is not equal to the Kelvin radius. This method is preferred because two parameters from the adsorption isotherm are used (volume and surface) rather than as in other idealizations where only one parameter (relative pressure) is used.

When the effect of the thickness of the multilayer film is to be considered, then models of either cylindrical or parallel plates must be assumed and a differently structured curve will result, dependent on the model chosen.

The Kiselev equation can be rewritten in terms of volume adsorbed, V_a, or condensed liquid volume, V_c (*see* Section 2.2.4.3):

$$S_p = \frac{-RT}{\gamma V_L} \ln \frac{P}{P_s} \int dV_a \tag{2.41}$$

which gives the volume–surface radius of the porous core as:

$$r_{vs} = \frac{V_a}{S_p} = \frac{\gamma V_L}{RT \ln (P_s/P)} \tag{2.42}$$

Comparison with the Kelvin equation shows that $r_{vs} = \frac{1}{2} r_K$ and is equivalent to the radius evaluated from Cohan's equation.

2.2.4.5 Pore size distribution by BJH computer modelling

Barrett, Joyner and Halenda (BJH)[26] developed a method for estimating the distribution of pores in porous solids from the physical desorption isotherms for nitrogen at $-195\,°C$. It is assumed that at any relative pressure between 0 and 1 all pores with a radius *larger* than a specific value, r_K as determined by Kelvin's equation, will contain an adsorbed layer of thickness t on the walls while all pores *smaller* than a specific value, r_K will be filled with capillary condensed liquid. Barrett, Joyner and Halenda thus assumed that a correction was necessary to account for this thickness of the multilayer film remaining on the walls of the pores after the inner capillary volume had emptied at a given pressure.

The total volume of adsorbate lost from a pore in a desorption step, ΔV_p, is therefore made up of the volume from the capillary desorption ΔV_c plus the volume from the decrease in the multilayer film in the pores, ΔV_f. The problem facing Barrett, Joyner and Halenda was to find ΔV_f which comes *not* from a flat surface but from a curved pore surface which is varying as a function of pressure.

The final equation developed by Barrett *et al.*, which they solved using a computational method, was:

$$\Delta V_p = R_n(\Delta V_c - C\,\Delta t\,\sum S_p) \tag{2.43}$$

where

ΔV_p is the actual volume of pores emptied in the desorption step

$R_n = r_p^2/(\bar{r}_k + t)^2$ where r_p is the actual pore radius, $\bar{r}_k$ is the mean capillary condensate radius, and t is the decrease in multilayer thickness

$\sum S_p$ is the sum of the surface area over all the desorption steps which for cylindrical pores is $S_p = 2\Delta V_p/r_p$

C is a correction factor to allow for the change in curvature of the pore wall as the range of pore size changes. Appropriate values are shown in *Table 2.8*

Table 2.8. Correction factors for the equation of Barrett *et al.*[26]

Pore radius range, r_p/nm	0—4.15	4.15—8.30	5.80—10.0	7.90—20.0
C	0.75	0.80	0.85	0.9

The correction factor corrected the assumption made by Oulton[88] for fine porous material that the thickness of the wall film remained constant for the entire pressure region over which desorption occurred.

Cranston and Inkley[29] made the BJH method more precise by applying a correction factor to each of the calculated pore sizes and extended the range of pore radii to less than 1.5 nm.

Innes[27] stated that for all practical work, where relative values are of importance, it is probably unimportant whether the adsorption or desorption branch of the isotherm is used as long as comparisons are made on a consistent basis. The desorption branch usually yields a narrower distribution of sizes but neither branch represents true equilibrium in all cases.

Roberts[89] has improved the procedure for estimating the distribution of pore volume and area of porous materials, by transforming the desorption isotherm into liquid volume units and using the concept of simultaneous

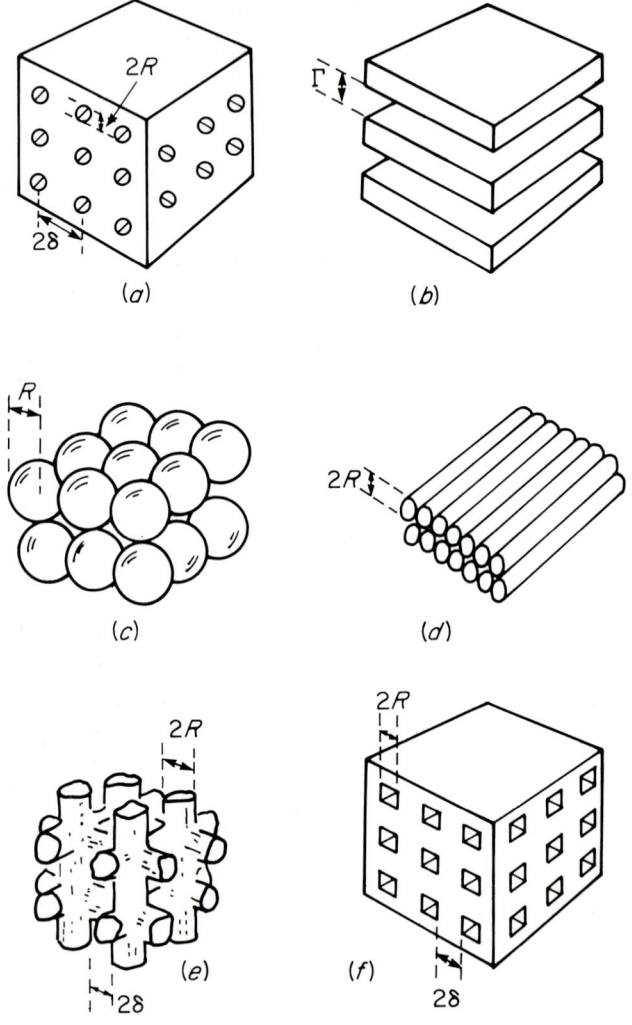

Figure 2.9.   Some pore systems. (a) Non-intersecting cylindrical capillaries. (b) Equally spaced plates. (c) Packed spheres of equal size. (d) Packed cylindrical rods of equal size. (e) Intersecting cylindrical rods. (f) Intersecting square capillaries (after Everett[91])

capillary condensation and multilayer adsorption. The method is applicable to both cylindrical and parallel plate pore models and also for adsorption and desorption branches of the isotherm.

Dollimore and Heal[90] used the equation of Halsey[77] (equation 2.32) to calculate t as a function of P/P_s which thus ignores the correction factor for varying capillary curvature.

2.2.4.6 *Appraisal of the computational methods of pore size analysis*

The application of the Kelvin equation usually assumed circular pores where:

$$r_{vs} = \frac{r_K}{2} = \frac{V_L \gamma}{RT \ln (P_s/P)} \cos \phi$$

but in reality for non-circular pores the Kelvin equation calculates the volume surface capillary ratio dV_a/dS_p. Only for cylindrical pores does the volume of capillary per unit length become πr_K^2 which then divided by surface area per unit length $2\pi r_K$ gives $r_K/2$. The Kelvin radius is thus twice the volume surface ratio r_{vs}. If the pores in a solid are not cylindrical the Kelvin radius r_K may not describe the actual pore dimension and comparisons with other methods of pore size analysis like mercury intrusion may not coincide. Everett[91] showed that for not-too-complicated pore networks, and ignoring multilayer thickness, the relationship between r_K and the actual pore size R in simple networks can be calculated from network geometry. Thus for non-intersecting cylindrical capillaries (*Figure 2.9*) $r_K = R$, but for intersecting square capillaries of width $2R$ and distance 2δ apart (*Figure 2.9f*) the function of r_K is equivalent to $R[1 + (R/3\delta)]$. If the distance between square capillaries is a third of the actual pore dimension ($\delta = R/3$) then the Kelvin radius is twice the actual width. For parallel fissures where Γ is equal to the distance apart of the walls, $\Gamma = r_K$. These variations arise because there is a difference between the pressure on the two sides of a meniscus, ΔP_1 and the free energy of the liquid ΔG can be related thermodynamically to the capillarity equation of Young and Laplace. Thus, from

$$\Delta G = -S\Delta T + V\Delta P$$

the relationships

$$\left(\frac{\Delta G}{\Delta P}\right)_T = V \text{ at constant temperature}$$

$$\left(\frac{\Delta G}{\Delta T}\right)_P = -S \text{ at constant pressure}$$

occur, where V and S are the volume and entropy terms of the system.

The function ΔP, by the general capillary equation of Young[92] and Laplace[93] is, for a capillary of radius r

$$\Delta P = \gamma \left(\frac{1}{r_1} + \frac{1}{r_2} \right)$$

where r_1 and r_2 are the two radii of curvature of a curved surface. Thus from the equation of the change in free energy for isothermal expansion:

$$\Delta G = RT \ln (P_s/P); \quad (\Delta G/\Delta P)_T = V$$

and

$$\gamma V \left(\frac{1}{r_1} + \frac{1}{r_2} \right) = RT \ln \left(\frac{P_s}{P} \right) \tag{2.44}$$

For a capillary meniscus which is hemispherical $r_1 = r_2 = r_K$. If however capillary condensation occurs between plates one radius is equivalent to infinity.

For spheres of radius R_s packed cubically $r_K = 0.613R_s$ or packed rhombohedrally $r_K = 0.229R_s$. The relationship between r_K and R_s can be compared with the radius of a sphere inscribed within the cavity, R_{in}, produced by spheres packed either cubically or octahedrally which give[94] $R_{in} = 0.732R_s$ and $R_{in} = 0.225R_s$ respectively (*see Table 2.1*). The comparison is not however valid for the radius of a circle which is inscribed within the throats of connecting cavities[12].

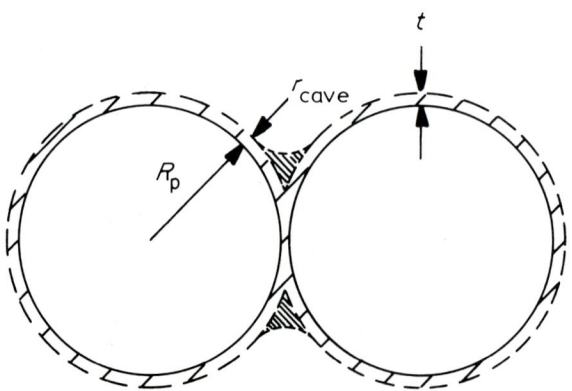

Figure 2.10.   Adsorption on two spherical particles in contact (after Wade, ref. 97) R_p = Radius of solid particle, r_{cave} = concave radius, r_{vex} = convex radius and t = statistical adsorbed layer

Dollimore and Heal[95] assumed, in the derivation of the volume of adsorbate obtained in a system of packed spheres, that the adsorption process could be regarded as:

1. Multilayer adsorption on to the sphere surface
2. Pendular ring condensation at points of contact between spheres
3. Condensation in the cavities between spheres

This was a similar approach initially reported by Kiselev, Karnaukhov and Aristov[96] and Wade[17, 97]. The saddle-shaped meniscus or torus from two spherical particles in contact (*Figure 2.10*) can be related to the radius of the solid particles, R_p and the multilayer thickness t. The concave radius of the meniscus is then r_{cave} and the convex radius r_{vex} is:

$$\frac{\sqrt{(R_p + t + r_{cave})^2 - R_p^2} - r_{cave} + R_p + t}{2} \tag{2.45}$$

Substitution of the two radii of curvature r_{cave} and r_{vex} into the Kelvin equation expressed in terms of radii of curvature (equation 2.44) then gives:

$$\ln\left(\frac{P}{P_s}\right) = 2.303 \frac{\gamma V}{RT}\left[\frac{1}{r_{cave}} - 2/(\sqrt{(R_p + t + r_{cave})^2 - R_p^2} - r_{cave} + R_p + t)\right] \tag{2.46}$$

This total amount of adsorbate adsorbed at a given relative pressure is thus dependent on the terms t, R_p and r_{cave} as well as the coordination number of the packed spheres.

Wade expressed equation 2.46 in terms of surface area and determined experimentally, for spherical particles of alumina, the amount of nitrogen adsorbed. For high specific surface area uncompressed samples (50 m^2 g^{-1}) the coordination number assumed in the successive approximation calculations had an overall value of 4. On compression the coordination number reached 12, the maximum value for hexagonal close packing of spheres.

Karnaukhov and Kiselev[98] showed that capillary condensation occurred in two stages:

1. Around the points of particle contact
2. When the saddle-shaped menisci joined together

In their view the coordination number determines the total pore volume rather than the particle radius. This approach could lead to new methods of pore size and shape estimation[99]. Ihm and Ruckenstein[100] compared various methods for the measurement of pore size distributions with the measurement made by electron microscopy of parallel cylindrical pores in alumina foils. The transmission electron microscopy value could then be used to verify directly the quality of the results obtained by adsorption measurement and the capillary condensation process.

Nitrogen adsorption isotherms at 77 K were used to compute pore size distributions from the following equations for capillary condensation in open cylindrical pores:

(a) Kelvin equation, usually applied to the desorption branch:

$$RT \ln \left(\frac{P_s}{P}\right) = \frac{2\gamma V_L}{(r_p - t)} \cos \phi$$

(b)   Cohan equation, usually applied to adsorption branch:

$$RT \ln \left(\frac{P_s}{P}\right) = \frac{\gamma V_L}{(r_p - t)} \cos \phi$$

(c)   Broekhoff and de Boer[84] equation:

$$RT \ln \left(\frac{P_s}{P}\right) = \frac{\gamma V_L}{(r_p - t)} + F(t)$$

where $F(t)$ can be obtained from the t-curve. The various t-curve equations are given in equations (2.32) and (2.33)

(d)   Nicholson equation[101]

$$RT \ln \left(\frac{P_s}{P}\right) = \frac{\gamma V_L}{(r_p - t)} + F(t) + V_L P_s \left(1 - \frac{P}{P_s}\right)$$

The thickness of the adsorbed layer on homogenous flat surfaces can be calculated from the following equations:

(a)   Lippens, Linsen and de Boer equation

$$t = 0.354 \left(\frac{V_a}{V_m}\right) (\text{nm})$$

(b)   Halsey equation

$$t = 0.43 \left(\frac{5}{\ln (P_s/P)}\right)^{1/3} (\text{nm})$$

(c)   Broekhoff and de Boer equations
(i)   for $t < 1.0$ nm

$$F(t) = 2.303 \, RT \left(\frac{13.99}{t^2}\right) - 0.034$$

(ii)   for $t > 1.0$ nm

$$F(t) = 2.303 \, RT \left(\frac{16.11}{t^2}\right) - 0.1682 \times \exp (-0.1137t)$$

From comparison of the computed pore sizes with those obtained by transmission electron microscopy (TEM) Ihn and Ruckenstein found that only the Broekhoff–de Boer equation[84] for the desorption branch showed a close agreement with TEM measurements. The correction term introduced by Nicholson[101] $V_L P_s [1 - (P/P_s)]$ had negligible effect. The probability of the validity of the model to show agreement between adsorption data and TEM measurements decreased in the order, Broekhoff–de Boer (desorption branch) > Kelvin ≫ Broekhoff–de Boer (adsorption branch) > Cohan.

In a similar study on the pore structure of synthetic chrysotile by electroscopy, nitrogen adsorption and mercury penetration, Scholten, Beers and Kiel[102] and de Witt and Scholten[103] showed that the pore size distribution calculated by the Kelvin equation corrected according to Broekhoff and de Boer's equation was in better agreement with the electron microscope pore size distribution than the pore size distribution calculated from the classical Kelvin equation.

Comparison of the pore size distributions of iron–chromium oxide, compressed aerosil and loosely packed and compressed chrysotile obtained from the Kelvin equation corrected by the method of Broekhoff and de Boer with those obtained from mercury penetration showed no overlap, mercury penetration giving distributions 30—40 per cent smaller than those from capillary condensation. The non-validity of the Washburn (Young–Laplace) equation was attributed to the factors of contact angle, surface tension and applicability of Washburn relationship to narrow capillaries.

In view of the uncertainty of the Kelvin equation with regard to the variation between adsorbed and bulk parameters, together with the variation of the *t*-curve with C_{BET} values, excessive refinement of the pore size distribution model has no warranty. This has been substantiated by Dollimore and Heal[95] who compared the pore size distributions obtained from models which assumed that the layers of adsorbed molecules were in the form of a torus and showed that there was little significant difference between these and capillary models. The work of Ihm and Ruckenstein together with that of de Witt and Scholten showed that, if the Kelvin equation was corrected by using the Broekhoff–de Boer correction, cylindrical pore size distributions obtained by nitrogen adsorption agreed with measurements of pore size obtained from electron microscopy.

In the calculation of pore size from adsorption isotherms it is not usual to continue calculations into the region of relative pressures below 0.3 which corresponds to a pore diameter of 1.5 nm. Many solids contain pores finer than this but owing to the uncertainty of the Kelvin equation the validity of pore sizes less than 1.5 nm is unsure.

2.2.4.7   *Multilayer adsorption on porous surfaces*

The disagreement seen with the initial relationships of relative pressure and the thickness of the multilayers of the adsorption film (*t*-curve) obtained for non-porous materials, from the work of Shull, Pierce, Cranston and Inkley and Lippens and de Boer with experimental values on porous materials has resulted in the re-evaluation and modification of the *t*-curves in which the apparent effect of the multilayer thickness is related to the structure and curvature of pores in porous material.

This interest, in knowing the way in which the number of adsorbed layers builds up with increase in relative pressure on porous surfaces, has led to the appearance of numerous modifications of the ideal non-porous *t*-curve. Mingle and Smith[81] as well as Butt[83] have derived expressions correlating the multilayer thickness with both the C_{BET} value and the mean pore radius of the adsorbent.

Mingle and Smith suggested that the contants in the Halsey equation could be related to the property characteristics of the adsorbent. The constant a' (equation 2.47) represents the deviation from ideal monolayer behaviour as a function of pore radius and suggests that monolayer thickness changes as the average interior surface becomes more curved. The exponent, $1/n$ is correlated with the relative pressure at monolayer coverage which in turn can be related to the C_{BET} value and the heat of adsorption on adsorbed layer thickness. The resultant equation became:

$$\frac{t}{0.4132} = a'\left(\frac{\log_{10}\left(\frac{P}{P_s}\right)_m}{\log_{10}\left(\frac{P}{P_s}\right)}\right)^{1/n} \tag{2.47}$$

where (P/P_s) and $(P/P_s)_m$ are relative pressures at surface and monolayer coverage respectively and the value of 0.4132 nm is the thickness of an ideal monolayer estimated from Lennard-Jones force potential function. Estimation of the values of a' and n from a number of porous catalysts using least-squares analysis gave:

$$a' = 0.126(\bar{r})^{-0.085}$$

where $\bar{r}$ is $2V_p/S_{BET}$ and $n = 1.42(1 + 0.00212C_{BET})$

The final equation in terms of $\bar{r}$, C_{BET} and relative pressure then became:

$$t = 0.519(\bar{r})^{-0.085}\left\{\frac{\log_{10}[(1 - C_{BET}^{\frac{1}{2}})/(1 - C_{BET})]}{\log_{10}(P/P_s)}\right\}^{1/[1.42(1 + 0.00212C_{BET})]}$$

$$\tag{2.48}$$

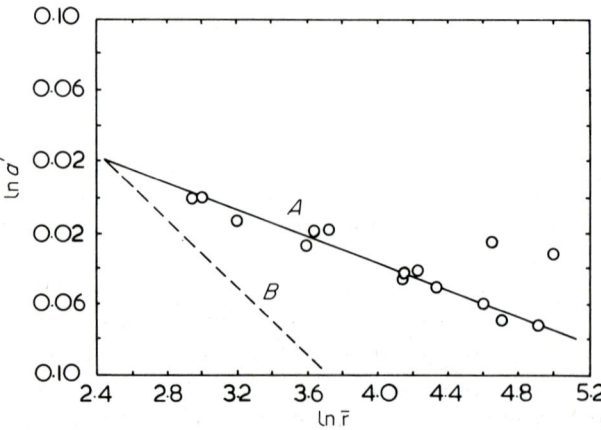

Figure 2.11.   Relation of the constant a' to average pore radius. Curve A, Butt (ref. 83). Curve B, Mingle and Smith (ref. 81)

Butt[83] found, using a modified Halsey equation, that the constants a' and $1/n$ could be expressed as:

$$\ln a' = -(3.65 \times 10^{-2}) \ln \bar{r} + 0.11 \qquad (2.49)$$

$$\ln (1/n) = 8.38(P/P_s)_m - 1.40 \qquad (2.50)$$

when 15 samples of porous material which encompassed a large range of surface areas $(17.1—609 \text{ m}^2 \text{ g}^{-1})$, pore radii $(2.0—14.5 \text{ nm})$ and pore system geometry (types A, B and E of de Boer classification together with slit-shaped pores; *see* Section 2.2.4.8) were analysed (*Figures 2.11 and 2.12*).

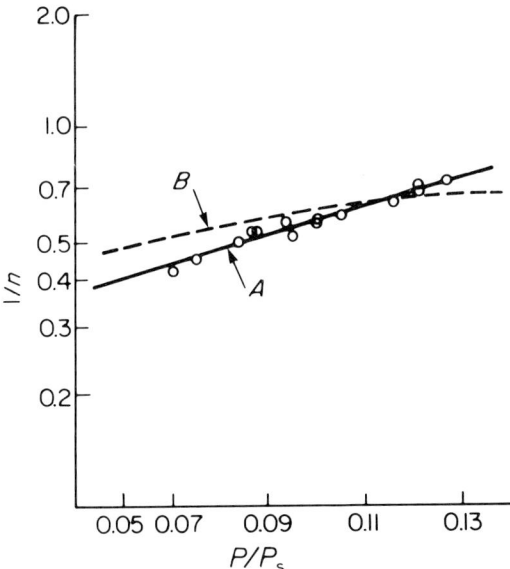

Figure 2.12.   Exponent of the Halsey equation as a function of relative pressure at monolayer coverage. Curve A, Butt (ref. 83). Curve B, Mingle and Smith (ref. 81)

Kiselev and Karnaukhov[85] attempted a correlation between a non-porous *t*-curve and a porous *t*-curve by allowing the value of *t* to vary with curvature of the walls of cylindrical pores.

Broekhoff and de Boer[84] derived expressions for the equilibrium thickness in pores and Viswanathan, Srinsvasan and Sastri[104] proposed a correlation between the monolayer saturation volume, V_m, and the C_{BET} value.

Girgis[86], in an attempt to provide a single multilayer curve for the multilayer thickness of nitrogen on non-ideal, porous, heterogeneous surfaces, used silica gel as the porous adsorbent and five computational equations to obtain an empirical equation for multilayer build up. The five equations used by Girgis to obtain *t* in addition to his own equation were:

(i) Halsey equation, where σ is the molecular dimension of the adsorbate

$$\frac{t}{\sigma} = \left(\frac{5}{\ln \left(P_s/P\right)}\right)^{1/3}$$

(ii) The t-curve of Cranston and Inkley

$$t = 1.7 - 1.368\left(0.96 - \frac{P}{P_s}\right)^{0.388} \text{nm}$$

(iii) The expression of Mingle and Smith

$$\frac{t}{0.4132} = a'\left[\frac{\log \left(\dfrac{P}{P_s}\right)_m}{\log \left(\dfrac{P}{P_s}\right)}\right]^{1/n} \tag{2.51}$$

(iv) Butt modification

$$N = \frac{t}{\sigma} = a' \left[\ln x_m/\ln x\right]^{1/n} \tag{2.52}$$

$$\sigma = 0.354 \text{ nm}$$

where N is the number of adsorbed layers

(v) t-Curve of Lippens, Linsen and de Boer

$$t = \frac{M_w V_{sp}}{22414} \times \frac{V_a}{S} \times 10^{-3} \text{ nm}$$

or

$$t = 0.354\left(\frac{V_a}{V_m}\right) \text{nm}$$

where M_w = molecular weight of adsorbate, V_{sp} = specific volume of adsorbate in $cm^3 g^{-1}$, V_a = adsorbed volume of adsorbate in cm^3 of gas at STP g^{-1} of adsorbent.

(vi) Girgis

$$t = 0.615 - 0.225 \ln \left(\ln \left(P_s/P\right)\right) + 1.05(P/P_s)^4 \text{ nm} \tag{2.53}$$

To show the deviation between various computer models and t-curve estimates Girgis plotted the ratio of S_{cum}/S_{BET} of the surface area computed by the BJH, CI and Roberts computational procedure with either the t-curve of Mingle and Smith or the Girgis empirical equation, for 26 silica gels with a range of surface area from 178 to 814 $m^2 g^{-1}$ against the BET surface area (*Figure 2.13*). Girgis's empirical equation (2.53) for use in estimating the multilayer thickness on a wide range of porous solids only shows small deviation, below a relative pressure of 0.4, with either the

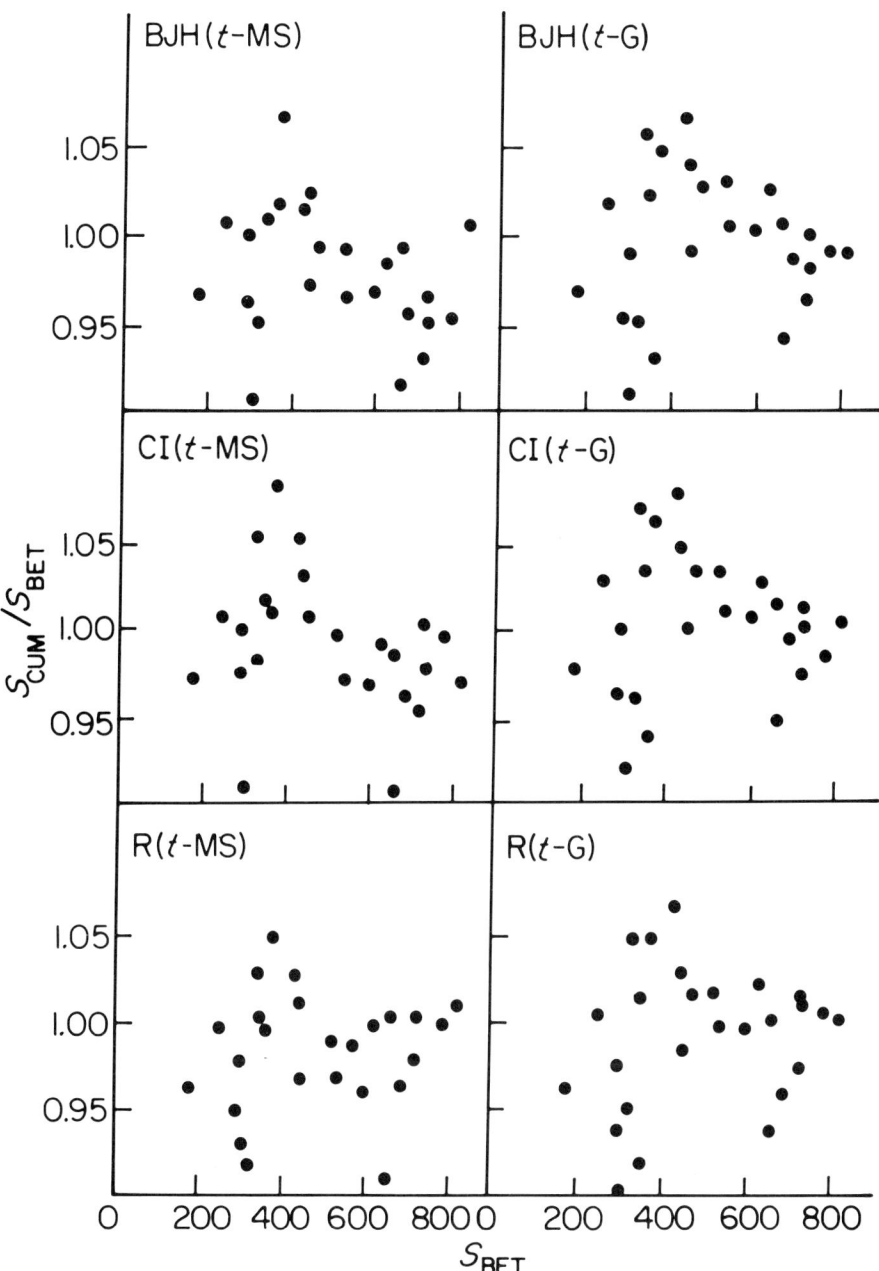

Figure 2.13. Correlation between the cumulative surface area S_{CUM} and S_{BET} for silica gel employing the procedures of Barrett, Joyner and Halenda (BJH), Cranston and Inkley (CI) and Roberts (R), with the t estimates of Mingle and Smith (t–MS) and the empirical equation of Girgis (t–G)

Halsey equation or the Cranston–Inkley correlation for non-porous material. At relative pressures above 0.4 deviation is, however, considerable, which may be explained from the observation and proposal of Derjaguin[105] that the curvature of the liquid–gas meniscus present in narrow pores is dependent not only upon the relative vapour pressure but also upon the magnitude of the interaction between solid pore walls and the capillary condensed liquid.

A statistical mechanical formulation of multilayer adsorption has been put forward by Steele and Ross[106] and reviewed by Steele[107]. This approach does not however describe thermodynamically the equilibrium between gas and adsorbent. Physical adsorption can be related to the van der Waals forces by the Lennard-Jones potential[108,109]. Ternan[74] deduced, from thermodynamics, an adsorption t-curve where the adsorbent was treated as being non-homogeneous and the gas–solid interaction could be represented in terms of dispersion forces. The equation developed agreed well with experimental t-curves for nitrogen obtained by the workers Lippens and de Boer, Pierce, Harris and Sing, Cranston and Inkley and Shull.

$$\frac{P}{P_s} = \exp\left\{\frac{-4\pi N\varepsilon\sigma^3\rho_s}{3RT}\left[\left(\frac{1}{2}\right)\left(\frac{\sigma}{t}\right)^3 - \left(\frac{1}{15}\right)\left(\frac{\sigma}{t}\right)^9\right] - \frac{V_L}{RT}(P - P_s)\right\} \quad (2.54)$$

where

N	= Avogadro's number
R	= universal gas constant
T	= absolute temperature
P and P_s	= adsorption and saturation pressure respectively
V_L	= molar volume of bulk liquid phase
ρ_s	= solid density in atom cm^{-3} obtained from the solid density of adsorbent
ε	= Lennard-Jones energy parameter = $(\varepsilon_G \varepsilon_S)^{1/2}$, where ε_G and ε_S are energy parameters of gas and solid phase
σ	= Lennard-Jones distance parameter = $(\sigma_G + \sigma_S)/2$
t	= thickness of adsorbed layer

Nicholson[101] compared the thermodynamic equations of Gibbs, Frenkel, Halsey and Hill (FHH)[110] and Cohan to predict the process of adsorption and filling of mesopores. The Cohan equation (equation 2.26) gave the poorest and most variable prediction although the correction for adsorption introduced by Broekhoff and de Boer[84] made a marked improvement. Equations based on the FHH slab theory, where the adsorbate is treated as a slab of fluid in the field of the adsorbent, are more suitable for the description of adsorption. The work indicated that it was unlikely that any thermodynamically based treatment could properly represent the adsorption process in the mesopore size range. It was also argued that the adsorption branch is preferable, for pore analysis, to the desorption branch because of the freedom of pore blockage during hysteresis.

In the analysis of pores it is best to use *t*-curves which correspond to the heat of adsorption of the adsorbent solid to be analysed. The choice of the correct *t*-curve is not however too critical in the case of mesopores and macropores because *t*-curves do not differ widely from each other and the pore sizes calculated from computational methods constitute only a part of the pore radius. This is not applicable for micropore analysis which is heavily dependent on the choice of *t*-curves. The Halsey equation (2.32) may

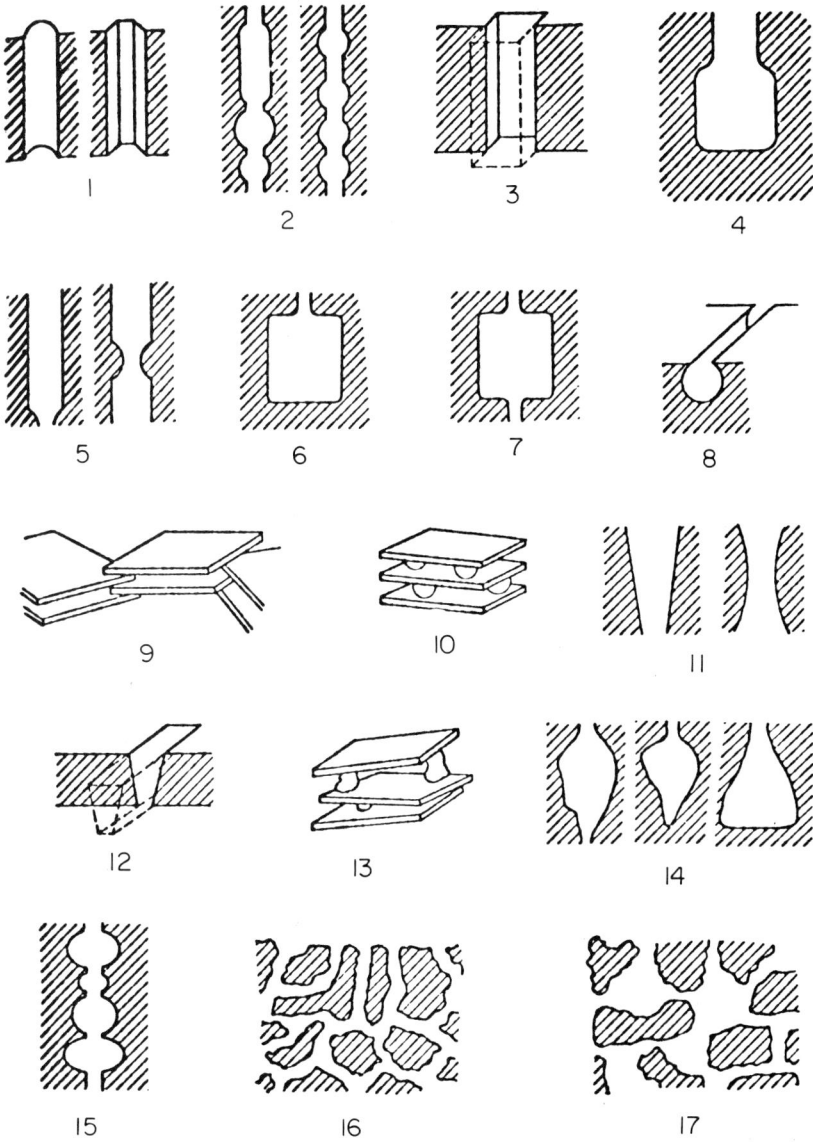

Figure 2.14.   The fifteen pore types and the two interparticle voids as classified by de Boer (see Section 2.2.4.8)

therefore be used as a convenient representation of a standard adsorption isotherm or *t*-curve.

2.2.4.8   Pore shape correlations

The relationship between the form of the hysteresis loop of the Type IV isotherm and the capillary pore shape has been classified by de Boer. When pores are consistent in shape and relatively uniform in size, de Boer classified 15 pore shape groups and two shape groups from the space between particles (*Figure 2.14*) into five major categories dependent upon the shape of the hysteresis loop (*Figure 2.15* and *Table 2.9*):

TYPE A   The adsorption and desorption curves are steep at intermediate

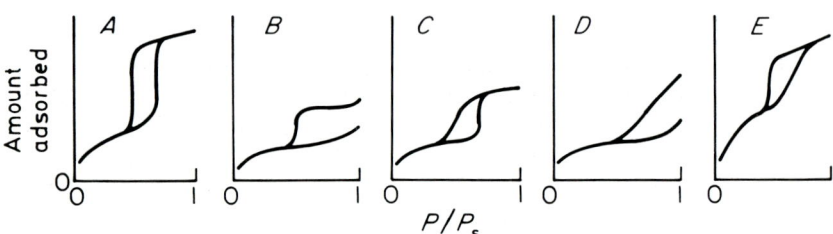

Figure 2.15.   Five types of hysteresis loop

Table 2.9.   Hysteresis loop correlation

Pore shape (Figure 2.14)	Hysteresis loop type (Figure 2.15)				
	A	B	C	D	E
1	×	—	—	—	—
2	×[a]	—	—	—	—
3	×[b]	—	—	—	—
4	×[c]	—	×	—	×
5	×[c]	—	×	—	×
6	×[d]	×	×	—	×
7	×[d]	×	×	—	×
8	×	—	×	—	×
9	—	×	—	×	—
10	—	×	—	×	—
11	—	—	×[e]	—	—
12	—	—	×	—	—
13	—	—	—	×	—
14	—	—	—	—	×[f]
15	—	—	—	—	×
16	×	—	—	—	—
17	—	—	—	—	×

[a] When $r_w \leqslant 2r_n$.   [b] When $1 \gg a$.   [c] When $r_n < r_w < 2r_n$.   [d] When $r_w > 2r_n$.   [e] When $R \leqslant 2r$.
[f] When r_w, r'_w and r''_w are large relative to r_n. r_w signifies the widest dimension and r_n the narrowest.

relative pressure. This occurs with tubular capillaries; tubular capillaries with two different dimensions; wide and narrow necked 'ink bottle' pores; rectangular cross-section pore and trough shaped pores (pore types 1—8 and 16, *Figure 2.14*).

TYPE B   The adsorption curve is steep at saturation and the desorption curve steep at intermediate pressure. This occurs with open slit shaped capillaries with parallel walls and capillaries with wide bodies and narrow necks (pore types 6, 7, 9 and 10, *Figure 2.14*).

TYPE C   The adsorption branch is steep at intermediate relative pressures with the desorption branch sloping at intermediate relative pressure. This occurs with heterogeneous distribution of pores; tapered and wedge shaped capillaries (pore types 4—8 and 11 and 12, *Figure 2.14*).

TYPE D   The adsorption curve is steep at saturation pressure while the desorption curve is sloping. This occurs with a heterogeneous distribution of pores; open slit shaped capillaries with parallel walls (pore types 9, 10 and 13, *Figure 2.14*).

TYPE E   The adsorption curve is sloping and the desorption curve is steep at intermediate relative pressure. This occurs with similar shapes to Type A; and with multidimensional and varying size along the capillary length (pore types 4—8, 14, 15 and 17, *Figure 2.14*).

Quite a few pore shapes do not give rise to hysteresis loops, such as tubular capillaries closed or nearly closed at one end and slit or cracklike pores closed at one edge. It is thus possible to calculate pores present in materials when none exist and *vice versa* and therefore conclude, from the absence of hysteresis, that no pores are present when in fact pores exist. Pore shape is a difficult feature to characterize but a simple criterion is the ratio of surface area to pore volume which from geometry is:

$$\beta = \frac{1}{r} \times \frac{2V}{A} \tag{2.55}$$

where A and V are the surface area and volume of pores respectively, r an average pore size and β a pore shape factor.

2.2.5   Type I, BDDT classification, isotherm

When a solid or a compact contains capillaries with dimensions approximately the same order of magnitude as the adsorbate dimensions, the adsorptive behaviour of the solid is such that the potential fields on opposite walls of these narrow capillaries overlap. The filling of these spaces—

micropores—at low pressure results in a Type I isotherm. In most cases a Type I isotherm has no hysteresis although a narrow hysteresis has been found in some adsorbate–adsorbent systems (nitrogen and Saran charcoal: Culver and Heath[111]; butane and carbon: Tyson[112]). The assumption with Type I adsorption isotherms is that the adsorbed layer on the solid walls build up to a monolayer which is reached at the plateau of the isotherm. Calculation of the surface area can be obtained from the linear form of the Langmuir equation:

$$\frac{P}{P_s n^s} = \frac{1}{n_m^s b} + \frac{P}{P_s n_m^s} \tag{2.56}$$

The determination of the slope or possibly extrapolation of the plateau region can be used to give n_m^s and the surface area of the solid. The surface area calculated from equation 2.56 may however give too high a surface area value. Culver and Heath[111] evaluated the surface area of charcoal from a Type I isotherm with nitrogen and found a surface area of 3130 m^2 g^{-1}. Since a carbon atom on graphite has an area of 0.07 nm^2 the number of gram-atoms in the charcoal sample is then of the order of

$$\frac{3.13 \times 10^3 \text{ m}^2 \text{ g}^{-1}}{0.07 \times 10^{-8} \text{ m}^2} \approx 4.4 \times 10^{22}$$

The number of carbon atoms, however, in a mole, from Avogadro's number, is

$$\frac{6.03 \times 10^{23}}{12} \approx 5.0 \times 10^{22}$$

Thus every carbon atom would have to adsorb approximately one atom of nitrogen gas to achieve the calculated surface area from the Langmuir equation. This would suggest a very open structure of the charcoal incompatible with the known structure and strength of Saran charcoal. Evidence with regard to the non-validity of surface areas calculated from the Langmuir equation has been shown from surface measurements on non-activated and activated charcoal and also by the application of the Gurvitsch rule.

As has already been shown, an assembly of non-porous particles usually gives a Type II isotherm and a Type IV isotherm after compaction. This has been substantiated by Carman and Raal[113] when nitrogen adsorption isotherms were measured on Linde silica powder. The initial part of the Type IV isotherm corresponded to the same monolayer region as the Type II isotherm. Compaction diminished the interparticulate spaces within the assembly of particles to a size in which capillary condensation could occur. This phenomenon usually occurs above relative pressures of 0.4.

Kiselev[114], on the compaction of carbon black and with hexane as the adsorbate, showed that although hysteresis was obtained with compacted material the pores produced by compaction did not influence the adsorption in the monolayer region because the parts of the adsorption isotherm preceding the commencement of the hysteresis loop were identical. The surface area calculated from the two isotherms was also identical. This was also seen with the work of Zwietering[115] on spherically shaped particles and with that of Wade[97,116] using alumina sphericles. The total pore volume, however, progressively decreased as compaction pressure increased. In some cases[113] the isotherm up to the relative pressures 0.3 to 0.4 did not remain the same for compressed and uncompressed material, the adsorption isotherm curve being depressed in the monolayer region owing to the loss of surface area by adhesion.

Barrer and Strachan[117] showed for carbon powders at a porosity, ε, of unity (uncompacted) and $\varepsilon = 0.64$ and 0.37 that the volumes of the monolayer and multilayer regions decreased with a decrease in porosity. The isotherm at $\varepsilon = 0.64$ showed a hysteresis loop while at $\varepsilon = 0.37$ an isotherm similar to Type I, which can be obtained with microporous material, ensued.

2.2.5.1 Lippens and de Boer micropore method

Valuable information about the specific surface area, size and shape of pores can be obtained by plotting the experimental volumes of adsorbed nitrogen, V_a, as a function of the statistical thickness, t, of the adsorbed layer as shown by Lippens and de Boer[78].

By plotting the experimental values of V_a (cm^3/g S.T.P.) for an unknown sample as a function of the Lippens and de Boer t-curve (t in nanometre) for non-porous solids a straight line passing through the origin will be produced as long as the multilayers are formed unhindered.

The slope of this straight line is a measure of the surface area S_t (m^2 g^{-1}):

$$S_t = 1.545 \, \frac{V_a}{t} \tag{2.57}$$

In most cases the first part of the V_a/t curve is a straight line through the origin but at higher relative pressure and thus higher t values deviations from this line may occur.

Three cases are readily distinguishable:

Case A

> The surface is freely accessible and the multilayer can form unhindered on all parts of the surface. The plot of V_a against t is a straight line.

Case B

> At certain pressures capillary condensation can occur in pores of certain shapes and dimensions. The porous material then adsorbs more adsorbate than the volume of the multilayer adsorbed on a non-porous solid. This then causes an increase in the slope of the V_a/t plot.

Case C

Some types of pore (slit shaped pores or large holes) cannot be filled by capillary condensation except at very high relative pressures. Large pores are filled at relative pressures near unity and slit shaped pores are suddenly filled by adsorbed layers on both parallel walls. With such pores the surface area is no longer accessible and the V_a/t slope will be smaller corresponding to the surface area available.

2.2.5.2   Micropore analysis method (MP method)

The 'MP method' is based on the t-curve method which is a plot of the statistical thickness of the adsorbed film against P/P_s for non-porous adsorbents. Several investigators have offered t-curves for nitrogen besides de Boer. Cranston and Inkley, for example, derived a composite curve from 15 non-porous materials. The t-curve of Cranston and Inkley at $P/P_s < 0.70$ is higher than de Boer's because the adsorbents used by the former had higher heats of adsorption than those used by the latter. It is essential for the analysis of micropores that a reference t-curve be chosen which has an adsorption coefficient value similar to that of the unknown structure.

The isotherm is converted to a V_a/t plot, using the de Boer t-curve if C_{BET} is of the order 130, by taking volumes from the smoothed isotherm curve at intervals of 0.05 starting with $P/P_s = 0.08$, which is equivalent to $t = 0.35$ nm.

The surface areas of the pore walls, for different pore groups, are obtained from the differences in the slopes of the straight lines (equation 2.57) at different pore widths. The volume of the group of pores being measured, ΔV_p, is obtained from:

$$\Delta V_p = 10^3 (S_{t1} - S_{t2}) \left(\frac{t_2 + t_1}{2} \right) (cm^3\ g^{-1}) \tag{2.58}$$

where S_{t1} and S_{t2} are the surface areas obtained from the slope of the V_a/t curves 1 and 2, and t_1 and t_2 are the thickness of film in the narrowest width pores of the group.

2.2.5.3   Dubinin, Radushkevich and Kaganer (DKR) equation

From the Polanyi theory of equipotential surfaces[118] the adsorption potential, ε^P, at a surface can be expressed as:

$$\varepsilon^P = RT\ \ln\left(\frac{P_s}{P} \right) \tag{2.59}$$

Dubinin and his co-workers[119] advanced the view that the adsorption space could be expressed as a Gaussian function of this adsorption potential such that:

$$W = f(\varepsilon_0) = W_0 \exp\left[-k(\varepsilon^P/\beta)^2\right] \tag{2.60}$$

where W is the space volume filled with adsorbate in the vicinity of a solid surface, W_0 is the total volume of micropores, ε^P is the adsorption potential, β is Dubinin's affinity coefficient between polar forces which occur between solid and adsorbate molecules and k is a factor which characterizes the pore size distribution within the solid.

The parameter W can be obtained from the amount adsorbed at a pressure P, which when expressed in grams of adsorbate per gram of solid, x, becomes x/ρ where ρ is the liquid density of adsorbate. Equation 2.60 then becomes:

$$\frac{x}{\rho} = W_0 \exp\left[-\frac{k}{\beta^2}\left(2.303\ RT \log_{10}\frac{P_s}{P}\right)^2\right] \tag{2.61}$$

or

$$\log_{10} x = \log_{10}(W_0\rho) - 2.303\frac{k}{\beta^2}(RT)^2\left(\log_{10}\frac{P_s}{P}\right)^2 \tag{2.62}$$

A plot of $\log_{10} x$ against $[\log_{10}(P_s/P)]^2$ should then give a straight line of slope $(2.303k/\beta^2)(RT)^2$ and intercept $\log_{10}(W_0\rho)$ where W_0 is then regarded as the micropore volume. The above equation has been found[119] to be valid over the relative pressure range 1.0×10^{-5} to 0.2 for microporous solids.

Kaganer[120] modified equation 2.62 to calculate the specific surface area from Type I isotherms because it was assumed that the distribution potential was related to the number of adsorption sites on the surface of the solid.

$$\log_{10} x = \log_{10} x_m - D\left(\log_{10}\frac{P_s}{P}\right)^2 \tag{2.63}$$

where D is $2.303k_1 R^2 T^2$ and the intercept is equivalent to the logarithm of the monolayer capacity rather than the logarithm of the pore volume.

2.2.5.4 Theory of volume filling of micropores (TVFM)

Dubinin[121] in describing the physical adsorption of gases and vapours in micropores proposed, with the aid of a computer program to correct the experimental isotherm, a two-term adsorption equation for the volume filling of micropores. This proposed theory has made it possible to classify total volume adsorbed into two porous volumes, one related to micropores of size less than 0.6—0.7 nm and the other to supermicropores (ultramicropores) of sizes in the range 0.6—0.7 to 1.5—1.6 nm.

The DKR equation was written generally as:

$$a = \frac{W_0}{V^*}\exp\left(-\frac{A}{\beta E_0}\right)^n \tag{2.64}$$

where

a = equilibrium adsorption at relative pressure P/P_s
W_0 = limiting volume of adsorption surface or micropore volume
V^* = molar volume of adsorbate at P/P_s equal to unity (variable with temperature and adsorptive)
A = changes in Gibbs free adsorption energy

$$A = -\Delta G = RT \ln (P_s/P)$$

β = affinity constant comparing standard with experimental values
E_0 = adsorption energy of standard vapour
n = generally equal to 2 for micropores in carbonaceous material but may range from 1.5 to 2.5 and even 3 for solids with fine pores

From the TVFM adsorption equation Dubinin has postulated that two experimental parameters can be determined from one isotherm, namely the micropore volume, W_0, and the adsorption energy, E_0.

These parameters may not, however, be single valued micropore parameters but may be descriptive for either a relatively narrow or a broadly distributed adsorption space volume of micropores.

Ramsay and Avery[122] have shown that on compaction of spherical particles of silica with a narrow size distribution, a Type II nitrogen adsorption isotherm was observed at zero compaction pressure while a Type IV isotherm was observed at 0.77 GN m^{-2} with a Type I isotherm at 1.16 and 1.54 GN m^{-2} (*Figure* 2.2). Analysis of the uptake of nitrogen and comparison of the calculated monolayer from either the BET or DKR equations indicated that the C_{BET} coefficient increased with increase in compaction. The C_{BET} increase reflects an increase in the adsorption energies from uncompacted or plane surfaces to compacted material. The monolayer capacities obtained from either the BET or the DKR equation are in close agreement, which supports the process of surface coverage in micropores initially proposed by Kaganer, rather than TVFM.

Thus, although Type I isotherms are invariably obtained by microporous adsorbents, the close agreement seen between the monolayer capacities from the BET and DKR equations (*Table 2.10*) favours the possibility of

Table 2.10. Physical adsorption data from N_2 isotherms on the compaction of spherical silica particles

Compaction pressure/ GN m^{-2}	Surface area/m^2 g^{-1}	C_{BET}	Total pore volume/cm^3 g^{-1}	Monolayer capacity/mmol g^{-1}	
				DKR eqn.	BET eqn.
0	739	73	—	7.64	7.59
0.77	346	184	0.222	3.48	3.55
1.16	268	601	0.152	2.87	2.76
1.54	213	$> 10^3$	0.109	2.15	2.19

surface coverage and not necessarily the volume filling mechanism pro-posed by Dubinin. The onset of volume filling, being dependent upon the size and type of adsorbate molecule and adsorbent, will not occur when D, the diameter of the pore width, is greater than 3.0d where d is the molecular diameter of the adsorbate[123,124]. For nitrogen, which on glass has a hard-sphere closest approach diameter of 0.40 nm, a pore size of less than $3 \times 0.4 = 1.2$ nm will be expected to fill by volume filling but with nitrogen on charcoal where the hard-sphere closest approach diameter d is 0.52 nm, pores of greater than $3 \times 0.54 = 1.6$ nm will not be filled. Knowledge of the monolayer capacities from BET and DKR can thus provide a distinction between the two processes of micropore coverage.

2.3   Mercury penetration

The method for evaluating the surface properties of a solid by mercury porosimetry is to force mercury into the pores within a solid or powdered substance. The technique of mercury intrusion is based on the principles of capillarity and the size of pore which fluids can enter.

For fluids like mercury, which have a contact angle greater than 90 degrees with respect to a solid, pressure is required to force liquid into the pores of a solid or voids of a compact.

2.3.1   Young–Laplace equation

The technique of mercury intrusion, based on the principles of capillarity to determine pore size, has been used for a variety of solids, including cements, clay, aluminium oxides, coal, coke, bricks, limestone and membrane filters. When mercury is forced into a porous solid, energy F is expended in creating an intruding mercury surface which can be expressed for iso-thermal conditions as:

$$dF_1 = \gamma_{LS}\, dS_{LS} + \gamma_{GS}\, dS_{GS} + \gamma_{LG}\, dS_{LG} \tag{2.65}$$

where γ_{LS}, γ_{GS} and γ_{LG} are the free surface energy per unit area of the mercury–solid interface, gas–solid interface and mercury liquid–vapour interface respectively, while S_{LS}, S_{GS} and S_{LG} are surface area of solid in contact with mercury, surface area of gas–solid interface and surface area of mercury in contact with its vapour respectively.

The energy needed to force an infinitesimal volume, dV, of mercury into a porous solid under reversible isothermal conditions is:

$$dF_2 = P\, dV \tag{2.64}$$

where P is the intrusion pressure.

Equating dF_1 and dF_2 and neglecting the change of the surface area of mercury menisci, S_{LG}, together with the relationship of Young, Laplace and Dupré that:

$$\gamma_{LS} - \gamma_{GS} = -\gamma_{LG} \cos \phi \tag{2.65}$$

where ϕ is the contact angle between solid and mercury, we obtain

$$P \, dV = -dS_{LS} \gamma_{LG} \cos \phi \tag{2.66}$$

where $dS_{LG} = -dS_{GS}$.

For pores of a circular cross-section open at both ends or closed at one end and with radius r the relationship between volume and surface area of mercury within a circular cross-section becomes:

$$\frac{dV}{dS_{LG}} = \frac{r}{2}$$

Substitution back into equation 2.66 gives:

$$P = \frac{-2\gamma_{LG} \cos \phi}{r} \tag{2.67}$$

For slit-shaped pores or porous substances with microfissures:

$$\frac{dV}{dS_{LS}} = \frac{d_c}{2}$$

Thus equation 2.66 transforms for slit-shaped pores of diameter d_c into

$$P = \frac{2\gamma \cos \phi}{d_c} \tag{2.68}$$

Equations 2.67 and 2.68 are the basis for the evaluation of pore sizes or void sizes using a Mercury Penetration Porosimeter where the amount of mercury forced into a porous solid is a function of pressure.

The size of a pore or void that mercury will penetrate at one atmosphere (1.06×10^6 dyne cm^{-2} = 1.06×10^5 N m^{-2}) is 7 micrometres, assuming a contact angle ϕ of 140 degrees and a surface free energy, γ, of 480 erg cm^{-2} (0.48 N m^{-1}) at 20 °C.

$$r = \frac{-2\gamma \cos \phi}{P} = \frac{2 \times 480}{1.06 \times 10^6} \cos 140° = 7 \times 10^{-4} \text{ cm} = 7 \ \mu\text{m}$$

2.3.2   Pore size distribution by mercury penetration

A mercury porosimeter, for the measurement of the cumulative pore-void size distribution, must generate and withstand high pressures while the measurement of the quantity of mercury forced into the pore-void space is achieved. A variety of porosimeter designs are now commercially available, since the initial work of Ritter and Drake[125], which cover a wide range of pore-void sizes.

In all cases porosimeters must have means for the removal of entrapped gases from the system and test specimen because unless the specimen is thoroughly evacuated prior to mercury intrusion, the compressibility of the entrapped gases can lead to considerable error.

The raw data from mercury penetration experimentation consists of the pressure applied and an indication of the penetration volume in terms of a count number. Each count number corresponds to a specific volume of mercury, which is of the order of 7.5×10^{-4} cm^3 per count, but varies with sample cell and the internal diameter of the cell and the pitch of the probe thread.

Compression of mercury as well as the compression of glass, begins at approximately 10^4 p.s.i.a., so a correction should be made by performing a blank run on an empty sample cell.

If it is assumed that the angle of contact is 130 degrees and the surface tension, γ, is 474 dyne cm^{-1} (erg cm^{-2}; 0.474 N m^{-1}), equation 2.67 transforms to:

$$r = \frac{-2 \times 474 \times (-0.6428)}{P_c \times 6.89 \times 10^4} = \frac{88.5}{P_c}$$

where P_c is the pressure in p.s.i.a. and r in micrometres.

In the valuation of mercury pore-void size distributions a number of factors must be considered:

1. A wide range of values have been found for the contact angle between mercury and various solid surfaces which range between 116 and 160 degrees (Rootare and Nyce[126], Scholten[127], Gmelin[128], and Scholten, Beers and Kiel[102]).
2. Variation of surface tension with temperature. Roberts[129] has found that the temperature coefficient of the surface tension of mercury, $(d\gamma/dT)$, is -0.21 erg cm^{-2} K^{-1}. The value of surface tension has been found by Young[130] to be 485 ± 5 erg cm^{-2} (0.485 N m^{-1}) at 20 °C which is in agreement with the value of Roberts of 487 erg cm^{-2} (0.487 N m^{-1}) at 20 °C for pure mercury. The literature values of surface tension show, however, a spread of approximately 100 erg cm^{-2} (Kernaghan[131], 432.2; Cook[132], 512 erg cm^{-2}). A deviation of 473—485 erg cm^{-2} can give a deviation of approximately 2 per cent in pore size which is not however as great as the deviation due to variation in the contact angle.
3. The compressibilities of materials, K, range from 10^{-11} to 10^{-12} cm^2 dyne^{-1}, while mercury and glass have compressibilities of 3.55×10^{-12} and 2.6×10^{-12} cm^2 dyne^{-1} respectively. Lea and Maskell[133] have shown that correction to the penetration volume, obtained from the raw porosimetry data, is essential, in addition to the blank correction, for non-porous compressible material such as nylon ($K = 19.0 \times 10^{-12}$ cm^2 dyne^{-1}) for pressures in excess of 5000 p.s.i.a.

Compression of powders and the packing of nearly spherical particles often results in the development of a pore system within the mercury intrusion sample cell. Particle size determinations can be achieved from the 'breakthrough pressure' in terms of the porosity of packed powder and the contact angle. Kruger[134], Frevel and Kressley[135] and Mayer and Stowe[136, 137] related the breakthrough pressure P_b, of mercury in the access openings in a collection of non-porous uniform solid spheres as:

$$P_b = \gamma_{LG}(L'/A)/r_s \qquad (2.69)$$

where r_s was the radius of the spheres. The function (L'/A) for all packings of spheres varying between a three dimensional cubic packing and a hexagonal close packed assembly was calculated by Mayer and Stowe (Tables II and III in their paper[136, 137]) over the porosity range 47.64—25.95 per cent. Stanley-Wood[138], using the relationship of Mayer and Stowe, investigated the penetration of mercury into packed assemblies of particles to indicate the presence and measure the shape of the size distribution of micrometre and sub-micrometre porous and non-porous powders.

With non-porous spheres mercury intrusion evaluated a similarly shaped distribution of particle sizes to that determined by other more conventional particle size techniques. The radius of sphere measured was not, however, the surface radius predicted by the Mayer and Stowe model.

With microporous irregularly shaped particles, the mercury intrusion method detected both micrometre and sub-micrometre particles while with mesoporous or macroporous material little correlation existed between the evaluated mercury radius and other particle size radii.

Internal surface area of porous solids can be calculated from mercury penetration independent of the BET method. Equation 2.66 can be rewritten as:

$$S_{LS} = -\frac{1}{\gamma \cos \phi} \int_{V_0}^{V} P \, dV \qquad (2.70)$$

Rootare and Spencer[139] used equation (2.70) to write a computer program in FORTRAN to calculate the total pore area, pore volume and pore-void size distribution of porous solids from mercury penetration experimental data. They found good agreement between the pore area of carbon Spheron 6 (a reference adsorbent for surface area measurements) determined by mercury (108.9 m^2 g^{-1}) and nitrogen adsorption (110 m^2 g^{-1}).

The relationship between pore size distributions determined by nitrogen desorption has been in good agreement with that determined by mercury penetration methods for solids which have a low pore volume[140,141]. With alumina, silica, silica aluminas and catalysts of less than 1.2 cm^3 g^{-1} nitrogen pore volume, Brown and Lard[142] also found agreement between nitrogen and mercury pore size distributions. However, with high pore volume silica (1.7—2.8 cm^3 g^{-1}), larger discrepancies were obtained between nitrogen and mercury pore size distributions. These discrepancies were attributed to the compression of silica from the exterior and interior of the particle. As mercury entered large crevices and pores the pore wall of the silica collapsed forming smaller pores. As mercury filled the voids resulting from the compression, an exaggerated macroporosity ensued. Brown and Lard concluded that the effects of compression were so severe that pronounced discrepancies occurred between nitrogen adsorption and mercury intrusion surface areas and pore size distributions could not be used for certain materials.

Zwietering and Koks[143] compared pore size distributions from an iron oxide–chromium oxide catalyst, obtained by nitrogen capillary condensation and mercury intrusion, and found agreement in the distributions using the classical Kelvin equation for nitrogen condensation and the Washburn[144] relationship for mercury intrusion. Zwietering *et al.* compressed non-porous spherical particles of Aerosil with a mean particle diameter of 0.15 nm and found that there was only fair agreement between the two methods. This degradation in agreement was attributed to the use of a circular cross-section pore model as opposed to the more realistic model of packed spheres. De Wit and Scholten[103] re-analysed the powders of Zwietering but used the Kelvin equation corrected according to Broekhoff instead of the classical Kelvin equation. In both cases the mode of the pore size distribution from nitrogen capillary condensation give approximately a 30 per cent larger radius than that from mercury intrusion.

The comparison of nitrogen and mercury pore size distributions on uncompressed and compressed (3500 kg cm^{-2}) porous (hollow tubed) powder[103] showed that the interparticle voidage distribution from mercury did not coincide with that obtained from nitrogen adsorption. The distribution peak of the intraparticle pores obtained from mercury penetration also gave a smaller value than that found by electron microscopy and nitrogen adsorption. The nitrogen adsorption distribution, from the corrected Broekhoff–Kelvin equation, was in agreement with the distribution obtained by electron microscopy.

2.4 Application of nitrogen isotherms Types II and IV and mercury intrusion to compacted solids

2.4.1 Nitrogen adsorption isotherms

Comparison of the nitrogen adsorption isotherms from six powders compressed over the pressure range 0—1480 GN m^{-2} showed[145] that Type II isotherm occurred with the uncompressed powder, with little or no hysteresis. When particles were forced together, the change to a Type IV isotherm indicated the formation of interparticle voids in the mesopore range. The powders chosen varied in particle shape and degree of deformability, so that the degree of interparticle adhesion and variation in voidage could be ascertained.

The greatest reduction (≈ 40 per cent) in surface area occurred with spherical silica particles in the size range 1—8 nm, and with irregularly shaped, flake like, mica particles of thickness 10—40 nm and breadth 50—

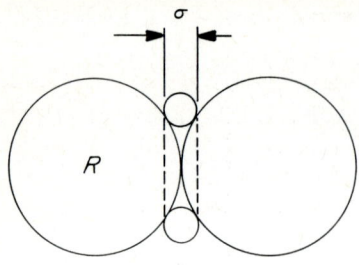

Figure 2.16.   *Diagram of two contacting spheres showing the decrease in accessible surface to a molecule of diameter σ*

3000 nm. The loss of surface of the silica spheres was attributed to the reduced accessibility of surface close to the region of contact of neighbouring particles. The accessible surface of contacting particles can be judged from the fact that when two spheres of radius R are in contact the total surface accessible to an adsorbant molecule of diameter σ is reduced (*Figure 2.16*). From the work of Ramsay and Avery[12] the accessible surface, S, can be taken as the difference between the total geometrical surface S_{geom} and the surface of the obstructed zone, which for spherical segments of height $\sigma/2$ becomes:

$$S = 4\pi R^2 - \pi R\sigma \tag{2.71}$$

Equation 2.71 can be rewritten in terms of number of contacts n as:

$$S = 1 - \frac{n\sigma}{4R} S_{geom} \tag{2.72}$$

A plot of co-ordination number, n, obtained from the packing of spheres *versus* the surface area, can be used to evaluate the term $\sigma/4R$. For spherical silica and zirconia[12] of 2.0 nm radius and with the adsorbate nitrogen of molecular dimensions 0.43 nm, the slope of equation 2.72 gave values of 0.057 and 0.063 respectively compared with a theoretical value of 0.053 (*Figure 2.17*). The work of Gregg and Langford[145] on silica and alumina showed that the value from the graph of co-ordination number against surface area was 0.06 and 0.014 respectively, compared with a theoretical value of 0.012. The softness and deformability of these two powders can be judged from Moh's hardness scale where silica has a value of 7 compared with alumina which has a value of 12. The harder material tended to conform to the concept of accessible area while silica did not.

Halloysite particles which are hollow in character with mean outside and inside diameters of 50 and 20 nm respectively, showed a remarkable resistance to crushing and reduction in surface area (6 per cent) on compaction, while kaolin, a flaky material of 30—60 nm thickness and 300—1000 nm breadth, on compaction showed an increase in surface area (≈ 2 per cent). The increase in surface area of kaolin was attributed to its disintegration under compaction pressures and the production of numerous interparticle voids of irregular shape caused by the generation of irregular shaped kaolin fragments which did not pack together like mica particles.

The effect of compaction on the surface topography of particles can be

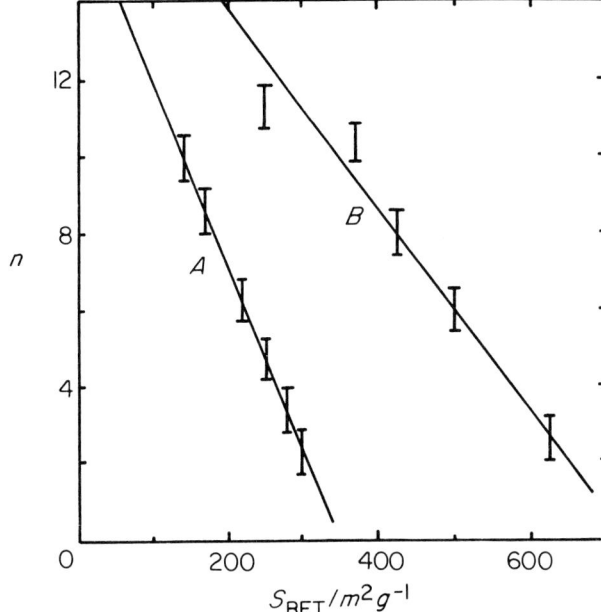

Figure 2.17. Relationship between specific surface area of compacts and the coordination number of particles; A, zirconia; B, silica

analysed by modifications of either the *t*-plot or the α_s plot (Section 2.2.5.1). A more direct method, however, is that of comparing the isotherm at various relative pressures of the compacted material with the isotherm from uncompacted powder at the same relative pressures, and then the calculation of a ratio (*f* values) from the adsorption values (*Figure 2.18*). Variation from the ratio value of unity indicates changes, due to compaction, from the original uncompacted structure of the solid[145].

Pore size analysis using either cylindrical or parallel plate models can be used to measure the pore size distribution in compacted materials. Gregg and Langford used the pore model of Roberts[89] and the *t*-correction based on the isotherm of the uncompacted powders. Silica and alumina showed the typical change from non-porosity to macroporosity and capillary condensation associated with Type II and IV isotherms. Silica uncompacted showed a characteristic isotherm of macroporosity which on compaction and use of the *f* ratio values suggested the formation of mesopores of a slit-like nature. Uncompacted illite powder had a small amount of mesoporosity in the size range 6.8—12.2 nm, which on compaction increased in volume indicative of a broadening of the distribution. Kaolin powder, mesoporous in the uncompacted state, on compaction showed an enhancement in adsorption without hysteresis. This is possibly due to capillary condensation in wedge shaped pores. Uncompacted halloysite contained, as

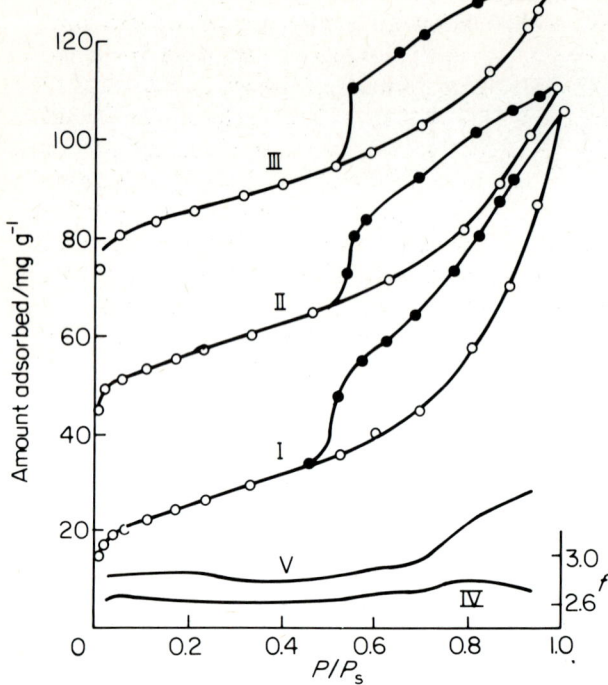

Figure 2.18. *Compaction of hallyosite powder*
  I   *Sorption isotherm of nitrogen at 77K on uncompacted powder*
  II   *Sorption isotherm of nitrogen at 77K with a compacting pressure of* 495 GN m^{-2}; *this curve is displaced upward by* 30 mg g^{-1}
  III   *Sorption isotherm of nitrogen at 77K with a compacting pressure of* 1480 GN m^{-2}; *this curve is displaced upward by* 60 mg g^{-1}. *Open circles, adsorption; full circles, desorption*
  IV   *Plots of f against P/P$_s$ based on uncompacted mica as reference with a compacting pressure of* 1480 GN m^{-2}
  V   *As IV, but for uncompacted powder*

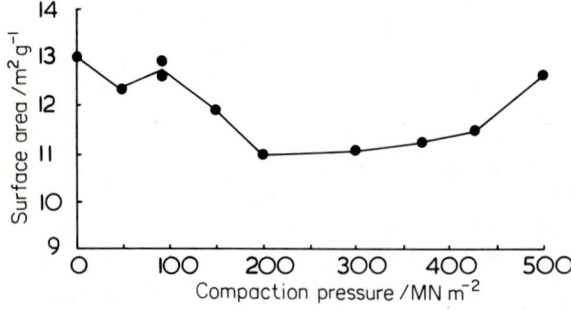

Figure 2.19. **BET** *surface area from compacts of magnesium oxide*

expected with hollow tubes, a high degree of mesoporosity, which on compaction showed a progressive elimination of the interparticulate mesopores.

Stanley-Wood[146] measured the surface area and pore-void size distribution of uncompacted and compacted porous magnesium oxide over the
compaction pressure range 35—490 MN m^{-2}. The surface area initially
decreased from an uncompacted surface area of 13.2 m^2 g^{-1} to
10.9 m^2 g^{-1} at a compaction pressure of 280 MN m^{-2} and then increased
from the minimum value to 12.3 m^2 g^{-1} at higher pressure (*Figure 2.19*).
The increase in surface area was attributed to the fracture of magnesium
oxide particles. Pore size analysis by use of the Lippens, Linsen and de Boer
t-curve showed that the uncompacted powder was microporous while the
compacts, by BJH pore size analysis, contained pore-voids in the range
1.34—5.17 nm together with unaffected micropores. Direct comparison of

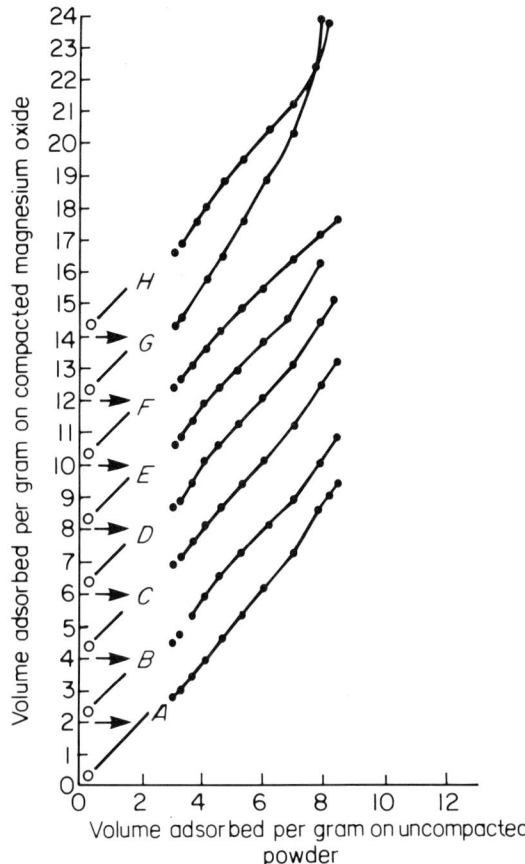

*Figure 2.20. Compaction-uncompaction curves for
magnesium oxide. The compaction pressure is as
follows: A, 34.9; B, 70.0; C, 140.3; D, 210.4;
E, 280.6; F, 350.7; G, 420.9; H, 490.6 MN m^{-2}.
The axis is displaced upwards for each compaction
pressure for clarity*

the adsorption isotherms obtained by nitrogen at 77 K from compacted materials to that of the reference uncompacted material was achieved by the graphical plot of the volume adsorbed on uncompacted magnesium oxide at specific relative pressure to that adsorbed on compacted material at the same relative pressure. Deviation from unit slope indicated changes in the solid due to compaction (*Figure 2.20*). The initial reduction in surface area was due to either deformation of the particles or pore closure. The closure of pore-voids was substantiated by the disappearance of certain pore-void sizes within the compacts. In an extension of this work, Stanley-Wood and Johansson[147] showed that for the porous materials magnesium oxide, bentonite and magnesium trisilicate, the surface area decreased with the increase in compaction pressure over the range 17.5—490 MN m^{-2}. The interparticle voidage was calculated from the subtraction of the sum of the intraparticle volume obtained by BJH computer analysis V_{BJH} and the reciprocal solid density $1/\rho_s$ from the reciprocal compact density $1/\rho_c$:

$$\text{Interparticle voidage} = \frac{1}{\rho_c} - \left(V_{BJH} + \frac{1}{\rho_s} \right)$$

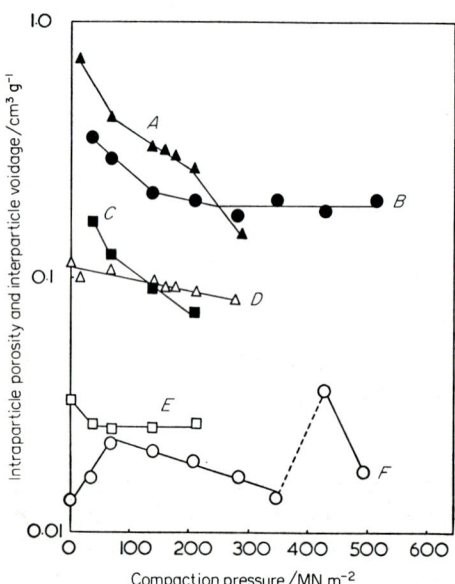

Figure 2.21.   Variation of intraparticle porosity and interparticle voidage with compaction pressure
   A   *Interparticle voidage in magnesium trisilicate*
   B   *Interparticle voidage in magnesium oxide*
   C   *Interparticle voidage in bentonite*
   D   *Intraparticle porosity in magnesium trisilicate*
   E   *Intraparticle porosity in bentonite*
   F   *Intraparticle porosity in magnesium oxide*

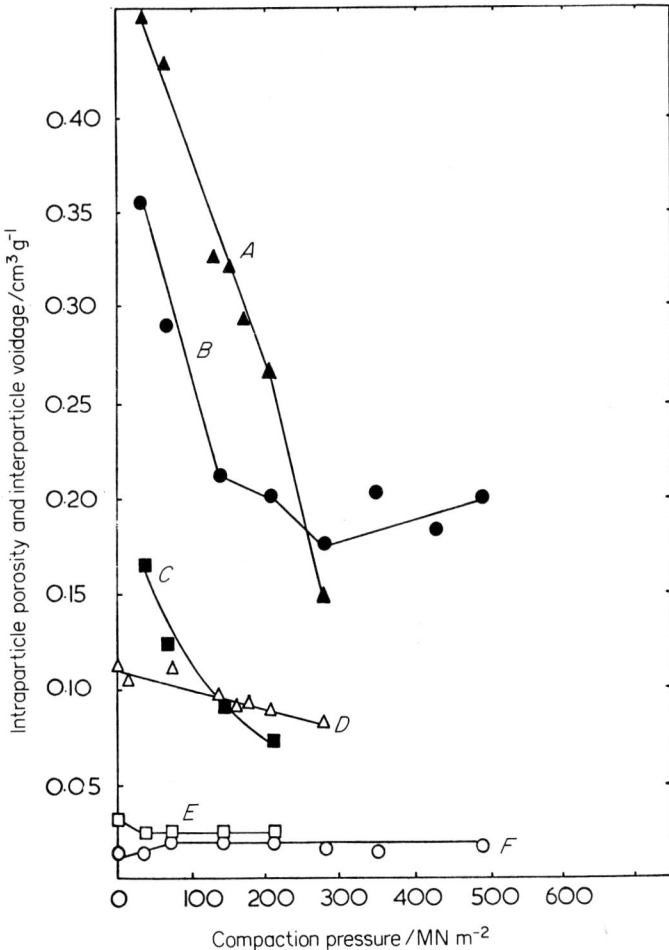

Figure 2.22. Variation of particle porosity and compact voidage with compaction pressure. For Key, see Figure 2.21

In the case of magnesium trisilicate and bentonite both the interparticle and intraparticle porosity decreased with compaction pressure. With magnesium oxide the intraparticle porosity initially increased and then decreased while the interparticle porosity remained constant over the compaction pressure range 17.5—140 MN m^{-2} (*Figures 2.21 and 2.22*). The contact area between particles of the three solids evaluated by the simple method of calculating half the difference between the surface area before and after compaction, was compared with the work of Hardman and Lilley[21] on sodium chloride, which is known to deform plastically. A linear relationship between contact area and compaction pressure occurs with sodium chloride and this was also observed with compacts produced from magnesium oxide and magnesium trisilicate in the middle range of com-

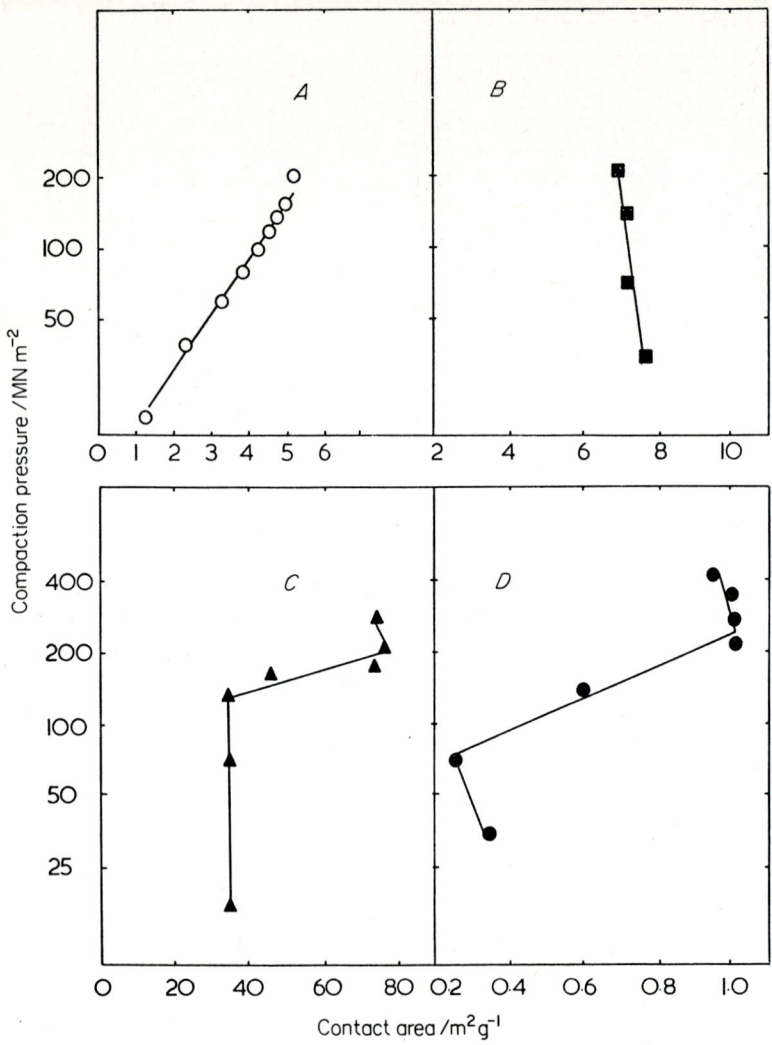

Figure 2.23.   The effect of compaction pressure on particle–particle contact area for A, *sodium chloride;* B, *bentonite;* C, *magnesium trisilicate;* D, *magnesium oxide*

paction pressures (*Figure 2.23*). At higher compaction pressures contact area decreased and was attributed to particle fracture. Bentonite showed no plastic particle–particle contact. A similar relationship has been shown by Gregory[148] with Lady Windsor coal, in which the projected area, obtained by microscopical examination (*Figure 2.24*), remained constant initially with increase in compaction pressure (region *a—b*) and then increased rapidly over a short intermediate pressure region (region *c—d*) to return eventually to a region (region *e*) where little variation in particle area occurred with compaction pressure increase. This phenomenon is described as micro-squashing.

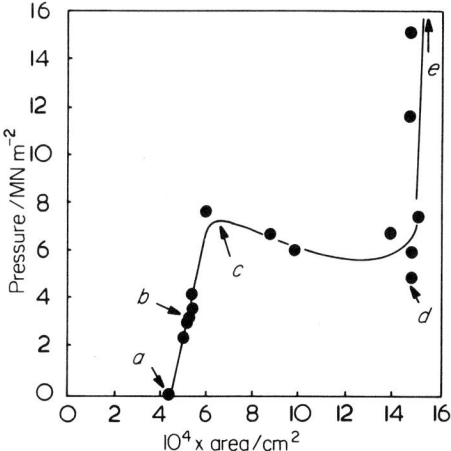

Figure 2.24.   Microsquashing of Lady Windsor coal

Strömgren, Åstrom and Easterling[149] studied the effect of interparticle contact area on the strength of cold pressed compacts by application of the work of Easterling and Thölen[150] who derived a simple relationship between the theoretical strength, σ_t, of a three dimensional metal powder compact, the work of adhesion across interparticle boundaries, W, and the mean particle radius, R, of a non-plastically deformed powder:

$$\sigma_t \simeq 3 \frac{W}{R} \tag{2.73}$$

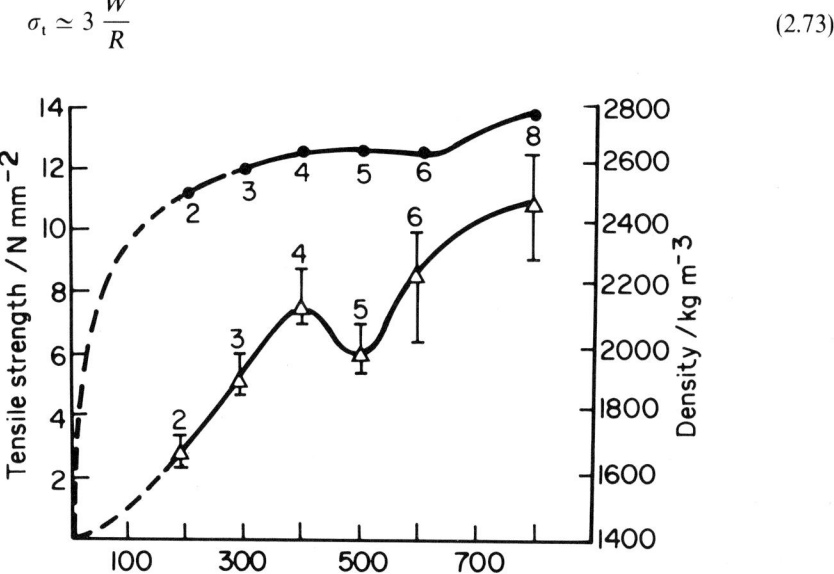

Figure 2.25.   Changes in tensile strength (triangles) and density (circles) of aluminium powder as a function of cold-compaction pressure

Because the work of adhesion is related to the geometry of contact between clean equisized metal particles, equation 2.73 can be transformed to:

$$W = \frac{4}{9\pi}(1 - v^2)E\,\frac{a^3}{R^2} \tag{2.74}$$

where v is the Poisson ratio, E is the modulus of elasticity and a is the radius of grain boundary between two spherical particles of radius R. Combination of equations 2.73 and 2.74 shows that σ_t is proportional to a^3 and the tensile strength of a green compact is dependent only on the grain boundary area development relative to the effective mean particle size. *Figure 2.25* shows the relationship of the tensile strength and density of aluminium compacts to compaction pressure while *Figure 2.26* shows

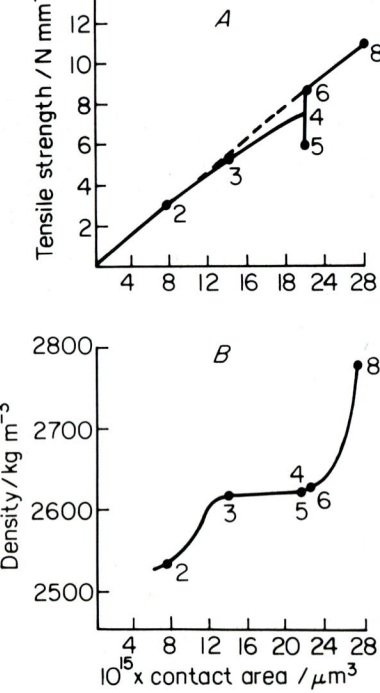

Figure 2.26.   The change in contact surface between deformed particles (expressed as a^3) as a function of A, tensile strength and B, density of the green compacts

the relationship of tensile strength and density to the change in contact surface. Since the relationship between density and compaction pressure is almost constant over the compaction pressure range 200—800 N mm^{-2} the relationship between density and contact area seen by Strömgren *et al.* is similar to the work of Gregory[148] and Stanley-Wood and Johansson[147] for non-metal powders.

Morrison and Richmond[151] used an analogue of the sphere flattening of Hertz elasticity for the flattening of a regular assembly of rigid, perfectly plastic spheres, compressed uniaxially. They computed the ratio of surface displacement to sphere radius, R, and the ratio of contact area to the square of sphere radius, R^2, and found the relationship:

$$P_A = 2.93 Y \qquad (2.75)$$

where P_A is the contact pressure and Y the yield stress and deduced that

$$a = 8.45 R^2 \left(1 - \frac{h}{D}\right) \qquad (2.76)$$

where a is the contact area with D the initial sphere diameter and h the distance between contact surfaces measured along the uniaxial loading direction.

The range of applicability of these equations to describe the variation of bulk density with compacting pressure is, however, restricted to the latter part of the initial region of the relationship of $\ln [1/(1 - D)]$ against compression pressure curves defined by Hewitt *et al.* and to cubic or hexagonal close packed assembly of spheres.

Some materials on compaction undergo fragmentation rather than plastic deformation and microsquashing. The material dicalcium phosphate (DCP) is known to undergo fragmentation when compacted[152]. Stanley-Wood and Shubair[153,154] compacted dicalcium phosphate with various concentrations of starch mucilage as binder and determined the surface areas and porosities of this granulated powder before and after compaction (*Figure 2.27*). They distinguished between the surface area contribution due to fragmentation and that due to bonding by measurement of the nitrogen

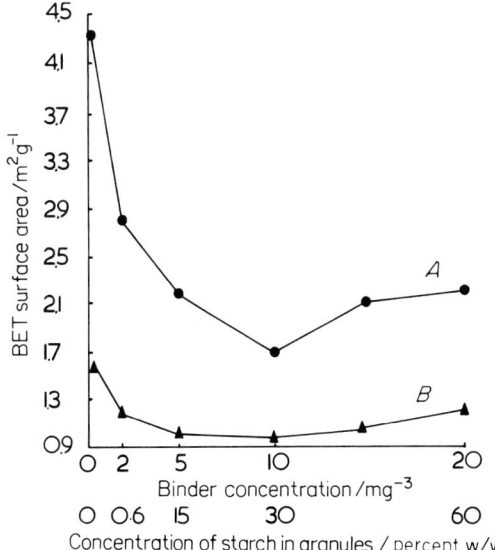

Figure 2.27.   Effect of compressional force on surface area of A, *compacts and* B, *granules*

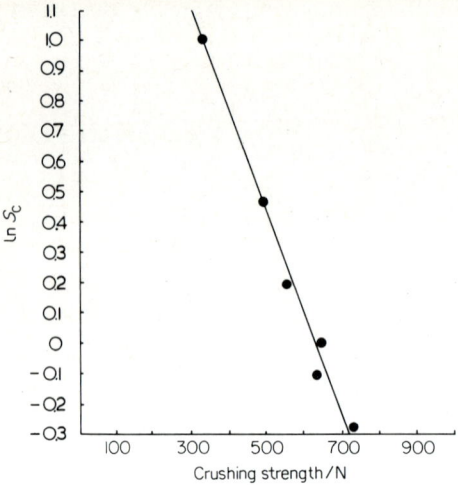

Figure 2.28.   Relationship between S_c and crushing strength of compacts. Extrapolation leads to zero crushing strength at 22.50 m^2 g^{-1}

BET surface area and crushing strengths of the resultant compacts produced at various compaction pressures. Extrapolation of the logarithmic relationship between measured BET surface area of compacts, S_c, and the crushing strength of compacts, C_f, to zero crushing strength gave a surface area generated solely by the application of force (*Figure 2.28*).

The surface area lost due to bond formation, S_b, is then:

$$S_b = S_f - S_c \tag{2.77}$$

where S_f is the surface area due to fragmentation of particles or granules obtained from the extrapolated value of the S_c *versus* C_f graph (*Figure 2.28*). The relationship between surface area of bonding to crushing force could then be expressed (*Figure 2.29*) as:

$$\frac{C_f}{S_b} = \frac{C_f}{S_{b_0}} + \frac{k}{S_{b0}} \tag{2.78}$$

where S_{b_0} was in agreement with the extrapolated value of *Figure 2.28* and is the maximum surface area of dicalcium phosphate powder that can be achieved by breaking and k is a constant.

Measurement of the compact pore volume by mercury intrusion also showed a similar relationship with the crushing strengths of compacts:

$$\frac{C_f}{V_{Hg}} = \frac{C_f}{V_{Hg_0}} + \frac{b}{V_{Hg_0}} \tag{2.79}$$

where V_{Hg} is the mercury intrusion pore volume, V_{Hg_0} is the minimum value of the mercury pore volume due to bonding, and b is a constant. The linear relationship between bond area and compact mercury intrusion volume

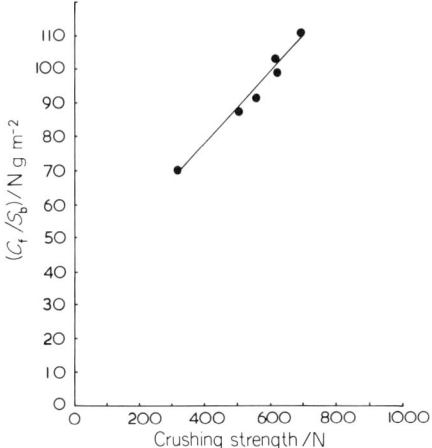

Figure 2.29.   Function of crushing strength and bond area related to compact strength

deduced from equations 2.78 and 2.79 has been verified experimentally[154] and is similar to the relationship between surface area of compacts and coordination number of particles seen with spherical particles of silicon and zirconia (*Figure 2.17*) when the area of contact is taken as a function of the coordination number (the higher the contact area, the more contact points and the higher the coordination number), and the volume of mercury penetrated is a function of the surface area of voids in compacts.

Tyler, Hambleton and Hockey[155] compressed Aerosil silica powder (RA2 and UA3) into discs at various compaction pressures and found that mechanical pressing and hydrothermal treatment increased the effective microporosity of the samples and enhanced adsorptive properties of the new surface. The RA2 silica powder yielded a mean macropore size from the Kelvin equation of 9.0 nm. When compressed at 5 and 10 tons in^{-2} (RA2/5 and RA2/10 respectively), the adsorption of nitrogen was enhanced and a mean pore size of 4.0 nm was obtained for both discs, the shape of the pores producing a mixture of Type A and E pores according to the classification of de Boer[156] (Section 2.2.4.8).

In addition to pore size analysis by the Kelvin equation, the nitrogen isotherms obtained were analysed by the de Boer *t*-curve technique and both RA2/5 and RA2/10 showed microporosity. Comparison was also made, with regard to the changes due to compaction, by a modified *t*-curve technique which used the uncompacted RA2 powder as a reference solid and compared the adsorption data obtained from compressed RA2/5 and RA2/10 with that of uncompressed RA2. This modified *t*-curve technique indicated that changes occurred in both microporosity and mesoporosity regions on compaction (*Figure 2.30*). The UA3 uncompacted powder showed no hysteresis and no microporosity. On compaction, however, to 5 tons in^{-2} the UA3 powder produced a Type A macroporous hysteresis

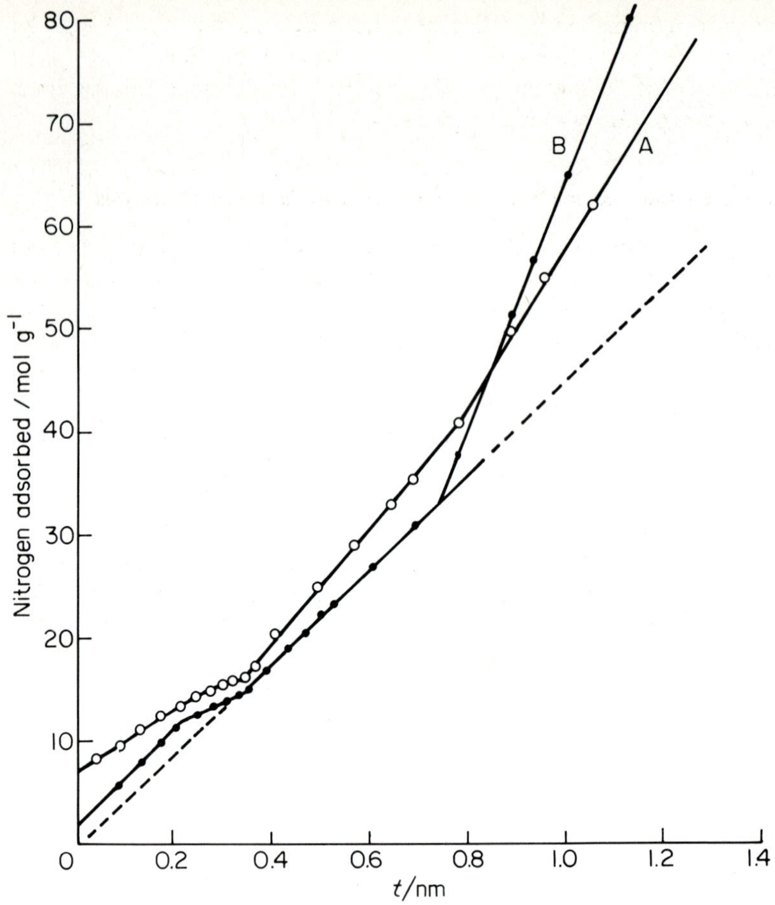

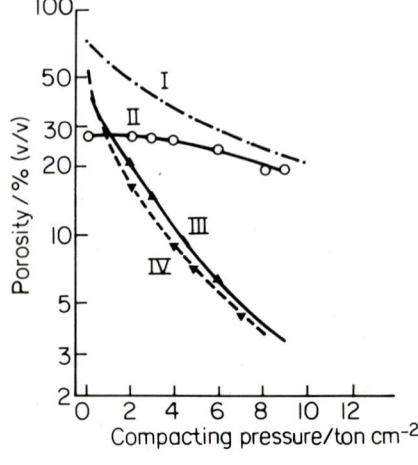

Figure 2.30. The t-plots for Aerosil powders RA2/5 (curve A) and RA2/10 (curve B) with RA2 powder (broken line) as reference solid

Figure 2.31. Change of total porosity (curve I), porosity inside powder particles (curve II), and porosity in between powder particles (curve III) for an iron powder obtained from pure magnetite by reduction at 500 °C in hydrogen, which had a particle size 24—100 mesh and a high degree of internal particle porosity. Curve IV shows total porosity of an electrolytic iron powder of particle size 40—150 mesh, having practically no internal particle porosity

isotherm with a mean pore radius of 8.0 nm but no microporosity. The technique of pore structure analysis by gas adsorption thus collaborated information gained on the change of structure from other techniques, such as heats of adsorption and infrared spectra.

2.4.2   Porosity measurement by mercury intrusion and lineal analysis

Bockstiegel[157] compacted magnetite of particle size range 44—417 μm (mean 63 μm) over the compaction pressure range 1.25—10 ton cm^{-2}. By microscopical lineal analysis of the etched samples, Bockstiegel showed that the internal particle porosity decreased differently from the interparticle porosity (voidage) of compacts (*Figure 2.31*). Increase in the compaction pressure caused individual pores to be eliminated from the compact in order of size. The larger sizes were eliminated before the smaller sizes. The largest size in a compact, D_{max}, when plotted on a logarithmic scale against the compaction pressure, P_A, revealed that the slope of the curve (d ln D_{max}/dP_A), became flatter with increase in compaction pressure. It was also found that not only the largest spaces between particles (interparticle voids) shrink under pressure but also simultaneously the smaller spaces inside particles (intraparticle pores) followed the same phenomenon although to

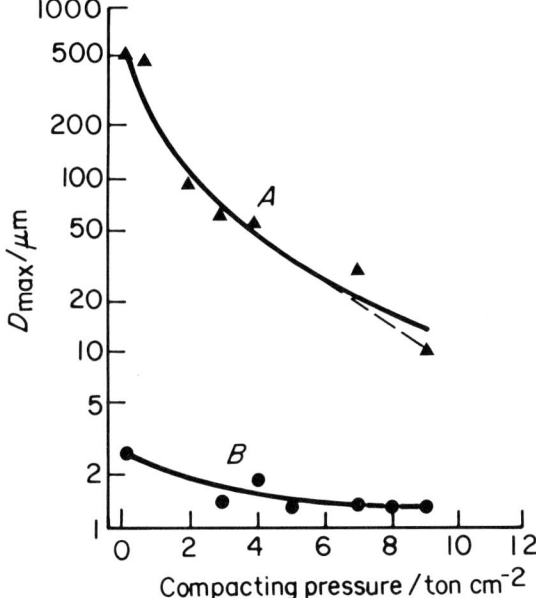

Figure 2.32.   Change of largest pore diameter, D_{max}, inside (curve A) and in between (curve B) powder particles with increasing compacting pressure, for iron powder. Pore diameters are plotted on a logarithmic scale and compacting pressure on a linear scale

a minor extent (*Figure 2.32*). The accumulated porosity, $u(D)$ arising from pores smaller than a given diameter, D, can be mathematically described by:

$$u(D) = u(D_R)\,erfc\,\ln\left(\frac{D}{D_R}\right) \tag{2.80}$$

where *erfc* is the complementary error function and D_R is an arbitrary size less than D_{max}.

Stanley-Wood[158] and Stanley-Wood and Johansson[159] showed that for compacts of magnesium trisilicate, dicalcium phosphate (DCP) and a mixture of dicalcium phosphate and Avicel (DCPA) that the intraparticle porosities in compacts can be measured by nitrogen adsorption and the

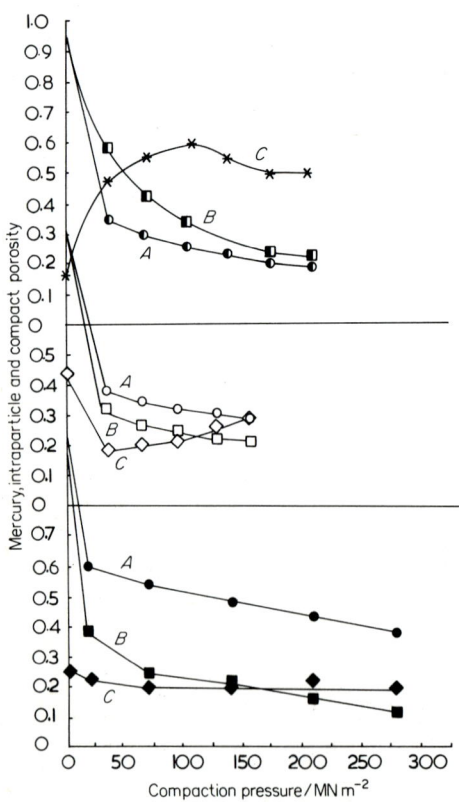

Figure 2.33.   Variation of porosity with compaction for a mixture of dicalcium phosphate and Avicel (top), dicalcium phosphate (middle) and magnesium trisilicate (bottom). A, compact porosity; B, mercury porosity; C, intraparticle porosity (÷10)

interparticle voidage measured by mercury intrusion. The intraparticle porosity of the fragmentable DCP produced a different relationship with compaction pressure from the interparticle voids. The intraparticle porosity increased with compaction instead of decreasing as seen with the interparticle voidage. With DCPA, owing to the presence of Avicel, a glidant, the increase in intraparticle porosity was less with increase in compaction pressure (*Figure 2.33*). The cumulative interparticle voidage distribution, ε_v, of different void sizes determined by mercury penetration for all three powders (*Figure 2.34*) showed that in the void size range 1.0—100 μm each compaction maintained its unique voidage. In the void size range below 1.0 μm, however, the void size distribution for compacts of magnesium trisilicate, DCP and DCPA converged on to three individual single lines. This convergence indicated that the decrease in voidage is not solely dependent on compaction pressure: diminution of void sizes and intraparticle porosity occurred in an orderly manner dependent on size, which is in agreement with that seen by the lineal analysis of Bockstiegel. The shrinkage and obliteration of the larger (1.0—100 μm) voids to a specific size D_{max} showed, however, a variation which, with the powders examined over the compaction pressure range used, could be expressed in a simpler mathematical relationship than that of Bockstiegel:

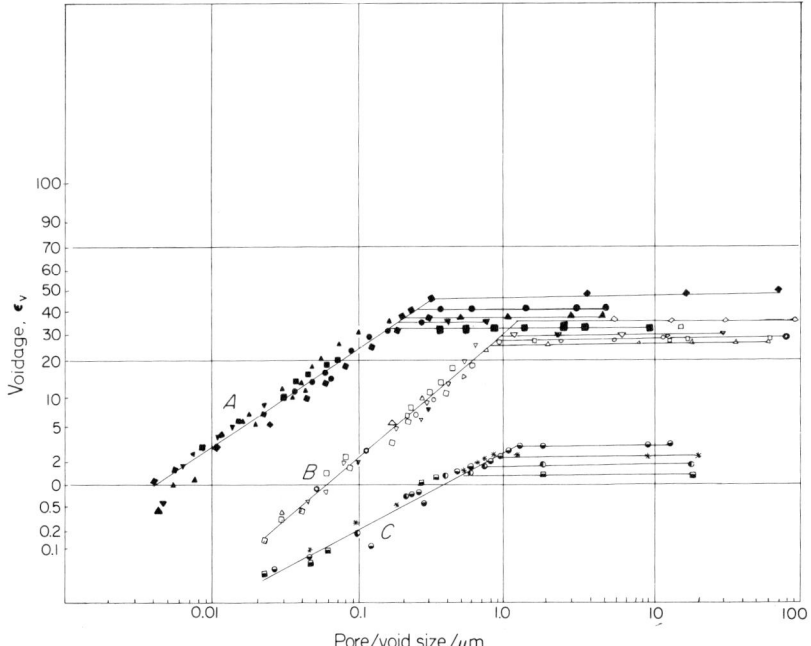

Figure 2.34. Cumulative interparticle voidage distribution for A, *magnesium trisilicate* (◆ 17.5, ● 70, ▲ 140, ▼ 210, ■ 280 MN m^{-2}); B, *dicalcium phosphate* (◇ 35, ▽ 65, ○ 97, □ 130, △ 160 MN m^{-2}); *and* C, *a mixture of dicalcium phosphate and Avicel* (◒ 35, ∗ 70, ◐ 140, ▣ 210 MN m^{-2})

$$\frac{P_A}{D_{max}} = \frac{P_A}{D_{max}^0} + \frac{1}{ab}$$

where a and b are constants and D_{max}^0 is the lowest void size produced from these materials under compaction conditions in the range 12—280 MN m^{-2}.

References

1. SMAKULA, A. and KALNAJAS, J., *Phys. Rev.* **99**, 1737 (1955)
2. SMAKULA, A., KALNAJAS, J. and SILS, V., *Phys. Rev.* **99**, 1747 (1955)
3. ADAM, N. K. A., '*Physical Chemistry*', Clarendon Press, Oxford (1956)
4. GRATON, L. C. and FRASER, H. J., *J. Geol.* **43**, 785 (1935)
5. MAGGS, F. A. P., SCHWAKE, P. H. and WILLIAMS, J. H., *Nature* **186**, 956 (1960)
6. BUDWORTH, D. W., *J. Br. Ceram. Soc.* **1**, 448 (1964)
7. FURNAS, C. C., *Ind. Eng. Chem.* **23**, 1052 (1931)
8. FRASER, H. J., *J. Geol.* **43**, 910 (1935)
9. WHITE, H. E. and WALTON, S. F., *J. Am. chem. Soc.* **20**, 155 (1937)
10. McGEARY, R. K., *J. Am. Ceram. Soc.* **44**, 513 (1961)
11. AYER, J. and SOPPET, F., *J. Am. Ceram. Soc.* **48**, 19 (1965)
12. RAMSAY, J. D. F. and AVERY, R. G., '*Proceedings of First International Conference on Compaction and Consolidation, Brighton*', Powder Advisory Centre, London (1972)
13. SHULL, C. G., *J. Am. chem. Soc.* **70**, 1405 (1948)
14. PIERCE, C., *J. phys. Chem.* **57**, 64, 149 (1953)
15. BRUNAUER, S., DEMING, L. S., DEMING, W. S. and TELLER, E. J., *J. Am. chem. Soc.* **62**, 1723 (1940)
16. CARMAN, P. C. and RAAL, E. A. *Proc. R. Soc. A* **209**, 59 (1951)
17. WADE, W. H., *J. phys. Chem., Ithaca* **69**, 322 (1955)
18. BARRER, R. M. and STRACHAN, E., *Proc. R. Soc. A* **231**, 52 (1955)
19. BOCKSTIEGEL, G., *Int. J. Powder Met.* **3**, 29 (1967)
20. GREGG, S. J. and POPE, M. I., *Fuel, Lond.* **39**, 301 (1960)
21. HARDMAN, J. S. and LILLEY, B. A., *Proc. R. Soc. A* **333**, 183 (1973)
22. COOPER, A. R. and EATON, L. E., *J. Am. chem. Soc.* **45**, 97 (1962)
23. BOWDEN, F. P. and TABOR, D., '*Friction and Lubrication of Solids*', Clarendon Press, Oxford, vol. 1 (1954), vol. 2 (1964)
24. JAMES, P., *Int. J. Powder Met.* **4**, 82 (1972)
25. SCHOFIELD, A. and WROTH, P., '*Critical State Soil Mechanics*', McGraw-Hill, London (1968)
26. BARRETT, E. P., JOYNER, L. G. and HALENDA, P. P., *J. Am. chem. Soc.* **73**, 373 (1951)
27. INNES, W. B., *Analyt. Chem.* **29**, 1069 (1957)
28. ANDERSON, R. B., *J. Catal.* **3**, 50 (1964)
29. CRANSTON, R. W. and INKLEY, F. A., *Adv. Catalysis* **9**, 143 (1957)
30. BRUNAUER, S., EMMETT, P. H. and TELLER, E. J., *J. Am. chem. Soc.* **60**, 309 (1935)
31. LIPPENS, B. C., LINSEN, B. G. and de BOER, J. H., *J. Catal.* **3**, 32 (1964)
32. de BOER, J. H. and LIPPENS, B. C., *J. Catal.* **3**, 38 (1964)
33. LIPPENS, B. C. and de BOER, J. H., *J. Catal.* **3**, 44 (1964)
34. de BOER, J. H., van den HEUVEL, A. and LIPPENS, B. C., *J. Catal.* **3**, 268 (1964)
35. MIKHAIL, R. S., BRUNAUER, S. and BODER, E. E., *J. Colloid Interface Sci.* **26**, 45 (1968)
36. DUBININ, M. M., *Progr. Surface Membrane Sci.* **9**, 1 (1970)
37. DUBININ, M. M., *J. Colloid Interface Sci.* **23**, 487 (1967)
38. KAGANER, *Zh. fiz. Khim.* **33**, 2208 (1959)
39. DUBININ, M. M., *Adv. Colloid Interface Sci.* **2**, 217 (1968)
40. DUBININ, M. M., PLAVNIK, G. H. and ZAVERINA, E. D., *Carbon* **2**, 261 (1964)
41. DUBININ, M. M., '*Proceedings of International Symposium on Surface Area Determination, Bristol*', Butterworths, London (1969)
42. URWIN, D., '*The Structure and Properties of Porous Materials*', Ed. Everett and Stone, Butterworths, London (1958)

43. ROUQUEROL, J., ROUQUEROL, F., PÉRÈS, C., GRILLET, Y. and BOUDELLAL, M., '*Characterization of Porous Solids*', Eds. Gregg, Sing and Stoeckli, Society for Chemical Industry, London (1979)
44. BALL, M. E. and NORWOOD, L. S., *J. chem. Soc. Faraday I*, **93**, 932 (1977)
45. JOHANSSON, M. E., M.Phil. Thesis, University of Bradford (1978)
46. LANGMUIR, I., *Phys. Rev.* **8**, 149 (1916)
47. LANGMUIR, I., *J. Am. chem. Soc.* **60**, 309 (1938)
48. BRUNAUER, S. and EMMETT, P. H., *J. Am. chem. Soc.* **59**, 1553 (1937)
49. HILL, T. L., EMMETT, P. H. and JOYNER, L. G., *J. Am. chem. Soc.* **73**, 5102 (1951)
50. SUPRYNOWICZ, Z., GORGAL, A. and WOJAK, J., *J. Chromat.* **148**, 151 (1978)
51. HUTTIG, G. F., *Mh. Chem.* **78**, 177 (1948)
52. BALEY, E. C., *Proc. R. Soc. A* **160**, 465 (1973)
53. LOPEZ-GONZALES, J. de D. and DEITZ, V. R., *J. phys. Chem.* **65**, 1112 (1961)
54. HARKIN, W. D. and JURA, G., *J. chem. Phys.* **11**, 430 (1943)
55. HARKIN, W. D. and JURA, G., *J. Am. chem. Soc.* **66**, 1366 (1944)
56. BARRER, R. M., MACKENZIE, N. and MacLEOD, D., *J. chem. Soc.* 1735 (1952)
57. BARRER, R. M., MACKENZIE, N. and MacLEOD, D., *J. chem. Soc.* 4184 (1954)
58. DOLE, M., *J. chem. Phys.* **16**, 25 (1948)
59. FOWLER, R. H. and GUGGENHEIM, E. A., '*Statistical Thermodynamics*', Cambridge University Press (1952)
60. ANDERSON, R. B., *J. Am. chem. Soc.* **68**, 686 (1946)
61. GREGG, S. J. and SING, K. S. W., '*Adsorption, Surface Area, and Porosity*', Academic Press, London (1967)
62. MacIVER, D. S. and EMMETT, P. H., *J. Am. chem. Soc.* **60**, 824 (1956)
63. ISIRIKYAN, A. A. and KISELEV, A. V. *J. phys. Chem.* **65**, 601 (1961)
64. HARRIS, M. R. and SING, K. S. W., *Chem. and Ind.* **66**, 487 (1962)
65. EMMETT, P. H., '*Advances in Colloid Science*', vol. 1, Interscience, New York (1942)
66. ANDERSON, R. B. and EMMETT, P. H., *J. appl. Phys.* **19**, 367, 370 (1948)
67. LIVINGSTON, H. K., *J. Colloid Sci.* **4**, 447 (1949)
68. McCLENNAN, A. L. and HARNSBERGER, H. F., *J. Colloid Interface Sci.* **23**, 577 (1967)
69. DEITZ, V. R. and TURNER, N. H., '*Proceedings of International Symposium on Surface Area Determination, Bristol*', Butterworths, London (1969)
70. HILL, T. L., *J. chem. Phys.* **14**, 263, 441 (1946); **17**, 520 (1949)
71. COHAN, L. H., *J. Am. chem. Soc.* **60**, 433 (1938)
72. PATRICK, W. A. and McGAVACK, J., *J. Am. chem. Soc.* **42**, 946 (1920)
73. ZSIGMONDY, R., *Z. anorg. Chem.* **71**, 356 (1911)
74. TERNAN, M., *J. Colloid Interface Sci.* **45**, 270 (1973)
75. WHEELER, A., '*Catalysis*', Ed. Emmett, P. H., vol. IIP, pp. 111–118, Reinhold, New York (1955)
76. WHEELER, A., '*Catalysis Symposia*', Gibson Island American Association for the Advancement of Science Conference, June 1945 (1946)
77. HALSEY, G., *J. chem. Phys.* **16**, 931 (1948)
78. LIPPENS, B. C. and de BOER, J. H., *J. Catal.* **4**, 319 (1965)
79. '*Proceedings of International Symposium on Surface Area Determination, Bristol*', Butterworths, London (1969)
80. LECLOUX, A., '*Proceedings of International Symposium on Surface Area Determination, Bristol*', p. 22, Butterworths, London (1969)
81. MINGLE, J. O. and SMITH, J. H., *Chem. Engng. Sci.* **16**, 31 (1960)
82. JOHNSON, M., *see* ref. 81
83. BUTT, J. B., *J. Catal.* **4**, 685 (1965)
84. BROEKHOFF, J. P. C. and de BOER, J. H., *J. Catal.* **9**, 8, 15 (1967); **10**, 153 (1968); **10**, 368, 377, 391 (1968)
85. KISELEV, A. V. and KARNAUKHOV, A. P., *Russian J. phys. Chem.* **34**, 1021 (1960)
86. GIRGIS, B. S., *J. appl. Chem. Biotechnol.* **23**, 19 (1973)
87. FOSTER, *see* ref. 61
88. OULTON, T. D., *J. phys. Colloid Chem.* **52**, 1296 (1948)
89. ROBERTS, B. F., *J. Colloid Interface Sci.* **23**, 266 (1967)
90. DOLLIMORE, D. and HEAL, G. R., *J. appl. Chem., Lond.* **14**, 109 (1964)
91. EVERETT, D. H., '*The Structure and Properties of Porous Materials*', Butterworths, London (1958)
92. YOUNG, T., '*Miscellaneous Works*', vol. I, p. 418, Murray, London (1855)

93. de LAPLACE, P. S., *'Mechanique Celeste'*, Paris (1806)
94. HERDAN, G., *'Small Particle Statistics'*, Butterworths, London (1960)
95. DOLLIMORE, D. and HEAL, G. R., *J. Colloid Interface Sci.* **42**, 233 (1973)
96. KISELEV, A. V., KARNAUKHOV, A. P. and ARISTOV, B. G., *Russian J. phys. Chem.* **36**, 1159 (1962)
97. WADE, W. H., *J. phys. Chem., Ithaca,* **68**, 1029 (1964)
98. KARNAUKHOV, A. P. and KISELEV, A. V., *Russian J. phys. Chem.* **31**, 2635 (1957)
99. ARISTOV, B. G., DAVYDOV, V. Y., KARNAUKHOV, A. P. and KISELEV, A. V., *Russian J. phys. Chem.* **36**, 1497 (1962)
100. IHM, S. K. and RUCKENSTEIN, E., *J. Colloid Interface Sci.* **61**, 146 (1977)
101. NICHOLSON, D., *J. chem. Soc. Faraday I*, 29 (1976)
102. SCHOLTEN, J. J. F., BEERS, A. M. and KIEL, A. M., *J. Catal.* **36**, 23 (1975)
103. de WITT, L. A. and SCHOLTEN, J. J. F., *J. Catal.* **36**, 30 (1975)
104. VISWANATHAN, B., SRINSVASAN, V. and SASTRI, M. V., *Indian J. Chem.* **9**, 1299 (1971)
105. DERJAGUIN, M., *'Proceedings of International Congress on Surface Activity, London'*, vol. 2, p. 153 (1957)
106. STEELE, W. A. and ROSS, M., *J. chem. Phys.* **33**, 464 (1960)
107. STEELE, W. A., *'The Solid–Gas Interface'*, Ed. Flood, E., vol. 1, pp. 353—366, Dekker, New York (1967)
108. BIRD, R. B., STEWART, W. E. and LIGHTFOOT, E. N., *'Transport Phenomena'*, Wiley, New York (1960)
109. HIRSCHFELDER, J. O., CURTISS, C. F. and BIRD, R. B., *'Molecular Theory of Gases and Liquids'*, Wiley, New York (1954)
110. STEELE, W. A., *'The Interaction of Gases with Solid Surfaces'*, Pergamon, Oxford (1974)
111. CULVER, R. U. and HEATH, N. S., *Trans. Faraday Soc.* **51**, 1596 (1955)
112. TYSON, R. F. S., Ph.D. Thesis, Exeter University
113. CARMAN, P. C. and RAAL, E. A., *Proc. R. Soc. A* **209**, 59 (1951)
114. KISELEV, A. V., *'Structure and Properties of Porous Materials'*, Eds. Everett and Stone, Butterworths, London (1958)
115. ZWIETERING, P., *'Proceedings of International Symposium on the Reactions of Solids'* (1956)
116. WADE, W. H., *J. phys. Chem.* **69**, 322 (1955)
117. BARRER, R. M. and STRACHAN, E. E., *Proc. R. Soc. A* **231**, 52 (1955)
118. POLANYI, M., *Verh. dt. phys. Ges.* **16**, 1012 (1914)
119. DUBININ, M. M., *'Proceedings of Conference on Industrial Carbons and Graphite'*, Society for Chemical Industry, London (1958)
120. KAGANER, M. G., *Zh. fiz. Khim.* **33**, 2202 (1959)
121. DUBININ, M. M., *'Characterization of Porous Solids'*, Eds. Gregg, Sing and Stoeckli, Society for Chemical Industry, London (1979)
122. RAMSAY, J. D. F. and AVERY, R. G., *J. Colloid Interface Sci.* **51**, 205 (1975)
123. STEELE, W. A. and HALSEY, G. D., *J. phys. Chem., Ithaca,* **59**, 57 (1955)
124. GURFEIN, N. S., DOBYCHIN, D. P. and KOPLIENKO, L. S., *Zh. fiz. Khim.* **44**, 741 (1970)
125. RITTER, H. L. and DRAKE, R. C., *Ind. Eng. Chem. analyt. Edn.* **17**, 787 (1945)
126. ROOTARE, H. M. and NYCE, A. C., *J. Powder Met.* **7**, 4 (1971)
127. SCHOLTEN, J. J. F., *'Porous Carbon Solids'*, Ed. Bond, R. L., Academic Press, London (1967)
128. *'Gmelins Handbuch der Anorganische Chemie'*, Ed. Deutsche Chemische Gesellschaft, Band 34, Teil A, p. 291
129. ROBERTS, N. K., *J. chem. Soc.* 1907 (1964)
130. YOUNG, T. F., *see ref. 3*
131. KERNAGHAN, M., *Phys. Rev.* **37**, 990 (1931)
132. COOK, S. G., *Phys. Rev.* **34**, 513 (1929)
133. LEA, J. A. and MASKELL, W. C., *Powder Technol.* **9**, 165 (1974)
134. KRUGER, S., *Trans. Faraday Soc.* **54**, 1758 (1958)
135. FREVEL, L. K. and KRESSLEY, L. J., *Analyt. Chem.* **35**, 1492 (1963)
136. MAYER, R. P. and STOWE, R. A., *J. Colloid Sci.* **20**, 893 (1965)
137. MAYER, R. P. and STOWE, R. A., *J. phys. Chem., Ithaca,* **70**, 3867 (1960)
138. STANLEY-WOOD, N. G., *Analyst* **104**, 97 (1979)
139. ROOTARE, H. M. and SPENCER, J., *Powder Technol.* **6**, 17 (1972)
140. JOYNER, L. G., BARRETT, E. P. and SKOLD, R., *J. Am. chem. Soc.* **73**, 3155 (1951)

141. BRUNAUER, S., *Chem. Engng. Prog. Symp. Ser.* **65**, 1 (1969)
142. BROWN, S. M. and LARD, E. W., *Powder Technol.* **9**, 187 (1974)
143. ZWIETERING, P. and KOKS, H. L. T., *Nature*, **173**, 688 (1954)
144. WASHBURN, R. W., *Proc. natn. Acad. Sci. U.S.A.* **7**, 115 (1921)
145. GREGG, S. J. and LANGFORD, J. F., *J. chem. Soc. Faraday*, 747 (1977)
146. STANLEY-WOOD, N. G., *Powder Technol.* **12**, 225 (1975)
147. STANLEY-WOOD, N. G. and JOHANSSON, M. E., *Arch. Pharm.* **117**, 568 (1978)
148. GREGORY, H. R., *Trans. Instn chem. Engrs* **40**, 241 (1962)
149. STRÖMGREN, M., ÅSTROM, H. and EASTERLING, K. E., *Powder Technol.* **16**, 155 (1973)
150. EASTERLING, K. E. and THÖLEN, A. R., *Powder Metall*, **16**, 112 (1973)
151. MORRISON, H. L. and RICHMOND, O., *Powder Technol.* **14**, 153 (1976)
152. de BOER, A. H., BOLHUIS, G. K. and LERK, C. F., *Powder Technol.* **20**, 75 (1978)
153. STANLEY-WOOD, N. G. and SHUBAIR, M. S., *J. Pharm. Pharmac.* **31**, 429 (1979)
154. STANLEY-WOOD, N. G. and SHUBAIR, M. S., *Powder Technol.* **25**, 57 (1979)
155. TYLER, A. J., HAMBLETON, F. H. and HOCKEY, J. A., *J. Catal.* **13**, 43 (1969)
156. de BOER, J. H., *'Structure and Properties of Porous Materials'*, Ed. Everett, D. H., Butterworths, London (1958)
157. BOCKSTIEGEL, G., *Int. J. Powder Met.* **2**(4), 13 (1966); **3**(1), 29 (1967)
158. STANLEY-WOOD, N. G., *'Third International Conference on Particle Size Analysis'*, p. 278 (1977) Heyden and The Royal Society of Chemistry, Analytical Division, Burlington House, London W1V 0BN
159. STANLEY-WOOD, N. G. and JOHANSSON, M. E., *Drug Dev. Ind. Pharm.* **1**, 69 (1978)

CHAPTER 3

Mixing of powders

N. Harnby

*Schools of Studies in Chemical Engineering and Powder Technology,
University of Bradford, Bradford*

3.1 Powder mixing

3.1.1 Mixture quality

When is a mixture well mixed? This fundamental question has to be asked
of all mixtures. The practical reply must be that the mixture is well mixed
when it is 'satisfactory' for its duty. A pigment is satisfactorily mixed into
neutral plastic chips when the resultant moulded product appears to have a
uniform colour. This is an arbitrary judgment of mixture quality as the
same mixture could well be 'unsatisfactory' if the moulded product was
thinner or was examined more closely. Another weakness of such a judgment
is that the manufacturer has no indication that a processing trend is about
to put his product off specification. For the manufacturer, quality control
demands that a quantitative assessment should be made of mixture quality.

Danckwerts[1] gives some helpful qualitative ideas on mixture quality.
Thus, the 'scale of segregation' of a mixture measures the size of regions of
unmixed or segregated material. The 'intensity of segregation' of a mixture
measures the amount of dilution of regions of unmixed material by other
mixture components. Evidently, the quality of a mixture is improved by
reducing the scale of segregation and reducing the intensity of segregation.
As the scale and intensity of segregation are reduced the mixture will pass
through a critical quality at which it is 'satisfactory' for its duty and beyond
which further mixing is unnecessary. The concept of a critical mixture quality was described by Danckwerts in terms of a 'scale of scrutiny' and
defined as 'the maximum size of the regions of segregation in the mixture
which would cause it to be regarded as imperfectly mixed'. Whilst this is an
imprecise definition it does concentrate attention on the importance of
identifying the volume, weight or area of mixture which should meet a
prescribed quality. Any mixture will be unsatisfactory if scrutinized closely
enough since the scale of scrutiny will then approach the scale of individual
particles. Thus in the case of the dispersion of pigment in plastic the scale of
scrutiny on the plastic surface will be a small area incapable of resolution
by the human eye. Microscopical examination of the plastic would reveal
colour variations which would require a further reduction in the scale of
scrutiny to give colour uniformity.

3.2 The mixing process

The mixture quality of a powder mixture will be improved if the scale and intensity of segregation are reduced. Unlike gaseous and miscible liquid systems this reduction process will not occur by molecular diffusion and work must be carried out on the system in order to generate interparticulate motion. The particles may be tumbled, kneaded, fluidized, sheared or scooped but the individual particles must be given the opportunity of relocating themselves within the mixture. The ability of particles to move freely is evidently an important characteristic and leads to a general division of powder mixtures into;

1. Free-flowing mixtures
2. Cohesive or non-free-flowing mixtures

A division between the two regimes is not easily defined. The bulk effect can be seen by rolling the mixture in a bottle. The free-flowing powder moves smoothly with well defined planes of movement whilst the cohesive powder exhibits 'stick–slip' motion with irregular surface characteristics. Particle size is probably the dominant influence on the type of flow regime. The gravitational force associated with a large particle is much larger than any restraining interparticulate forces, with the result that individual particles retain their freedom of movement. As the particle size decreases various interparticulate forces can potentially dominate and the particles attempt to retain a structured arrangement. The actual transition size will be a function of the nature of the interparticulate forces and of other particle characteristics, but in general terms it can be stated that particles of nominal diameter greater than 50 μm will tend to be free flowing whilst particles of nominal diameter less than 50 μm will tend to be cohesive. What effect has the flow property of a powder on the mixing process? Let us consider the two regimes in turn.

3.2.1 Free flowing mixtures

Both from the marketing and the processing point of view free flowing powders have many desirable features.

Because of the larger particle size they represent less of a dust hazard. They can be poured smoothly after transport whether it be from a bottle or a rail car. Within the process the flow of the powder is consistent and control of flow rates and product quality is simplified. With such attractions it is not surprising that in many process industries the trend is towards free flowing particulate products, even if this means adding an agglomerating operation to the process to increase the mean size of the particles.

The disadvantage of such mixtures is that they can be subject to segregation or 'unmixing' on a severe scale. The very freedom which allows a particle to move smoothly and independently of its neighbours gives it the freedom to move preferentially in a particular direction. Thus a small particle will percolate preferentially through a bed of moving particles. A large particle will be projected further than a fine particle given an equal initial velocity. That such preferences can result in a very large 'scale of segregation' can be seen in *Figures 3.1* and *3.2*. *Figure 3.1* illustrates the effect of pouring a mixture of multisized particles into a sectioned heap. The finer particles percolate preferentially through the moving inclined surface of the heap whilst the coarser particles preferentially tumble away from the pouring point. The rolling of particles on such an inclined plane is one of the chief causes of gross segregation in industrial applications and is also one of the main causes of the segregation shown in the sectioned V-Mixer of *Figure 3.2*. The V-Mixer is one of a large class of 'tumbler' mixers which find wide industrial application in batch processes. The rotation of the vessel provides an excellent free tumbling or percolation plane for the powder and extreme segregation can occur even with relatively small particle size differences. If batch mixing is a requirement for such free flowing systems then mixers should be chosen which limit the freedom of movement of individual particles and which tend instead to move changing groups of related particles from one location to another within the mixer. Examples of such mixers are the range of ribbon blenders and a variety of scooping blade machines. For a suitable process continuous mixing is an excellent way of avoiding gross segregation. Within the continuous mixer the quantities of mixture at the point of mixing are minimized and the potential scale of segregation is considerably reduced.

Even if a large mass of free flowing powder is 'satisfactorily' mixed great care has to be taken in subsequently using the mixture. Many storage and handling processes can destroy the mixture quality that has been so carefully created. Only when the mixture has been 'frozen' at its final point of usage can the mixture be regarded as safe. This is the point at which the particles lose their mobility or at which segregation is no longer important, and could be the formation of a tablet, the filling of a package or a point at which the powder is put into solution.

The advice for the process engineer must be to locate the mixing point as closely as possible to the point of usage and to avoid all bulk storage and transport between these points.

3.2.2   Cohesive mixtures

If for the free flowing mixture the art of the process engineer is often to restrict the freedom of movement of individual particles, then for the cohesive mixture the problem is reversed. The cohesive system has a natural structure which has to be repeatedly broken down in order to give individual particles within that structure an opportunity of relocating themselves. The selective grading of particles which produces a very large scale of segregation in free flowing mixtures is not possible and it is relatively easy

Figure 3.1.   *A mixture of multisized particles poured into a sectioned heap*

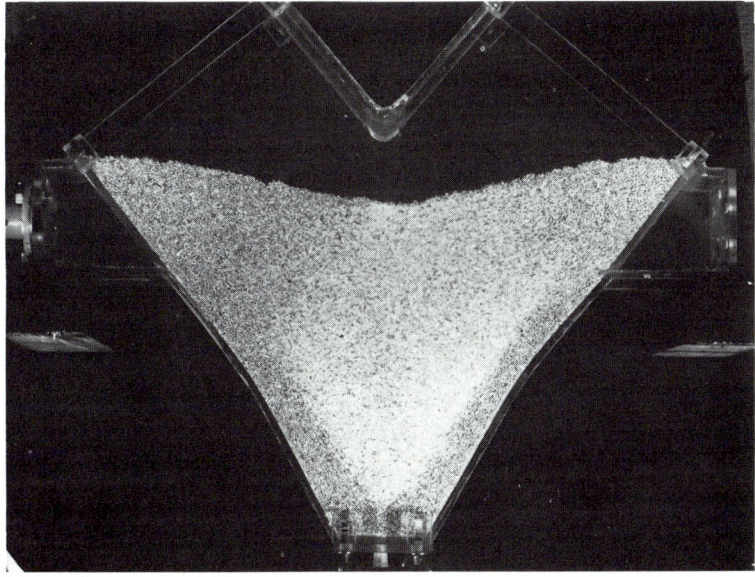

Figure 3.2.   *Segregation of multisized particles in a sectioned V-Mixer*

to produce a 'good' cohesive mixture quality in most industrial mixers. Problems of mixture quality can arise in processes which demand a very small scale of scrutiny for their final product. It is commonly found that whilst the scale of segregation of cohesive mixtures is small the intensity of segregation is very high and that little, if any, dilution of the initial ingredient has occurred. This is caused by small agglomerates of individual mixture ingredients retaining their structure throughout the mixing process. The strength of these agglomerates and the ability of different mixers to break them down to the scale of the individual particle is the central study of cohesive mixtures.

The nature and the strength of the interparticulate forces acting within the cohesive powder system will determine the ease, or difficulty, likely to be experienced in re-locating individual particles within a mixture and will also determine the final scale and intensity of segregation attainable within a given mixture. If solid 'bridge formation' between particles is excluded then likely bonding mechanisms are,

1. Due to moisture
2. Due to electrostatic charging
3. Due to van der Waals forces

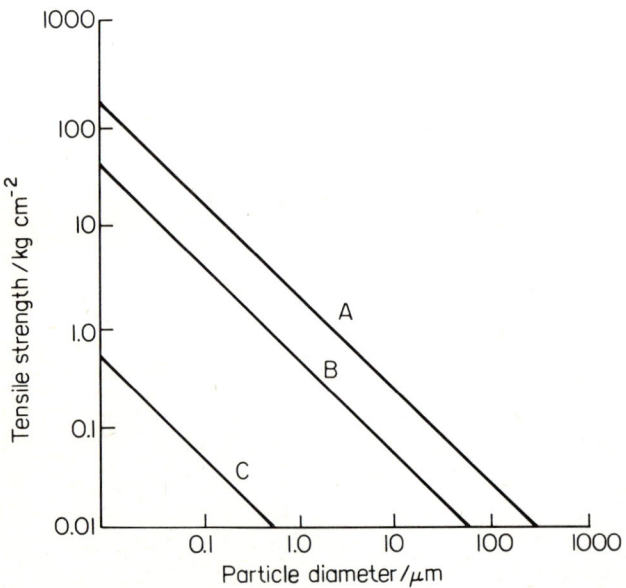

Figure 3.3.   Theoretical tensile strength of agglomerates (after Rumpf)
  A   *Liquid bridge or capillary*
  B   *Adsorbed moisture*
  C   *van der Waals*

Some appreciation of the relative magnitude of these forces is given in *Figure 3.3* where a theoretical interparticulate force for bulk material is plotted against particle size. If a tensile strength of 0.01 kg cm^{-2} is arbitrarily chosen as a 'negligible' powder strength than it can be seen from the graph that van der Waals forces will have a 'significant' cohesive effect only for particles of diameter less than 1 μm. When adsorbed surface moisture is present and valency forces exist at the points of contact then the limiting particle diameter is about 80 μm. At very high humidities, or for wet powder, surface moisture is free to flow over the particle surface and 'liquid bridges' are formed between particles providing cohesion due to surface tension. In this case particles of the order of 500 μm can exhibit cohesiveness. No corresponding curve for electrostatic forces is provided as the attraction is highly dependent on the powder, its surface characteristics and the presence of impurities.

Under most atmospheric conditions a 'dry' powder mixture would not have liquid bridging and the cohesive forces would be due to van der Waals forces, electrostatic forces or adsorbed moisture. The work of Rumpf[2] would tend to confirm the generalization that a powder will 'tend' to become cohesive in nature when its diameter is less than 50 μm.

The range of application of these forces is most important. For moisture bonding a physical contact between particles is required. The van der Waals forces are inversely proportional to the cube of the distance between attracted particles whilst electrostatic forces remain constant over macroscopic distances.

In practical mixing problems electrostatic forces, because of the unpredictability of their strength, are generally avoided rather than encouraged. Whilst the bonding provided could be useful in avoiding gross segregation, the quality of such a mixture would be as unpredictable as the strength of the electrostatic attraction. Additionally, the electrical charging can be a safety hazard.

Evidently, for adsorbed liquid and van der Waals bonding the degree of compaction of the powder will affect the distance between adjacent particles and also the contact area. This leads to the conclusion that greater compaction will result in more cohesive systems.

Within the mixing process the formation of a structured powder and perhaps agglomerates is not a bad feature as long as the mixing mechanism is capable of repeatedly breaking down such structures and agglomerates to the scale of individual particles. If the mixer is not capable of doing this then a small scale of segregation with a high intensity of segregation is likely to result. The most difficult powder mixtures to restructure repeatedly will have submicron constituents with some moisture present and will have been subjected to compaction at some stage. To mix such a system the 'mixer' is likely to have high shearing or impaction characteristics and could well be a particle comminuter rather than a conventional 'mixer'. Industrial mixtures seldom can be defined by a single particle size and frequently the size spectrum will span the size boundary between the free flowing and cohesive systems. Two special cases in this category will be mentioned:

(*a*)   A very wide particle size range. Such mixtures can exhibit the worst features of both the free flowing and the cohesive powder systems. The coarser particles and some of the fine particle agglomerates can be free flowing and produce a large scale of segregation whilst the bulk of the material remains cohesive. The mixture does not have the smooth flowing advantages of the coarser powder or the mixing advantages of the finer particles. Such 'fence-sitting' systems should be avoided if possible.

(*b*)   A mixture containing fine and coarse particles both of which have a small size spectrum. Without bonding such a system would strongly segregate. If carefully chosen the fine particles will preferentially bond to the coarser or 'parent' particles and can produce a mixture of a very high quality. This is the so-called coating process.

3.3   Quantitative assessment of mixture quality

If the scale of scrutiny for a mixture can be specified then the quantity of mixture to be sampled and analysed for uniformity of quality can be determined. The control of the quality of the mixture is then based on a statistical assessment of a selection of such samples[3]. It must be re-emphasized that such a quality assessment is only valid at that scale of scrutiny and should be scaled up or down with extreme caution.

The variance of a component in a number of samples is a measure of the consistency or quality of the product. The smaller the variance the better the mixture. Such a measure gives a graded assessment of mixture quality which enables process trends to be followed and problems to be predicted and avoided. The variance value alone gives no absolute assessment of mixture quality. To do this the experimentally determined value has to be related to the limiting mixture variance values of complete segregation and of the randomized mixture.

Generally, the variance of the randomized mixture is the more helpful datum of comparison as it normally represents the attainable mixture quality in an industrial process as it represents a random positioning of the components within the mixture with no segregation or preferential positioning.

The limiting random variance can be calculated for most industrial mixtures. The simplest case is that for a two-component mixture of equisized particles[4] where

$$\sigma_R^2 = \frac{p \times q}{B} \tag{3.1}$$

σ_R^2 is the random variance of the samples, p and q are the proportions of the two components and B the number of particles in the sample.

Similar, but more complex, relationships are available for multisized two component systems[5], for multisized multicomponent systems[6] and for some cases of agglomerating mixtures[7]. These random variance values represent an invaluable datum of comparison for experimental values. They represent the possible goal of a mixing process but do not take into account any of the shortcomings of the mixer or of a segregating mechanism in a real process.

The mixing process attempts to minimize the variance value between samples. As an ideal, σ_R^2, should be kept as small as possible. Inspection of equation 3.1 shows that the random variance can be reduced either by having one component in a very small proportion or by having a very large number of particles in each sample. Usually, the proportion of the components is fixed by the process and cannot be adjusted. The number of particles in the sample can be increased by either having a larger sample, or smaller particles, or both. If the weight of the samples is fixed then the only process freedom remaining is to reduce the size of the particles in the mixture. This will both reduce the attainable value of mixture variance and also move the mixture towards the cohesive flow region and possibly reduce the amount of gross segregation.

Inspection of the expressions for the random variance of more complex systems confirms the importance of comminution and especially the control of particle size if a consistent and high quality of mixture is to be maintained.

Whilst a random arrangement of particles is normally the attainable ideal for powder mixing it is possible, in certain circumstances, to have an ordered, or ideal, arrangement of particles and so approach a mixture variance of zero. Such mixtures are evidently of great interest to those industries requiring very high mixture standards[8].

References

1. DANCKWERTS, P. V., *Research, Lond.* **6**, 355 (1953)
2. RUMPF, H., *Chemie-Ingr-Tech.* **30**, 144 (1958)
3. HARNBY, N., *'Proceedings of Powtech 71'*, 1st International Conference on Powders and Bulk Granular Solids, p. 19
4. LACEY, P. M. C., *Trans. Inst. Chem. Engrs.* **26**, 331 (1954)
5. STANGE, K., *Chemie-Ingr-Tech.* **26**, 331 (1954)
6. STANGE, K., *Chemie-Ingr-Tech.* **35**, 580 (1963)
7. COELHO, M. and HARNBY, N., *'Second European Conference on Mixing, March 1977'*, Cambridge
8. HERSEY, J. A., *Powder Technol.* **11**, 41 (1976)

CHAPTER 4

Mechanisms of size enlargement

P. J. Lloyd

Department of Chemical Engineering, Loughborough University of Technology, Loughborough

4.1 Introduction

There are many situations where changing the size distribution of a powder is desirable or essential. Granulation is the generic name for particle size enlargement and is seen as the answer to different powder flow problems, to ensure better results when mixing difficult powders, to reduce a dust hazard problem, to ensure a uniform fill in a tabletting or die press machine or to obtain controlled release of chemicals in fertilizers and insecticides. These are just a few of the situations where granulation might be considered advantageous.

There are two principal methods of granulation: dry granulation and wet granulation. In the first, granules are made by compression and the strength of granulation depends on the sintered bonds between the particles. This method is the subject of another paper. In wet granulation a liquid, which is usually water, is used to make the preliminary bond between the particles and no direct compression is employed. Two main approaches are made. In the first, a controlled quantity of liquid is sprayed on to the dry powder. The liquid forms contacts between particles and a granule will grow. The second approach is that of forming droplets from a slurry or suspension of the powder and drying it. The excess of liquid is removed and the end result is somewhat similar to that of the first method.

This paper reviews the basic mechanisms and relates them to three common methods of granulation: pan granulation, fluidized bed granulation and spray drying or prilling.

4.2 Basic mechanisms

The basic mechanisms have been reviewed by Rumpf[1] in what may be considered a classic review. There are three types of force available:

1. Bonds due to free liquids
2. Bonds due to solid bridges
3. Surface forces

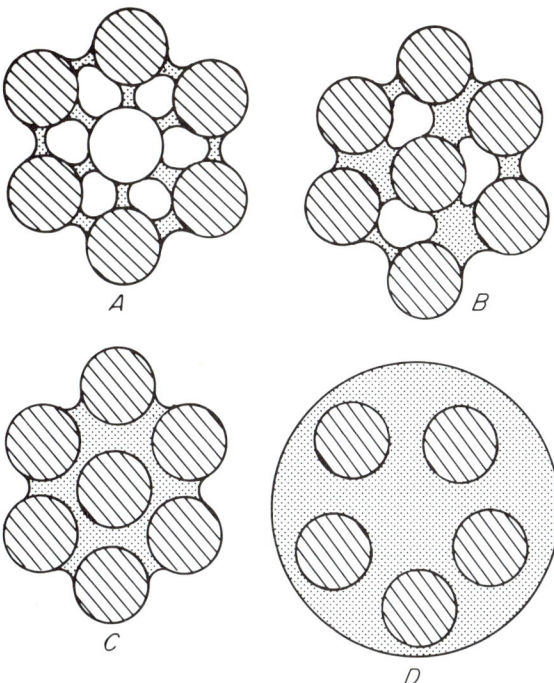

Figure 4.1. Types of liquid bonding in granules: A, *pendular;* B, *funicular;* C, *capillary;* D, *droplet*

4.2.1 Bonds due to free liquids

When a liquid is in contact with a solid the predominant force becomes that due to surface tension. Newitt and Conway-Jones[2] identified four types of liquid bonding depending on the proportional amount of liquid present. The four states are pendular, funicular, capillary and droplet (*Figure 4.1*).

In the pendular state, the liquid bridges between the particles are all separate and independent, being centred on points of contact between the particles. These forces are strong; the surface tension draws the particles together. If the proportion of liquid is increased the liquid can now freely move and the bonding force is less strong. At higher liquid contents all the interstices are filled with liquid and the strength of the granule is due to curvature of the liquid between particles on the outside. There is little internal strength in this case but if the liquid is removed by evaporation the particles are drawn together and can form a successfully strong granule. In the final state the particles are not in contact and the droplet has very little strength.

Rumpf[3] calculated the strength of the bond due to the pendular bridge and showed that the strength of the bond was not very dependent on the amount of moisture present until the funicular state was reached, but the resistance of the bond to rupture increased with increased liquid content because the particles could be pulled further apart without individual bridges breaking.

Thus it can be seen that spraying a liquid on to a dry powder will create pendular bridges which will form strong granules. The quantity of liquid used is not critical but too much should be avoided. It will be seen later that the size of the granules formed can be controlled by quantity of liquid added. On the other hand spraying a slurry or suspension will produce droplets which on drying will produce the capillary state and dense granules.

4.2.2   Bonds due to solid bridges

Although liquids form a very important part in the initial stages of granulation, it is not possible to rely on these bonds to form a strong granule. It is preferable to convert the bridge into a more permanent structure. There are a number of possibilities.

4.2.2.1   Crystalline bridges

If the powder is slightly soluble in the liquid used, consequent evaporation of the pendular ring will cause any dissolved solid to be thrown out of solution and since in the process the pendular ring will be shrinking a strong bridge will be formed.

4.2.2.2   Use of liquid binders

If the powder is not sufficiently soluble for crystalline bridges to be formed, another method is to add to the liquid a binder such that when the liquid is evaporated the binder effectively 'glues' the particles together. The binder may set solid or may remain as a high viscosity liquid.

4.2.2.3   Use of solid binders

An alternative to the liquid binder is the use of a finely ground solid such as cement which reacts with water. The solid binder is intimately mixed with the powder prior to granulation. On the addition of the liquid during the process the chemical reaction takes place and cements the particles together. This has been found to be particularly useful in the granulation of iron ore.

4.2.3   Surface forces

There exist between all solids molecular-derived forces known collectively as van der Waals forces. The magnitude of these is known for some materials and they are in general very small compared with those above. Furthermore, these forces are further weakened by surface contamination and adsorbed gases and so are unlikely to be significant in granulation.

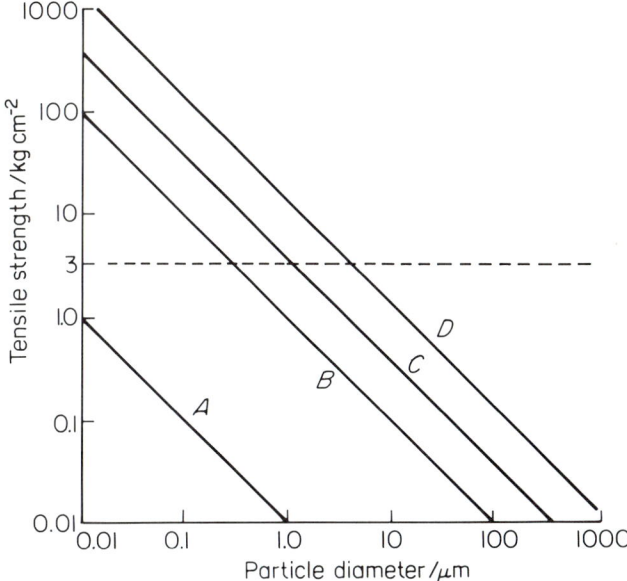

Figure 4.2.   Relative magnitude of bonding forces
  A   *van der Waals forces calculated at distance of* 3.3 nm
  B   *van der Waals forces in the presence of adsorbed liquid*
layers with zero interparticle distance
  C   *Pendular bridges between particles (water)*
  D   *Capillary bonding (water)*

The broken line represents the limit for crystalline salts

4.2.4 Summary on forces

Rumpf[1] produced *Figure 4.2* to illustrate the relative magnitude of the forces. As is seen, compression will produce the strongest binding forces. One point to note is that crystalline salts have an intrinsic strength of about 3 kg cm^{-2} and this will become the intrinsic strength of the bond.

4.3   The granulation process

The discussion of the basic mechanisms of granulation has shown how granules can be formed but granules formed in this way would be wet, small and weak and would break up. It is in the method by which they are produced that the granules are consolidated and grown to the required size. Because of this each method is discussed separately.

4.3.1 Granulation in rotating pans or drums

The most common granulator is a shallow cylindrical dish which is rotated about an inclined axis of which the angle of inclination is variable. If powder alone is placed in the rotating pan then it is most probable that small granules or nucleii will begin to form. These will be soft and will not

grow above a certain small size. If the powder does not nucleate sponta-
neously then it may be necessary to add small 'seed' granules continuously.
If the liquid to be used is now sprayed on to the powder, the particle surface
becomes wet and granules now begin to grow. There are many mechanisms
for growth and which mechanism is dominant will depend on many factors
such as the size, strength plasticity and surface wetting of the granules, the
degree of agitation and the amount of fines in the powder bed.

The basic mechanisms are:

1. Nucleii creation
2. Nucleii growth
3. Coalescence of nucleii
4. Crushing of smaller granules
5. Layering of fragments on to larger granules

Compression of the nucleii by continuing collisions causes the newly
created nucleii to move from the pendular state through the funicular state
to the capillary state where liquid is squeezed out on to the surface of the
granule. This liquid and other sprayed-on liquid enable the growing granule
to collect more fine particles on the outer surface and grow. This layered
outer surface is further consolidated by collision. Clearly, this process could
continue but other mechanisms now compete. The growing granule can
coalesce with the other small granules or nucleii and thus make step changes
in particle size, or the weaker granules can be broken by the continuing
collisions. The fragments are then layered on to the growing granule. Because
there are so many possible growth mechanisms control over the resultant

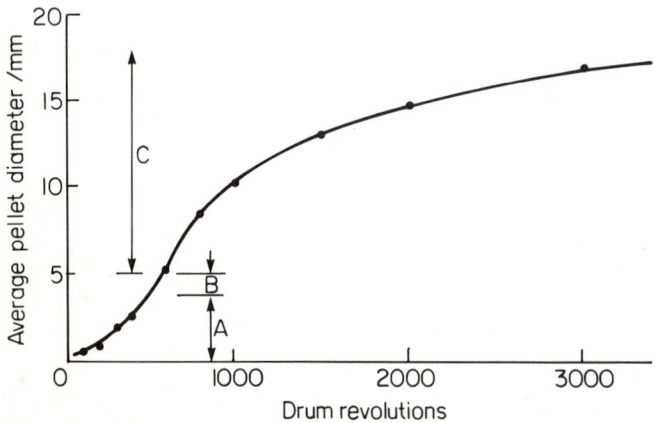

Figure 4.3.   *Growth stages in granulation of pulverized limestone
having a water content* 44.5 *per cent* (v/v)
  A   *Nuclei growth region*
  B   *Transition region*
  C   *Ball growth region*

size distribution is not easy. As the process is moisture-dependent the size and strength can be controlled by controlling the rate of liquid input and by the fineness of the spray, but even so some oversize range granules will result and these will have to be separated, crushed and returned to the granulator. Too fine granules can also be returned. In this way the size distribution is controlled. One reason why the pan is such a useful piece of equipment is that by adjustment of the inclination granules that achieve a desired size automatically leave the granulator.

Figure 4.3 shows the type of growth expected in a batch experimental granulator which shows the relative importance of the various regions.

The authors[4] modelled the process identifying three regions:

1. Nucleii growth where $dx/dt \propto x$
2. Transitional growth region where $dx/dt \propto 1/x$
3. Ball growth region where $dx/dt \propto 1/x^3$.

where x is the mean particle size and t the time.

4.3.2 Granulation in fluidized beds

Granulation in fluidized beds is a relatively new technique which has become very important because more control over particle size is possible and the drying stage takes place as part of the same operation. In a fluidized bed air, in this case preheated, passes through a porous bed support. If the velocity is high enough the powder on top of the support is fluidized, i.e., the packing of the particles is broken and the bed expands so that it has the appearance of a boiling liquid. The particles move randomly and make many collisions. If the binding liquid is sprayed on to the bed granulation takes place. Some consolidation will take place by particle collisions but the resulting granule will not be as dense as that formed in a pan granulator. Granules will grow by layering into the outer surface and as they become larger and therefore heavier they will tend to fall towards the bed support. As they slowly migrate downwards they are dried by the incoming air and the granules are then removed continuously from close to the bed support. Fresh powder is continuously fed in at the top. The granules formed by this method are of a narrow size range and are ideal for subsequent processing. A disadvantage is that fine particles are continually being removed by the exhaust air and it is necessary to recover these by use of cyclones and/or bag filters and return the fines to the fluidized bed.

4.3.2.1 *Design parameters*[5,6]

4.3.2.1.1 *Minimum bed height*

The unexpanded bed height should be in the range half to one third of bed diameter.

4.3.2.1.2   Bed expansion

An optimum has been suggested at 1.6 times the initial bed height.

4.3.2.1.3   Granulation dispersion

This is not found to be very important but some dispersion as a fine spray is necessary.

4.3.2.1.4   Position of spray head

This is not critical.

4.3.2.1.5   Granule size

The resulting granule size is the net result of growth and attrition. The size obtained is found to be a function of the ratio of the granulating liquid to powder flow rate when results similar to those in *Figure 4.4* are obtained.

4.3.2.1.6   Mechanical agitation

The use of an agitator has a number of advantages:

1. Large granules are broken down, thus giving closer control on size
2. Average mechanical strength is increased as weak granules are broken
3. Granule size not a function of feed rate
4. Necessary when powder is difficult to fluidize

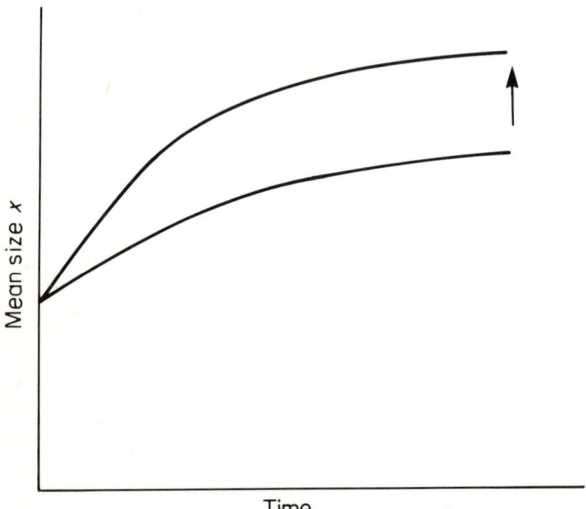

Figure 4.4   Growth of granule size with time; the ratio of
the granulating liquid to powder flow rate is indicated by
the arrow

4.3.3 Prilling or spray drying

The powder is suspended in a granulating liquid and is sprayed at the top of a high tower up which hot air is passed. The atomizing of the suspension creates droplets with the particles inside. The subsequent drying during falling creates the capillaric state and subsequently a granule the size of which is a function of the original drop. If the rising air is at very high temperature the liquid may be flash vaporized and the resulting granule shape is then a hollow near-sphere with the particles forming a crust.

It is a very important method for producing fertilizers, but the method is difficult to control, consumes large amounts of energy and causes pollution problems. Consequently, other methods are being explored.

4.3.4 Granulation by extrusion

Granules can also be formed by extruding a paste made up of the powder and granulating liquids through a coarse sieve. The granules produced in this way are produced immediately in the capillaric state and have to be dried to give some strength. This is a useful method for small scale and laboratory production but great care has to be taken to obtain just the correct consistency of the paste to obtain adequate granules.

Bibliography

KAPUR, P. C. and FUERSTENAU, D. W., *Ind. Eng. Chem. Process Des. Dev.* **5**, 5 (1966)
CAPES, C. E. and DANKWERTS, D. W., *Trans. Instn chem. Engrs.* **43**, T125 (1965)
MEISSNER, H. P., MICHAEL, A. S. and KAISER, R., *Ind. Eng. Chem. Process Des. Dev.* **5**, 10 (1966); **3**, 197 (1964)
KAPUR, P. C. and FUERSTENAU, D. W., *Ind. Eng. Chem. Process Des. Dev.* **8**, 56 (1969)
SHERRINGTON, P. J., *The Chem. Engr.* 201 (1968)
BUTENSKY and HYMAN, D., *Ind. Eng. Chem. Fundam.* **10**, 212 (1971)
ORMOS, Z., PATAKI, K. and CSUKAS, B., *Hung. J. Ind. Chem.* **3**, 631 (1975)

References

1. RUMPF, H., in '*Agglomeration*', Ed. Knepper, W. A., Interscience, New York (1962)
2. NEWITT, D. M. and CONWAY-JONES, J. H., *Trans. Instn chem. Engrs* **36**, 422 (1958)
3. RUMPF, H., *Chemie-Ingr.-Tech.* **30**, 144 (1958)
4. OUCHIYAMA, N. and TANAKA, T., *Ind. Eng. Chem. Process Des. Dev.* **13**, 383 (1974)
5. ORMOS, Z., PATAKI, K. and CSUKAS, B., *Hung. J. Ind. Chem.* **1**, 475 (1973)
6. SCOTT, M. W., LIEBERMAN, H. R., RANKELL, A. S. and BATTISTA, J. V., *J. pharm. Sci.*, **53**, 320 (1964)

CHAPTER 5

Flow and handling of solids; the design of solid handling plants

J. C. Williams
School of Studies in Powder Technology, University of Bradford, Bradford

5.1 Introduction

The handling and storage of particulate solids plays an important part in the processes of many industries. The design of such plants to ensure the satisfactory flow of material is very important since it can determine the profitability of the process. Yet the general standard of design is very much lower than would be tolerated in plants handling fluid materials. This is largely due to a failure to incorporate in the design the results of recent work on the flowability of particulate materials resulting in frequent stoppage of the process, involving loss of production and the costly use of staff to restore flow. In particular, problems arise owing to the difficulty of withdrawing material from a storage hopper without interruption and at the required rate, in spite of the fact that in the majority of cases the hopper could be designed to discharge continuously. Since this is the most important part of the plant from the point of view of flow, the method of designing storage hoppers will be treated in some detail.

5.2 Types of storage hopper

Examination of the flow pattern occurring when a hopper is emptying shows that there are two types of hopper: in a core flow hopper, flow is confined to a central core above the hopper outlet, the material outside this core being stationary; in a mass flow hopper, when discharge from the outlet is taking place every particle in the hopper is moving and in particular particles in contact with the wall are sliding on the wall.

Core flow hoppers have the following disadvantages:

1. For cohesive materials flow can be stopped either by the formation of a stable arch between the walls in the converging section of the hopper or by the development of a rathole, in which case the material above the outlet falls out leaving an empty vertical channel whose walls are strong enough to resist the stresses exerted on them by the remaining material.
2. If the hopper is operated so that it is periodically partially emptied and then refilled, some material will remain in the dead space for a long

time, not leaving the hopper until it is eventually emptied. This long residence time is not desirable for materials which change their properties with time.

3. When operating a hopper as in 2 fine particles may percolate into the dead space; when the hopper is eventually emptied the last material to come out will contain a very high proportion of fines.
4. The material in the dead space may become so densely compacted that it will not flow out of the hopper when complete emptying is required.
5. The flow rate and bulk density of the material leaving the hopper may be very variable.
6. Severe size segregation of the particles may be produced.
7. For fine powders the material in the hopper may become fluidized and the whole contents may pour out of the hopper in an uncontrolled manner, overloading the conveying system and spilling over on to the floor.

In a hopper designed for mass flow the following advantages can be obtained:

1. When materials are flowing out of the hopper stoppages due to the formation of arches or ratholes will not occur.
2. If flow is stopped for a short time, for example by closing the gate at the outlet, it will recommence on opening the gate.
3. If the material gains strength by time consolidation when flow is stopped the hopper can be designed to take account of this, so that flow will recommence on opening the gate.
4. Material comes out of the hopper in approximately the same order as it is put in, so that no material is kept in the bin for much more than the interval between fillings.
5. The flow rate and bulk density of the material emerging from the hopper will be more uniform than in the case of a core flow hopper. This is particularly important when feeding material on to a continuous weigh-belt feeder, since it reduces very considerably the amplitude of the fluctuations presented to the control system.
6. The effects of segregation on filling the hopper are considerably reduced by using a mass flow hopper, since some remixing occurs as the hopper empties.
7. Flooding of the material out of the hopper is much less likely to occur in a mass flow hopper.

The main disadvantages of a mass flow hopper are:

1. It will generally require more headroom, which may create difficulties in replacing a hopper in an existing plant.
2. In mass flow, particles slide in contact with the wall. For very abrasive materials this may lead to excessive wear, requiring the use of replace-

able liners. For foodstuffs and pharmaceutical products it may lead to unacceptable contamination unless a suitable wall material can be chosen.

A core flow hopper may be used for a free flowing material which never gains cohesive strength and for which segregation is not a problem, or for very abrasive materials which will wear away the walls of a mass flow hopper. In most other cases a mass flow hopper is preferred.

5.3   Measurement of the failure properties of a particulate solid

Before considering the method used to design a hopper for a given material so as to give mass flow we must define those properties of the particulate material which are relevant to this problem. It cannot be too strongly emphasized that the angle of repose, which some writers still refer to in this connection, is totally irrelevant to hopper design. What is important is to measure the way in which the shear strength of the material varies with the compressive stress acting on it, using some kind of shear cell. The equipment most generally used is the Jenike shear cell, shown in *Figure 5.1*, consisting of a circular base, ring and lid. The box is filled with the material to be tested and the lid is placed in position. By placing a weight on the load frame suspended from the point on the lid a compressive force is applied to the specimen. The ring and lid are pushed forward at a constant speed and the shear force needed to maintain this motion is measured by a load cell and displayed on a recorder. The recorder trace is a plot of shear stress against time; since the shear strain rate is constant it is also a graph of shear stress against shear strain. The recorder trace obtained will be of the form of one of the lines (*A, B, C*) shown in *Figure 5.2*.

The explanation of these traces is that, when the specimen is sheared, in general its volume will change until it reaches a critical value. If the specimen is initially tightly compacted it will expand on shearing; in that case work has to be done both against internal friction between the particles and

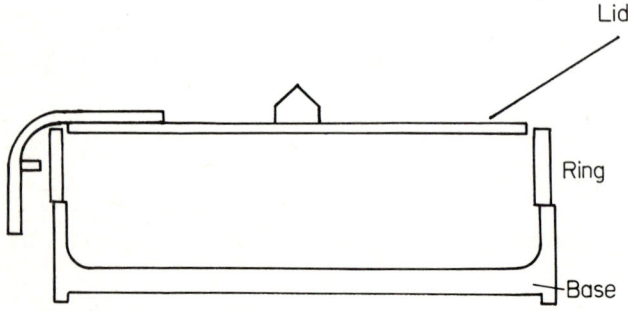

Figure 5.1.   The Jenike shear cell

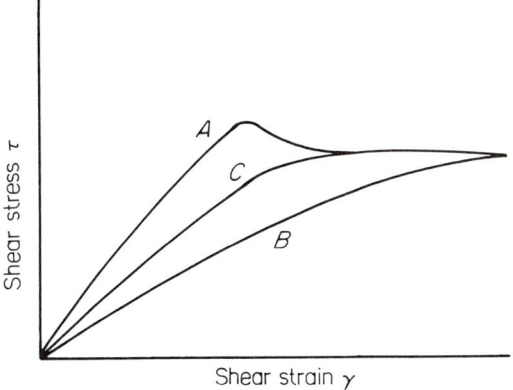

Figure 5.2. Types of stress—strain trace

also to raise the load on the lid. The latter component approaches zero as the volume of the specimen approaches its critical value so that the shear force needed to maintain the motion passes through a maximum, as shown in trace *A*. If the specimen is initially very loosely compacted it contracts on shearing and the load on the lid falls and some of its potential energy is recovered; the shear force necessary to maintain motion is therefore less than in the previous case and a trace of type *B* is obtained. If the specimen is initially at its critical volume to a first approximation no volume change occurs in shearing. The shear stress rises at first during the elastic deformation of the powder then remains at a constant value after plastic failure begins (trace *C*). This type of trace is of particular importance as it represents conditions in which all the work done by the shear force is used in overcoming internal friction in the powder.

For details of the test procedure the reader should consult references 1 and 2. A set of identical specimens is prepared and by trial and error the stresses applied during preparation are adjusted so that a trace of type *C*, representing failure without change in volume, will occur when the normal load on the specimen has some predetermined value referred to as the critical load. A number of specimens is then prepared by exactly the same method; each has applied to it a different load, less than the critical load, and in each case the shear force necessary to cause failure is determined. The traces obtained will be of type *A*, the peak giving the shear force needed to cause failure. A plot of shear force at failure against normal force is called the Jenike failure locus. For a cohesive powder this will be of the form shown in *Figure 5.3*. The applied forces have been divided by the cross sectional area of the shear cell to convert them into stresses. Every point on the failure locus represents conditions under which failure will occur; the point *E* represents the conditions required for failure without change in volume.

Further failure loci are obtained by changing the load applied to the specimens during preparation.

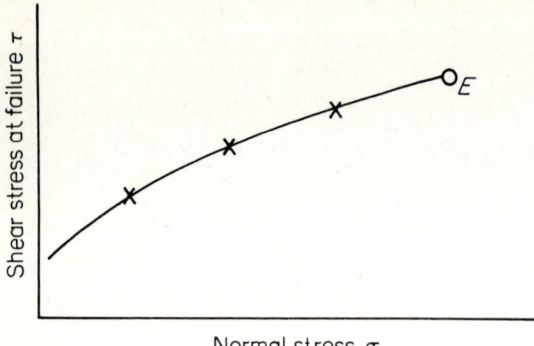

Figure 5.3.   The Jenike failure locus

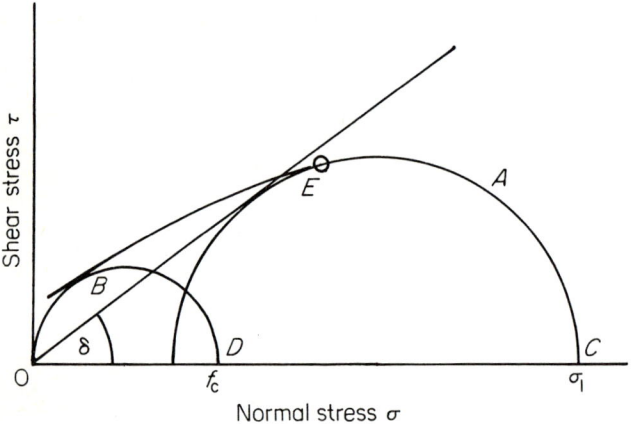

Figure 5.4.   Derivation of the failure function and effective angle of internal friction

If material is flowing in a hopper it can be assumed that it will have been sheared sufficiently to be at its critical bulk density. The stress conditions in the powder can therefore be represented by a Mohr circle, drawn with its centre on the horizontal axis, touching the failure locus tangentially at its end point E (*Figure 5.4*, circle A). The intercept C made by this circle on the normal stress axis gives the major principal stress (σ_1) acting in the powder. If flow should stop owing to the formation of an arch near the outlet the strength of the arch can be found by drawing a Mohr circle with its centre on the horizontal axis; the circle must pass through the origin and touch the failure locus tangentially (circle B in *Figure 5.4*). The intercept D gives the value of the major principal stress (f_c) needed to cause collapse of the arch. For this failure locus the intercepts f_c and σ_1 give the ultimate yield strength (f_c) of an arch formed in a material in which the major consolidating stress is σ_1. By repeating this procedure for each failure locus a graph can be drawn showing how the ultimate yield strength (f_c) varies

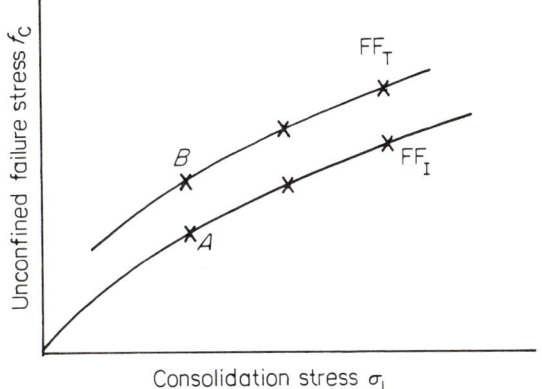

Figure 5.5.   Instantaneous (A) and time-consolidated (B)
failure functions

with the major consolidating stress (σ_1). This graph is called the Instantaneous Failure Function (FF_I) of the powder (*Figure 5.5*).

If a straight line is drawn from the origin tangential to the Mohr circle
passing through the end point of a failure locus (*Figure 5.4*) it can be shown
that the ratio of the major and minor principal stresses (σ_1/σ_3) is given by
the relation:

$$\frac{\sigma_1}{\sigma_3} = \frac{1 + \sin \delta}{1 - \sin \delta} \tag{5.1}$$

where δ is the angle between the tangent and the horizontal axis (*Figure
5.4*). The straight line is called the effective failure locus of the powder and
the angle δ is the effective angle of internal friction of the powder. It is
experimentally observed that the values of δ obtained from the different
failure loci for a powder are practically the same and a mean value can be
taken as the effective angle of internal friction.

In a mass flow hopper slip occurs between the particles and the hopper
wall. The angle of friction between the particles and the wall is therefore
important. It is measured by placing the ring from the shear cell on a plate
of material similar to the hopper wall. The ring is filled with the particulate
material and the lid is placed in position. The shear force needed to maintain uniform sliding motion is measured when different normal forces are
applied to the lid. The plot of shear force against normal force is a straight
line whose slope ϕ_w is the required angle of wall friction.

The three properties of the particulate material required for the design of
a mass flow hopper are therefore:

1.   The failure function (FF);
2.   The effective angle of internal friction of the material δ; and
3.   The angle of wall friction ϕ_w.

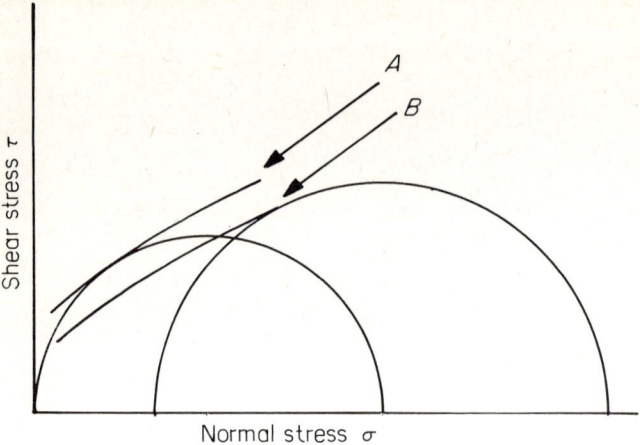

Figure 5.6.   Time-consolidated Jenike failure locus (A) *and instantaneous Jenike failure locus* (B) *for a time-consolidated powder*

Some particulate materials will gain strength if left in a hopper without motion. To ensure that flow will recommence after such a stoppage it is necessary to determine experimentally the amount by which a specimen gains in strength during storage and to obtain a failure function based on the results of tests made on specimens which have been time-consolidated. After an instantaneous yield locus has been obtained by the method described above further samples are prepared by the same treatment and are then left for an appropriate time under a consolidating stress σ_1 (*Figure 5.4*). The normal load on each specimen is then replaced by a smaller load (the same loads being used as in the tests on the instantaneous specimens) and the shear forces needed to cause failure determined. A time-consolidated yield locus is thus obtained (*Figure 5.6*) and the corresponding value of the unconfined yield stress (f_c) found by drawing the Mohr circle of type *B* in *Figure 5.4*. The value of the major consolidating stress (σ_1) is the same as that for the instantaneous yield locus. By repeating this procedure for the other yield loci a time consolidated failure function can be drawn (*Figure 5.5*).

This concludes the measurement of the relevant failure properties of a powder. The results are used to design a mass flow hopper.

5.4   Design of mass flow hoppers

Calculations of the stress distributions acting in a particulate material flowing in a converging channel have been made by Jenike[1]. He showed that the ratio of the major principal stress (σ_1), which determines the state of compaction and therefore the strength of the material, to the stress which will be applied to the surface of stoppages, such as an arch (A.S.) is a function only of δ, ϕ_w and θ, the wall slope. In a given system, where δ, ϕ_w and θ are constant, the ratio will be constant; it is referred to as the flow factor (ff) of the system.

$$\text{Flow factor (ff)} = \frac{\text{consolidating stress } (\sigma_1)}{\text{stress acting on arch (A.S.)}}$$

Jenike calculated the values of the flow factor for different conditions for three hopper types; conical, symmetric plane flow, and plane flow channels with one vertical wall. The results are presented in the form of flow factor charts in ref. 1.

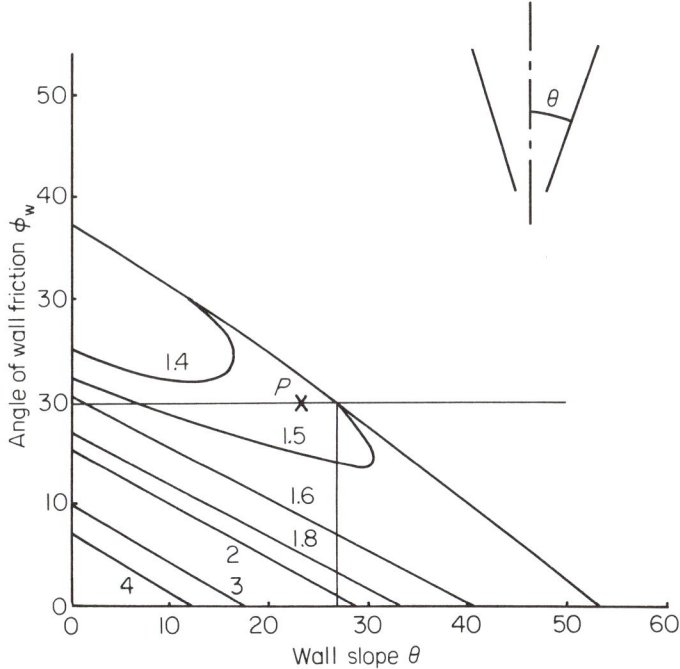

Figure 5.7.   *Flow factor chart for a conical hopper with $\delta = 40$ degrees*

To illustrate the use of a flow factor chart a typical example is shown in *Figure 5.7*. It applies to a powder of effective angle of internal friction 40 degrees in a conical hopper. First it is noted that the chart exists only within a defined area. If the angle of wall friction ϕ_w is known, an upper limit is set to the wall slope θ for which mass flow will occur (note that θ is the semi-included angle of the cone). For example, if $\phi_w = 20$ degrees mass flow will not occur if $\theta > 27$ degrees. In practice it is advisable to select a value for θ three degrees less than the limiting value, giving $\theta = 24$ degrees in this case. When θ has been selected this fixes an operating point on the chart, shown as *P* in *Figure 5.7*. The contour lines drawn on the chart connect points of equal flow factor; the flow factor for any point can therefore be obtained by interpolation. For the point *P* the flow factor is 1.46. This value is used to determine the minimum opening diameter to ensure mass flow.

The failure function for the powder, which was obtained from the shear cell tests, shows how the strength of an arch varies with the consolidating stress under which the arch was formed. The flow factor, obtained from the charts, gives the relation between the stress which is applied to the arch and the consolidating stress. By comparing the failure function and the flow factor we can therefore establish a criterion for conditions under which the arch will collapse and discharge from the hopper will occur. For the flow factor, if we plot applied stress against consolidating stress, the result will be a straight line through the origin of slope 1/ff. *Figure 5.8* shows the failure function and the flow factor plotted on the same graph, the two lines intersecting at the point *A*. In the region *0A*, for any value of the consolidating stress (*0C*) the stress applied to the arch (*CD*) is less than the stress required to break the arch (*CE*). A stable arch can therefore form and flow will stop. In the region *AB*, for any value of the consolidating stress (*0F*) the stress applied to an incipient arch (*FH*) will be greater than the stress required to break it (*FG*); an arch will therefore not form and flow will be continuous. The point *A* therefore represents the limiting stress conditions for flow to occur and gives a critical value of the applied stress (CAS) which must be exceeded. As the size of a hopper opening is increased the values of the applied stress and consolidating stress at the outlet increase so that there is a size of opening for which the stress applied to an arch reaches the critical value. From the results of shear cell tests for a powder *Figure 5.8*

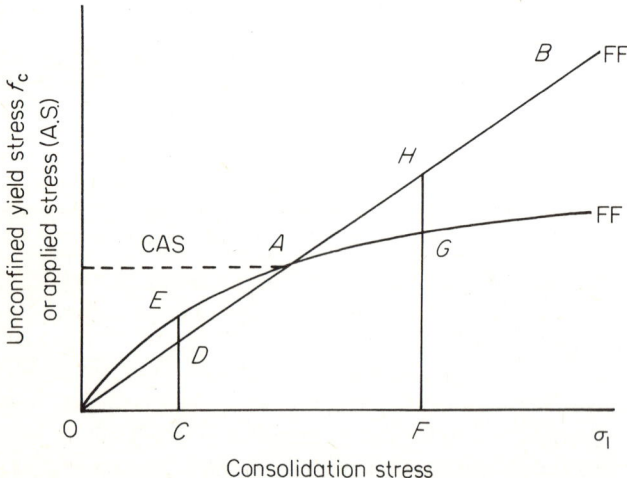

Figure 5.8.   Comparison of failure function (FF) and flow factor (ff). The broken line represents the critical applied stress (ACS)

can be drawn giving the value of the critical applied stress. For a conical hopper with a circular outlet the minimum outlet diameter (D) required for mass flow is given by:

$$D = H(\theta) \frac{CAS}{\gamma}$$

where

$$H(\theta) = 2 + \frac{\theta}{66}$$

and γ is the specific weight of the particulate material. The limiting values of wall slope and opening size for a mass flow hopper have therefore been determined.

The above method can also be used to design a symmetric plane flow hopper, that is, a wedge-shaped hopper with a rectangular slot outlet. The method gives the minimum slot width; the length of the slot must be at least three times its width.

In the case of a core flow hopper it is possible to calculate the size of opening necessary to prevent the occurrence of arching and piping. This will not give mass flow but it will ensure that uninterrupted flow will occur. This is particularly useful in modifying an existing core flow hopper by cutting back the converging section of the hopper to give a bigger opening.

5.5   Design of a plant for mass flow

In designing a plant to ensure the uninterrupted flow of particulate solids the first step is to design the hoppers satisfactorily, either by using mass flow hoppers or by making the outlets of core flow hoppers sufficiently wide to prevent the formation of stable arches or pipes. There are, however, other causes of stoppage and it is important that the whole plant should be designed with flow in mind; some of the problems to which attention should be given will now be discussed.

Even if mass flow hoppers have been properly designed it is possible to fill them in such a way that flow will not start when required. It must be remembered that a mass flow hopper has the following properties:

1. Once flow has started it will not stop.
2. If flow is stopped, for example by closing a gate, it will restart on opening the gate.

During flow, a stress distribution is set up in the powder so that failure will occur. If the hopper is filled by dropping material into it from a height, material near the outlet may be tightly compacted owing to impact pressure and the resulting stress distribution may be very different from that occurring during flow, so that on opening the gate flow will not begin. There are a number of ways of solving this problem.

1. A vibrator may be fitted to the hopper to be used only for starting flow after the hopper has been filled. It should be switched off as soon as flow has started. In every case where a vibrator is fitted to a hopper its switch should be interlocked with that opening the gate so that the vibrator cannot be switched on when the hopper outlet is closed.
2. If a discharge from the hopper, even at a very slow rate, can be maintained while the hopper is being filled, this will produce the required stress distribution near the outlet and prevent stoppages when the flow rate is increased to the required level.
3. If the hopper is never completely emptied the material near the outlet is protected from impact pressures and will flow out when required. The height of material left in the hopper should be at least twice the outlet width to be effective.

In emptying a hopper the device fitted should withdraw material from the whole of the hopper outlet. This is particularly important in the case of slot outlets. The screw conveyor, belt, air slide or other type of feeder must have an increasing capacity in the direction of flow, so that it draws material from the whole of the slot. In the case of a screw conveyor this can be achieved by using a tapered screw or a variable pitch screw.

When fitting a valve to the bottom of a hopper it is important to ensure that there are no projecting steps into the flow stream. There are some valves available in which the maximum aperture for flow is less than the hopper outlet size; this may prevent the development of mass flow even in a properly designed hopper, and with a cohesive powder may prevent flow.

The layout of a solids handling plant is very important and should be done by a designer with an understanding of the flow properties of his materials. A common fault when a hopper is feeding a machine or a conveyor belt is to leave a gap between the two, which is connected by a vertical pipe. I have seen a case where this pipe extended through the height of a storey in a building. This pipe was badly hammered, and was a constant cause of stoppage. Incidentally, wherever hammering has to be used it will eventually deform the hammered surface and make stoppages more frequent. In this case the trouble could have been avoided by anticipation of the problem at the layout stage. A short vertical connector may be acceptable if it is carefully designed; ideally it should be slightly diverging to reduce the effect of wall friction.

In designing chutes any changes in flow direction are danger points. The reason for this is that when a particulate solid flows round a corner it is subjected to shear and generally will expand. The increased pressure can lead to a blockage. The cure for this is to give room for expansion by increasing the cross-sectional area of the chute in the region of the bend.

The important principle I wish to emphasize is that if a plant is not to be subject to flow stoppages the whole plant must be designed for flow. The situation to be avoided is that in which units of equipment used to make up the plant are obtained from different suppliers and the arrangement for connecting these units is left to a draughtsman who is unaware of the flow problems that may arise.

References

1. JENIKE, A. W., '*Storage and flow of solids*', Bulletin 123, Utah Engineering Experiment Station, University of Utah, Salt Lake City, Utah, USA (1964)
2. WILLIAMS, J. C. and BIRKS, A. H., *Rheol. Acta* **4**(3), 170 (1965)

CHAPTER 6

Pharmaceutical granulation and compaction

B. Hunter

Imperial Chemical Industries plc, Pharmaceutical Division, Macclesfield, Cheshire

6.1 Introduction

An attempt is made in this chapter to identify some compaction problems related to pharmaceutical production, and to outline scientific and empirical solutions which have been developed to deal with them.

Before proceeding however it would be useful to consider why the pharmaceutical industry chooses to make the compacts called tablets, which represent the major part of its output. The compressed tablet evolved from the moulded 'pill' as a solid dosage form which is both elegant and convenient for the patient to carry and swallow. It maintains the drug in a dry, stable presentation, yet disintegrates rapidly into a finely divided form when ingested, thus ensuring rapid dissolution of the active agent. It is also easy and cheap to pack and transport, and can be produced on high-speed tabletting machines, with suitable emboss marks to effect product identification.

6.1.1 Compressor design

The speed of production has been an important factor in the development of pharmaceutical compressors. Single punch reciprocating machines were soon replaced by multistation rotary presses, which squeezed the punches between two rollers at the compression stage. A typical punch displacement sequence is shown in *Figure 6.1* and machines of this type can operate at turret speeds of up to 100 r.p.m. Indeed the larger multistation machines also include a precompression stage and perform two compression cycles each revolution. Such machines are capable of producing over 500 000 tablets per hour. The turret speed in this design is limited to 50—100 r.p.m. owing to the increasingly disruptive effect of centrifugal forces on powder flow and packing. Manesty Machines Ltd. have however recently developed a prototype 'radial' compressor which utilizes centrifugal force to fill the powder into the dies. This machine can operate satisfactorily at turret speeds up to 500 r.p.m. which means that the compression cycle involving die filling, weight adjustment, compression, relaxation, and ejection takes place in less than 120 ms.

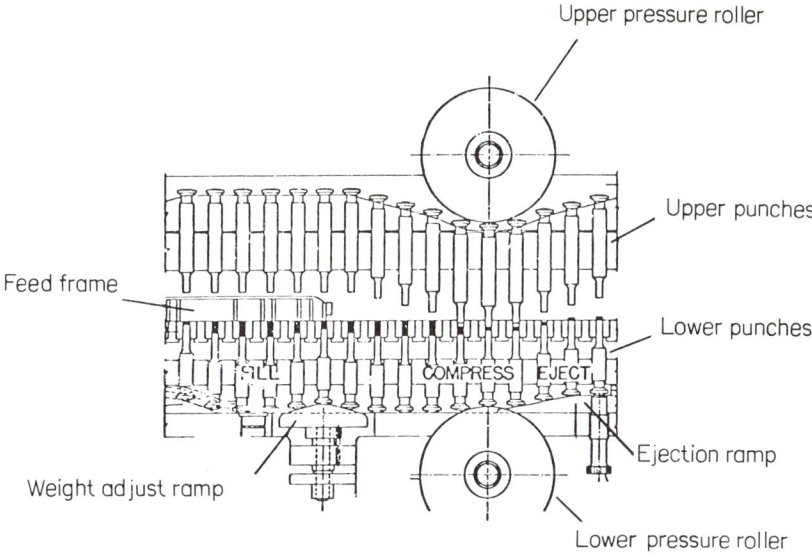

Upper pressure roller

Upper punches

Feed frame

Lower punches

COMPRESS EJECT

FILL

Ejection ramp

Weight adjust ramp

Lower pressure roller

Figure 6.1. Typical rotary tabletting machine

6.2 Theoretical considerations

6.2.1 Flow into the die

The need for accurate reproducible tablet weight control, with a coefficient of variation often less than 2 per cent, requires uniform flow and packing on the compressor die table. The problems are most acute with direct powder compaction. In these cases the use of shear cell data is invaluable in determining the likely behaviour of the powder system in the machine hopper or during die filling. The use of this technique was pioneered by Jenike[1] and this work has led to the development of scientfically designed mass flow hoppers. Complex force feeding systems have similarly been developed by compressor manufacturers to aid filling on the die table by preventing the formation of powder bridges and ensuring a uniform bulk density. The new 'radial' press described above utilizes centrifugal force for the same purpose.

The flow behaviour of a particular powder system can also be altered by changing its bulk properties. For instance, altering the particle size, size distribution, moisture content, etc. will alter the shear strength of a powder bed[2].

Another method used to improve flow is to add a proportion of fine powder to the system to present close contact of the main particles. Suitable additives are fumed silica and talc, which are added in small amounts to act as 'glidants'.

6.2.2   Powder compaction

The compression behaviour of a multiparticulate system is extremely difficult to characterize because there are several discrete mechanisms in operation which overlap and change as compression proceeds. Seelig and Wulff[3] identified three such regions. The initial mechanism involved particle rearrangement and packing down, followed by elastic and plastic deformation, and finally cold welding of the compact.

Despite this complexity there have been numerous attempts to characterize the compaction process such as that proposed by Kawakita and Lüdde[4]:

$$C = \frac{V_0 - V}{V_0} = \frac{abP_A}{1 + bP_A}$$

where C is the degree of compaction, V_0 the initial apparent volume, V the volume under applied pressure P_A, and both a and b are constants characteristic of the particular powder under consideration. This equation is useful because of its wide applicability in the field of powder compaction and its accurate results for many pharmaceutical systems.

The other most generally accepted equation is that due to Heckel[5] which has the form

$$\ln \frac{1}{1 - D} = KP_A + A$$

where D is the relative density at pressure P_A, K is equal to the reciprocal of the mean yield pressure, and A is a function of the original compact volume. This equation is of particular value as it provides information on the actual mechanisms of consolidation of the powder.

The validity of the Kawakita and Heckel relationships have been compared for pharmaceutical systems by Hersey and Rees[6] and again by Hersey, Cole and Rees[7].

6.2.3   Bonding mechanisms

Returning to the changing processes which occur during compression, the relative amounts of the different mechanisms will depend on the magnitude of the applied pressure. Bonding occurs whenever an area of true contact is established between two surfaces because the interfacial energy is always less than their surface energy. Johnson et al.[8], for instance, demonstrated that the tensile strength of the bond produced when a material with low elastic modulus such as gelatin was compressed was dependent on the surface energy. Similarly Easterling and Tholen[9], working with metal powders, showed that increasing surface contact due to plastic deformation during compression resulted in increased bond strength.

Hiestand[10] has extensively reviewed the mechanisms which occur with pharmaceutical powder systems. He considered the possibility of asperity melting at high compaction pressure, which has been proposed for other systems, but he concluded that plastic deformation was the major mechanism causing increased true contact area and bonding.

6.2.4   Decompression effects

The mechanisms governing powder bonding outlined above illustrate why increased compression pressure is required to form strong compacts but this action exacerbates the problems associated with decompression which can lead to lamination of the formed compact. Indeed the phenomenon of 'capping', where the tablet separates into two parts on ejection, is a common occurrence during pharmaceutical compression.

In a recent dissertation, Barton[11] reviewed the possible causes of lamination and capping in pharmaceutical systems. Capping can occur at two points in the compression cycle. First, fracture is possible during decompression of the compact following the maximum applied pressure. As the load on the punches is released, the tablet will expand axially owing, initially, to its normal elastic recovery and, secondly, to the Poisson's ratio effect of the residual radial and hoop stresses in the compact. The net axial strain ε_a in this circumstance is given by

$$\varepsilon_a = \frac{v}{E}\,(\sigma_r + \sigma_\theta)$$

where σ_r and σ_θ are the radial and hoop compressive stresses in the compact, and v and E are Poisson's ratio and Young's modulus for the material.

A pharmaceutical compact tends to behave like a brittle material and it is often unable to accommodate this imposed strain by plastic flow, in which case tensile failure results. A further factor is the differential expansion of adjoining areas of the compact due to the non-uniform distribution of stress which occurs on compaction and which was highlighted by Train[12]. This effect, illustrated in *Figure 6.2*, will also set up additional shear stresses in the compact.

A third possibility is related to the dynamic situation. If the load on the punches is released quickly enough, tensile unloading waves will propagate from both punch tips and travel towards one another, relieving the elastic compressive stress in the compact as they pass. However, at the instant when they meet, the two pulses will add and there will be a net tensile stress on the tablet cross-section at that point equal in magnitude to the original elastic axial compressive stress. Thus as the tensile strength of a pharmaceutical compact is very much less than the maximum applied pressure, under these circumstances, tensile failure of the tablet in a plane perpendicular to the axis of the die is probable especially as, owing to the still porous nature of the compact, there will be no shortage of sites from which a crack could propagate. Similar situations involving the sudden unloading of brittle materials have been investigated for prestressed concrete[13] and for plaster of Paris[14].

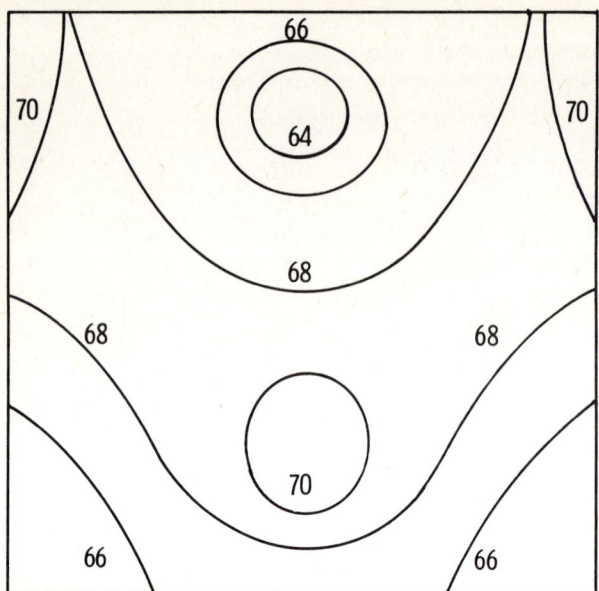

*Figure 6.2.   Apparent density distributions within a lubri-
cated compact, expressed as percentage of solid present
(from Train[12])*

The second stage of the cycle at which capping may occur is during
ejection of the compact from the die. That part of the compact which is
clear of the die will be free to recover elastically in the radial direction and
also, to a certain extent, in the axial direction as die-wall friction may have
previously prevented complete axial expansion. However, the lower part of
the compact remaining in the die will still be confined and this differential
strain distribution will lead to shear stresses being set up. If the original
stress in the compact, particularly in the radial direction, is high enough
then this will cause fracture along the top of the compact causing it to 'cap'.

Hiestand and his coworkers[10,15] have extensively studied the effects of
decompression for pharmaceutical systems and state that whether or not
fracture occurs during the shear deformation which accompanies decompression depends on the ability of the material to relieve the stresses by
plastic deformation without undergoing brittle fracture, and this ability is a
time-dependent phenomenon. Those materials that relieve stresses rapidly
are less likely to 'cap' or laminate. Hiestand has proposed several methods
to assess the readiness of materials to laminate, ranging from measuring the
speed at which the residual die wall pressure decays, to a 'brittle fracture
propensity' test which compares the tensile strength of a compact that
contains a built-in stress concentrator 'defect' with one that does not.

In earlier work Leigh, Carless and Burt[16] studied the axial and radial
pressure cycles of several pharmaceutical systems. The curves obtained were
compared with those of several ideal solid bodies, namely, a perfectly elastic

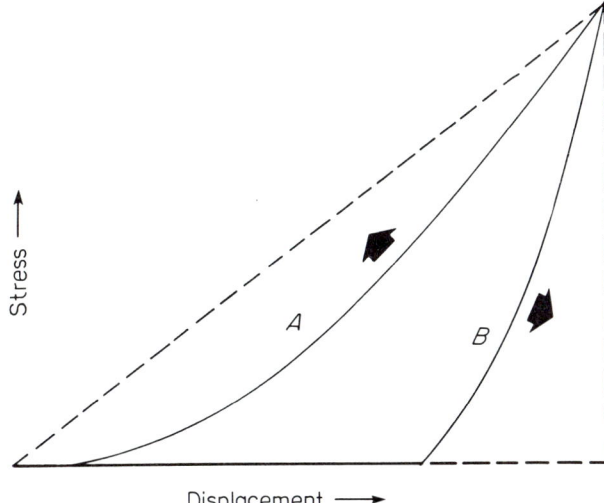

Figure 6.3.   Optimization of stress–displacement compression profile (from Guyot[17]). A, compression; B, decompression

body, a body with a constant yield stress in shear and a Mohr's body, i.e., a body in which there is a simple relationship between the shear yield stress and the normal stress on the plane of yield. The results indicated that materials which tend to laminate on decompression behave like a Mohr's body.

A simple method for characterizing materials has recently been proposed by Guyot[17], expanding the theory of Führer[18], where the upper punch stress is plotted against its displacement during the compression cycle. Ideally, a right angled triangle is obtained, indicating perfect plastic behaviour with none of the disruptive effects of elasticity. A real material takes the form shown in *Figure 6.3* and excipients should be selected to move the profile for the mixture towards the ideal as indicated by the arrows.

6.3   Powder preconditioning

It will be apparent that the flow, compression and relaxation properties of most simple fine powder systems as outlined above is not conducive to producing strong fault-free compacts on high speed tabletting machines. Therefore, to facilitate tablet production, the material fed to the compressor is normally preconditioned.

It is occasionally possible simply to blend the drug with a free flowing, readily compressible material, such as microcrystalline cellulose; but the

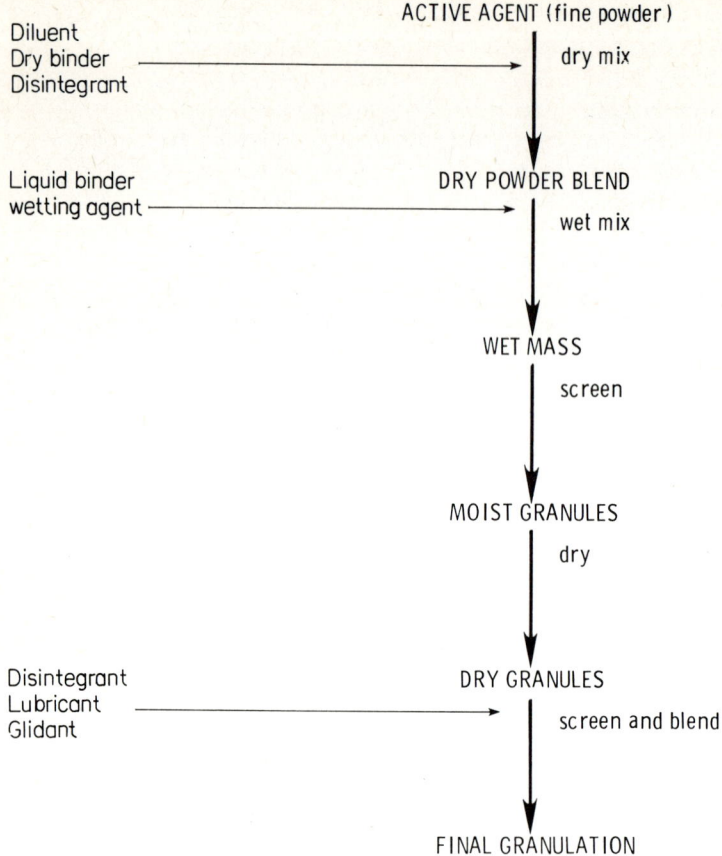

Figure 6.4.   *Typical pharmaceutical granulation process*

most universally accepted approach is to include a granulation stage prior to compression.

6.3.1   Wet granulation

The granulation process involves massing the powders with a liquid 'binder' to form moist granules which are then dried and finally force screened to produce a granulation with the required size distribution. A typical granulation process is shown in *Figure 6.4* which illustrates that the pharmaceutical tablet contains many ingredients (excipients) in addition to the active agent. The 'binders' are materials selected for their ability to form strong bonds even at high compression–decompression rates. The solid 'disintegrants' conversely enable the tablet to disrupt quickly when wetted, i.e., when ingested into the body. Other materials with specialized functions are similarly incorporated such as 'wetting agents', solid 'lubricants' and 'glidants'.

The process may appear overcomplicated, involving, as it does, many unit operations, but these may be telescoped by using specialized equipment and the advantages of a granular intermediate outweigh the processing difficulties. For instance, the ingredients are optimally mixed during the process, then maintained in the correct proportions within each dried granule, thus ensuring intra-tablet uniformity. The mean granule size and size distribution are fixed by the process conditions assuring optimal flow and packing characteristics with minimal dust generation. The granular intermediate also has improved compression properties because the process conditions and excipients are selected to achieve a friable granulation which deforms plastically to produce strong tablets with uniform surface appearance using relatively low compression pressure (approximately 200—300 MPa). Uniform distribution of the wetting agent and liquid binder also determines that the formed tablets have a hydrophilic pore structure, which allows rapid penetration of liquid and subsequent disintegration when ingested.

6.4   Compression scale-up

The granulation process facilitates the subsequent tabletting stage, but the extra stresses introduced during higher speed compression often still result in 'capping' occurring, which leads to a limitation on the compressor's production rate. This inadequacy is not altogether surprising as the formulator is often asked to develop a formulation containing a large proportion of a very fine, poorly compressible drug at a time when only a few kilograms of drug are available for his experiments. This makes the assessment of compression scale-up difficult because the high speed multistation rotary compressors which will eventually be used for tablet production require a minimum of a few kilograms of granules to run at all. Even when sufficient drug is available, assessment is hampered because the high speed machines are difficult to instrument and differ from one another in rates of compression–decompression, ejection, etc.

6.4.1   High-speed compression simulator

In an attempt to overcome some of these problems we have recently developed a compression simulator which can subject a powder sample to a complex high speed compression cycle but using a single station, this allowing individual tablets to be made in the same manner as on production machines.

The simulator consists of two independently operated, servo-controlled, hydraulic actuators which can move standard pharmaceutical punches within a die at velocities up to 400 mm s^{-1} to a positional accuracy of 5 μm. In addition the simulator can follow a force–time profile with the same relative accuracy up to a maximum compressive force of 55 kN. The two

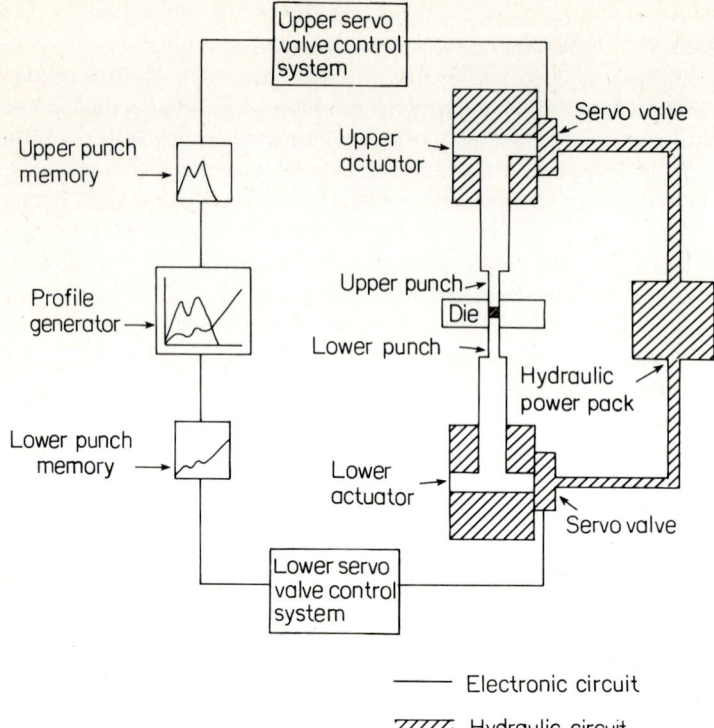

Figure 6.5. *Schematic diagram of the compression simulator designed for Imperial Chemical Industries plc*

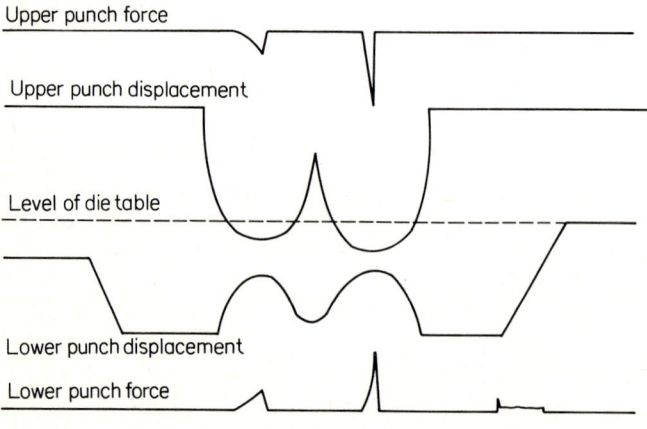

DIE FILL   PRECOMPRESSION   COMPRESSION   EJECTION

Figure 6.6. *Typical ultraviolet recorder traces*

actuators have separate control systems which can be programmed. The required upper and lower punch displacement–time profiles are converted to digital form by use of an ultrasonic graph plotter or a mini-computer. The two profiles are stored separately then fed out simultaneously to the two actuators at any rate between 0.001 and 30 Hz which gives a range of cycle times of 33.3 ms to 1000 s. *Figure 6.5* is a schematic diagram of the simulator.

Each actuator is instrumented with a precision load cell and displacement transducer which serve two purposes: first, to feed back force and positional data to the control system; and secondly to provide a permanent record of the force and displacement histories of each punch. For the latter purpose the outputs from the transducers are relayed to galvanometers in a multichannel ultraviolet recorder and, via A to D converters, to a mini-computer. They may also be monitored on peak-picking digital voltmeters. A typical recorder trace is reproduced in *Figure 6.6*.

The basic purpose of the simulator is to reproduce the compression cycles of production compressors and to investigate new machine developments; for instance it is able to follow the punch displacement profiles of the new 'radial' press described in Section 6.1.1.

6.4.1.1 Rate effects

A more fundamental use of the machine is to study rate effects for each part of the compression cycle. Fell and Newton[19] compressed crystalline and spray dried lactose and found that both forms offered a greater resistance to compaction at higher rates. Barton[11] used the I.C.I. Simulator to produce exponentially decreasing displacements for the two punches travelling towards each other during the compression stroke. He was thus able to compress powders at constant compression rates, and he investigated three such rates which compressed the powder bed to half its original volume in 700 s, 7 s and 70 ms respectively. The results for sodium chloride, potassium bromide and a direct compression sugar agreed with Fell and Newton, i.e., the materials compacted more readily at the lower rates. However, direct compression paracetamol consistently produced the opposite effect and this is currently being investigated.

The simulator has also been used to study the effect of recovery as discussed in Section 6.2.4. For instance some initial results for direct compression paracetamol lubricated with 0.5 per cent magnesium stearate are

Table 6.1. Effect of relaxation time on capping force

Relaxation time/ms	Capping force/kN
477	8.2
320	7.6
211	7.4
143	7.0

given in *Table 6.1*. The 'relaxation time' is the time interval between maximum compression and the beginning of ejection. All other parts of the cycles including the rates of compression, decompression and ejection were constant for the four tests. As the relaxation time was increased, the maximum force which could be applied at the compression stage without causing 'capping' also increased, thus allowing stronger tablets to be produced. This is probably due to the compact's being able to dissipate a larger proportion of the internal forces which remain after compression. Thus the rapid stress relaxation which could occur on ejection is diminished, reducing the tendency for the compact to 'cap'.

6.5   Formulation and process optimization

The product should be fully defined at a reasonably early stage in the total development timetable which spans several years. This is because details of the formulation and process have to be scrutinized by government bodies worldwide and then registered in each country, with supporting stability, bioavailability and clinical data, before the tablet can be sold in that country.

Some of the important factors governing process and excipient selection have already been outlined. The simulator is used to stress new formulations to ensure that they can withstand higher compression–decompression rates than they will ever meet in production and still produce satisfactory tablets with the required drug release properties.

The formulation development in conjunction with process design and scale-up must also ensure that the effect of material and process variation is minimized, particularly as the product will have to be manufactured in numerous overseas factories. Such a formulation is termed 'robust' and is the aim of every formulator, but inevitably there will be some properties of the starting materials and process which if allowed to vary unchecked would affect the final product. One solution is to control every possible variable within close limits, but this course, although it should ensure no reject batches, would be difficult and costly to maintain and many variables would be controlled which did not have a significant effect on the quality of the final product. The significant variables should therefore be predetermined before a new formulation is fully established in production, and only those variables need be subsequently controlled or monitored.

6.5.1   Computer optimization

The variables which could possibly affect the final product should therefore be identified during formulation development. These would probably include the surface activity, temperature, viscosity and volume of the liquid binder, and the relative proportion, size distribution, shape and moisture content of the starting materials, and of the intermediate granules. The effect of each of these independent variables on tablet properties can be determined individually or by using computer optimization techniques, such as one based on the mathematical model described by Box and Wilson[20]. This 'fine tuning' of the formulation should be completed before the process is scaled-up and will be the basis of the provisional physical specification for the ingredients and process. The general procedures for computer optimization of pharmaceutical systems have been discussed by Schwartz *et al.*[21].

6.5.2 Production monitoring

When the formulation is transferred to production scale there should be an establishment period during which the process is extensively monitored. The crystal form of the drug and other important properties of the system will have been defined but the other variables which were significant during the small scale experiments and data collected during the production scale granulation process should now be correlated with the physical properties of the tablets produced. At least 30 batches should be monitored during this establishment period and the dependent tablet properties correlated with the independent process and material variables using multiple regression analysis. The resultant relationship for each dependent variable should include only those independent variables which significantly improve the fit of the equation. The required limits for each dependent variable will have been selected and by substitution the accepted range for each significant independent variable can be calculated. The final physical specification for the formulation can now be set and the significant in-process variables routinely monitored. These data can subsequently be used to keep the process in check by cusum analysis or by simply using control charts to eliminate process 'drift' and thus reduce the risk of compression problems.

References

1. JENIKE, A. W., '*Storage and Flow of Solids*', Bulletin 123, Utah Engineering Experiment Station, University of Utah, Salt Lake City, Utah (1964)
2. PILPEL, N., *Adv. pharm. Sci.* **3**, 173 (1971)
3. SEELIG, R. P. and WULFF, J. *Trans. Am. Inst. Min. metall. Engrs*, **166**, 492 (1946)
4. KAWAKITA, K. and LÜDDE, K.-H., *Powder Technol.*, **4**, 61 (1970/71)
5. HECKEL, R. W., *Trans Am. Instn mech. Engrs* **221**, 671 (1961)
6. HERSEY, J. A. and REES, R. E., '*Conference on Particle Size Analysis*', Bradford (1970)
7. HERSEY, J. A., COLE, E. T. and REES, J. E., '*Proceedings of First International Conference on Compaction and Consolidation of Particulate Material*, Brighton', 165 (1972)
8. JOHNSON, K., KENDALL, K. and ROBERTS, A., *Proc. R. Soc. A.* **324**, 301 (1971)
9. EASTERLING, K. and THOLEN, A., *Acta metall.* **20**, 1001 (1972)
10. HIESTAND, E., '*Physical Processes of Tabletting*', International Conference on Powder Technology and Pharmacy, Basel, Switzerland (1978)
11. BARTON, D., M.Sc. Dissertation, University of Manchester Institute for Science and Technology (1978)
12. TRAIN, D., *J. Pharm. Pharmac.* **8**, 745 (1956)
13. RINEHART, J. S., '*International Symposium on Stress Wave Propagation in Materials*', Interscience, New York (1960)

14. LLOYD, R. B., M.Sc. Thesis, University of Manchester (1974)
15. HIESTAND, E., WELLS, J. E., PEOT, C. B. and OCHS, J. F., *J. pharm. Sci.* **66**, 510 (1977)
16. LEIGH, S., CARLESS, J. E. and BURT, B. W., *J. pharm. Sci.* **56**, 888 (1967)
17. GUYOT, J. C., '*Physical Measurements during Compression and their Practical Relevance to Formulation of Tablets*', International Conference on Powder Technology and Pharmacy, Basel, Switzerland (1978)
18. FÜHRER, C., *Pharm. Ind.* **25**, 674 (1963)
19. FELL, J. T. and NEWTON, J. M., *J. pharm. Sci.* **60**, 1866 (1971)
20. BOX, G. E. P. and WILSON, K. B., *J. R. statist. Soc. B* **13**, 1 (1951)
21. SCHWARTZ, J: B., FLAMHOLZ, J. R. and PRESS, R. H., *J. pharm. Sci.* **62**, 1165 (1973)

CHAPTER 7

Mechanisms of compaction

J. J. Benbow
Formerly Imperial Chemical Industries plc, Agricultural Division, Billingham, Cleveland

7.1 Introduction and scope

In the formation of catalysts and catalyst supports for use in the manufacture of gases, for example in the production of ammonia, a range of compacted catalysts each carrying out a particular reaction and made in several

Figure 7.1. Catalyst pellets

different shapes from a variety of materials is required. *Figure 7.1* shows some examples of these pellets. The compacting (or pelleting) requirements for catalysts are critical for three specific reasons:

1. The metal oxides used are very hard.
2. Catalysts must be simultaneously strong and porous.
3. Catalysts must also have a high internal surface area.

These needs make the approach to catalyst production somewhat special but most of the fundamental aspects of compaction are common to *all* compacted materials. The pelleting machines used by different industries are also basically similar. It is with these basic, common features of compaction that this chapter is concerned. The specific differences between products and compaction methods are described in other parts of the book.

Explanation of the mechanisms of compaction has been restricted to uniaxial compaction in a cylindrical punch and die: isostatic pressing is not described (*see* Chapter 12), but several mechanisms are the same in both axial and isostatic compaction. The principal steps in uniaxial die compaction are:

1. Flow of the correct amount of powder into the die.
2. Application of pressure.
3. Compaction, i.e., reduction of voids and increase in bulk density.
4. Bond formation.
5. Removal of pressure.
6. Ejection of compact.

The aim here is to describe in outline the principal mechanisms which occur during compaction, i.e., steps 2, 3, 4 and 5. In principle it should be possible to predict quantitatively the mechanisms from the physical and chemical

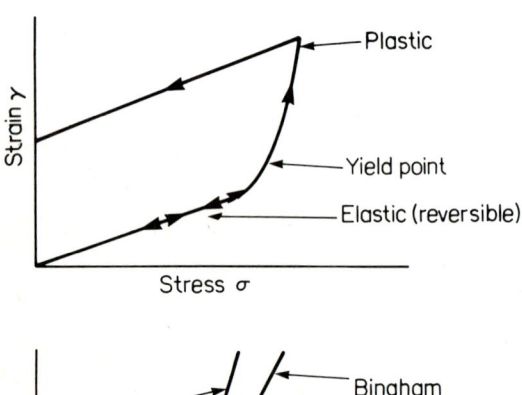

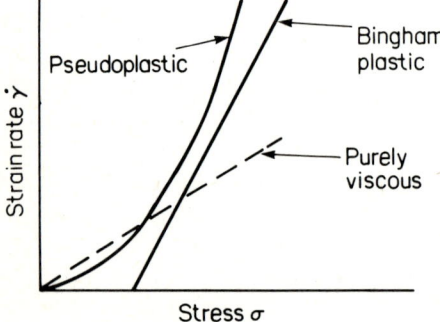

Figure 7.2.   *Typical strain and strain rate curves as functions of stress*

properties of the powders and the applied stresses. In practice, however, this would involve describing theoretically the shape of the particles, their size and spatial distribution, and then combining the effects of all the individual mechanical processes which take place simultaneously. This is not yet practicable.

It is possible to generate theories which fit the experimental data but to select the correct theory in this way is extremely difficult. What can be done is to make detailed experimental measurements during compaction or on products and then ensure that their interpretation is consistent with the expectations from basic physical and chemical laws and mechanical models.

In the following description experimental evidence is used rather than the relevant mathematical theories. The references quoted will allow theories to be studied in detail. Comprehensive accounts of compaction are given by Kirsop[1], Heckel[2], Onoda[3], James[4], Train[5] and by Hardman and Lilley[6]. Key references are given rather than a full literature survey. Each gives numerous historical references.

The powder feed has a size distribution determined by the type of pelleting machine to be used and the end use of the product. The initial powder density distribution is not necessarily uniform but will not significantly upset the subsequent compaction mechanisms. The actual powder density in the die will depend on the particle size distribution and on particle shape but for most powders it is likely to be about 50—60 per cent of the solid density. Thus the starting point is a bed of particles having a bulk density equal to 50—60 per cent of that of individual particles. After the application and removal of pressure the average density will have increased to about 80—90 per cent of the maximum.

Within a die there is a microcosm of physical and mechanical processes. These include powder flow, percolation, lubrication, friction, fracture, and elastic, viscous and plastic deformation, as well as the important operation of generating interparticle bonds. By these mechanisms a powder is transformed into a porous, coherent well defined shape in a fraction of a second.

The main effects which govern tabletting can be divided into material properties and machine conditions. These can be subdivided and amplified as follows.

1. Intrinsic material properties
    (a) Mechanical nature of the material to be tabletted, i.e., viscous, plastic or elastic.
    (b) Material properties, i.e., values of its viscosity, plasticity, hardness etc.
    (c) Properties and amounts of additives, lubricants and binders[7].
2. Particulate properties
    (a) Mean size and size distribution[8-14].
    (b) Shape[15].
    (c) Agglomerate porosity.
    (d) Moisture content.

3. Nature of applied load
    (a)   Axial compression, isostatic compression or shear.
    (b)   Rate of load application and removal. Punch pressure *versus* time or position *versus* time.
4. Die geometry
    (a)   Length.
    (b)   Diameter.
    (c)   Shape complexity, regions of different depth or thickness, etc.

The interaction between the material characteristics listed under 1 and 2 and the applied mechanical conditions in 3 and 4 determine the properties of the product.

In view of the variety of materials which are tabletted it is useful to describe briefly the main ways in which single phase substances react to an

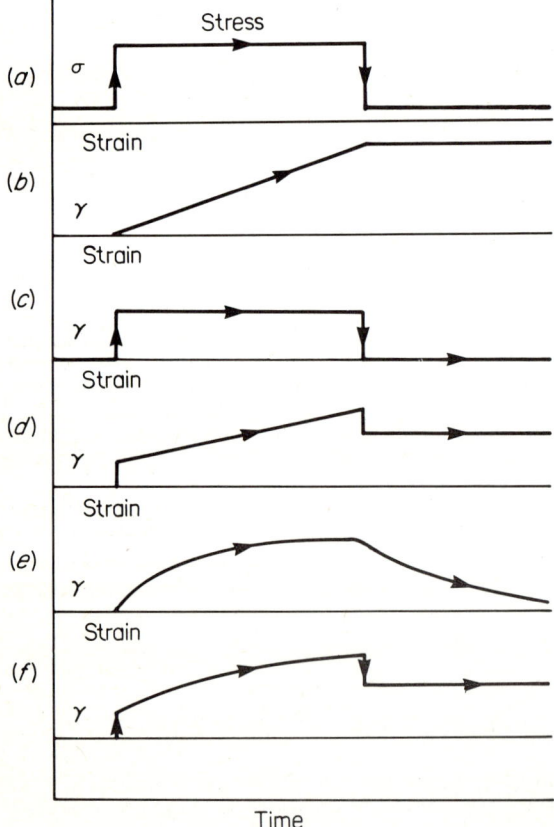

Figure 7.3.   *Plots of an idealized stress pulse σ* (a) *and the resultant strain γ in* (b) *a viscous liquid,* (c) *an elastic solid* (d) *an elastic liquid,* (e) *a viscoelastic solid and* (f) *a plastic solid when σ > yield stress*

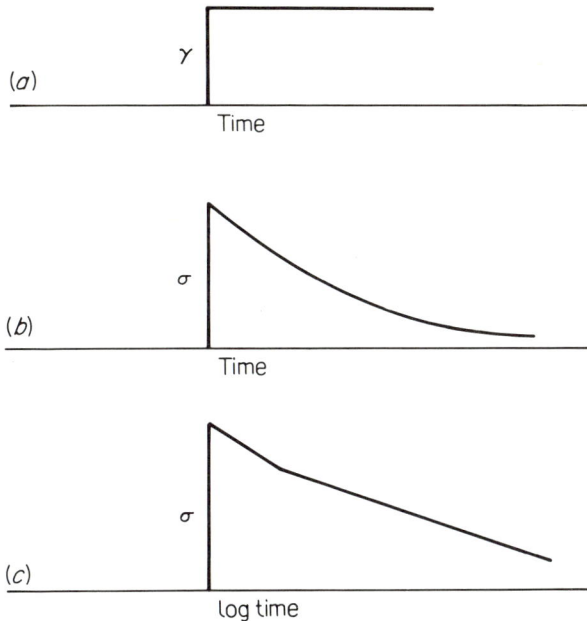

Figure 7.4.   Stress behaviour as a function of time for the step function strain. (a) The imposed strain γ. (b) The resultant stress σ in a Maxwell body where $\tau = \eta/G =$ the relaxation time. (C) The resultant stress σ for real materials (relaxation-time spectra)

applied mechanical condition. *Figure 7.2* shows some typical stress *versus* strain and stress *versus* strain rate curves which are plotted together with annotated descriptions. To emphasize the relationship between these properties and die compaction it is also informative to present the response as a function of time to an idealized stress pulse. *Figure 7.3* shows some typical types of behaviour. In *Figure 7.4* the stress relaxation of an ideal model solid is compared with that of a typical real material. Even though pelleting materials are complicated by internal voids in the feed powder and by the presence of lubricants, these fundamental types of deformation still describe their behaviour in dies.

7.2   Application of pressure and frictional effects

The load applied to the punch varies with time but depending on the type of press involved it will either have a prescribed position *versus* time characteristic, a pre-set force *versus* time characteristic, or a combination of both. For the present purpose it is however assumed that the applied force increases slowly and uniformly with time and is then removed in a similar time interval.

7.2.1 Stress distributions

The force on the punch is transmitted both to the wall and to the bed of the pellets[16]. Elementary theory predicts that the pressure exerted on the powder near the wall will decrease exponentially with distance from the moving punch owing to the die wall friction. The decrease depends on the ratio of axial to radial pressure and the average friction coefficient.

This theory has recently been superseded by some careful, elegant measurements by Strijbos *et al.*[17,18]. Their measurements on compacted iron (III) oxide with stearic acid lubricant show that the measured distribution of wall stress differs significantly from that predicted by simple theory. They found for an unlubricated powder that the stress normal to the wall was constant for roughly half the pellet length away from the moving punch and then fell *linearly*. For an L/D of 0.4 the total decrease was to about 80 per cent and for an L/D of 1.20 to 40 per cent of the maximum. For a lubricated system the values in the constant region were similar for different L/D ratios but the decreases near the lower punch were appreciably larger. The same team also measured the transmitted axial and radial pressures as a function of final pellet length.

The fractional decrease of transmitted axial pressure falls linearly with increasing height to diameter ratio (H/D) whereas the proportion of radial pressure transmitted is independent of the final aspect ratio.

From experiments with composite punches they deduced the distribution of normal stress across the upper and lower punches for both lubricated and non-lubricated systems.

The distribution on the upper punch was not affected by the presence of lubricant. In contrast the variation of stress across the lower punch was approximately parabolic and was considerably altered by the presence of lubricant.

7.2.2 Density distribution

Aketa *et al.*[19] have examined the effects of applied pressure on density distribution and have compared experimental results with theoretical predictions. A comparison between their calculated density distributions and the actual distributions showed a good overall quantitative agreement but the local areas of high and low density found experimentally along the pellet axis were not however predicted theoretically. Train[20] has also made theoretical predictions of the density distribution and compared them with measured results.

Comparison of the density distributions of compacts made from iron oxide[19], nickel powder[21,22], magnesium carbonate[23] and catalyst shows that the patterns are similar in several features, i.e., high densities at the upper edge as well as a low density region at the lower outer edge. It should be noted that for catalysts the non-uniform density distributions are small compared with the average density.

The high and low density regions at the upper and lower edge respectively are consistent with the pressure distributions described by Strijbos[18]. The high shear stresses at the outer edge of the die give rise to the dense regions. Correspondingly, the low pressures near the outer edge of the lower punch account for the drop in density in this region.

7.3   Particle rearrangement

The pressure distribution is not uniform at any time within the die, but at any point within the powder it can be assumed to be subjected locally to a steadily increasing compressive force. There will be some increase in density without major deformation as particles flow past each other into voids. A measure of this is the 15 per cent difference between poured and tapped bulk density. Vibrations will promote this percolation but sticky, rough or irregularly shaped particles will inhibit the tendency. Particle size distribution will have a critical effect on the densification which takes place. Williams has described the flow behaviour of powders in bulk in Chapter 5. Recently Bridgewater[24] has related percolation to movement in simple shear cells. Percolation velocities were found to be influenced by total shear strain and (perhaps more relevant to powder compaction) by particle diameter ratio. Lesser effects arise from particle density and elasticity.

As the applied force continues to increase, gravitational effects will become relatively less important. Further rearrangement can take place by processes which depend on the mechanical behaviour of the material.

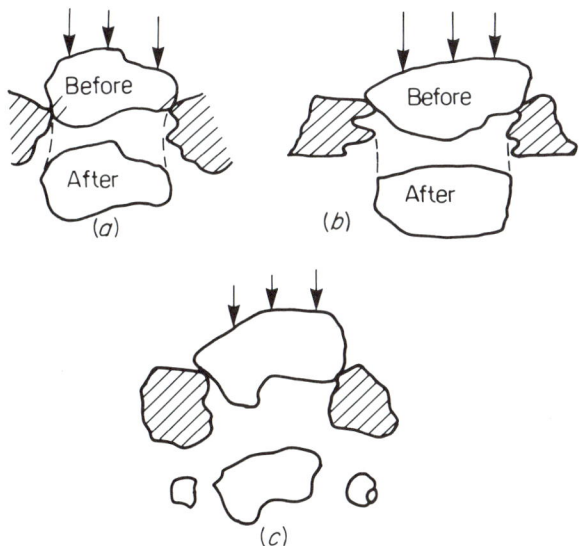

Figure 7.5.   (a) *Elastic,* (b) *viscous or plastic, and* (c) *brittle fracture deformation of particles before and after loading*

Change of relative positions may involve local elastic contraction in which a particle deforms to pass through a small gap and then recovers [*Figure 7.5(a)* and curve (*c*) of *Figure 7.3*]. Alternatively it may flow in a viscous or plastic manner and remain permanently changed in shape [*Figure 7.5(b)* and curves (*b*) and (*c*) of *Figure 7.3*]. Finally the asperities may be broken off or the particle may be shattered [*Figure 7.5(c)*] in which case the change is permanent and results in an increased surface area.

Hardman and Lilley[6] describe the compaction mechanisms which occur during the densification of sucrose, sodium chloride and coal. This comprehensive paper is illustrated by several stereo-scan electron micrographs.

In the case of agglomerated powders such as catalysts, where the powder particles are themselves made up from even smaller units, the amount of rearrangement will depend on the ability of the *agglomerate* to respond to the applied load. If the predensification is too low the final porosity will be high. On the other hand if the predensification is too high the load needed to produce a strong compound will be extremely high and machine damage may result.

7.4   Deformation without rearrangement

Eventually (some milliseconds later), when the processes so far mentioned are completed, the relative positions of individual particles will remain unaltered. Since, generally speaking, the prime purpose of compaction is to bind particles together into a chosen shape the subsequent mechanisms are crucial.

Compact strength depends on the product of bond strength per unit area and the area of intimate *permanent* interparticle contact. Elastic deformation, being recoverable, constitutes a disrupting influence rather than a strength-producing mechanism. The important mechanisms are those which impart *permanent* areas of contact, i.e., viscous and plastic deformation and perhaps fracture. This refers to deformation at the points of interparticle contact, not to the whole particles described in Section 7.3.

7.4.1   Viscous flow

If the material is viscous it will flow irreversibly but the local stress is reduced, as the area of contact increases, so that the amount of viscous flow decreases with time [*Figure 7.3(b)*]. Any highly viscous binder present will behave in the same way. If however the binder has a very low viscosity it will be forced out of the space between the two asperities and will tend to occupy the interparticle voids. The rheological properties of ceramic binders are discussed by Onoda[25]. If the material is purely elastic the area of contact will disappear when the tabletting load is removed [*Figure 7.3(c)*]. Purely viscoelastic solids can be compacted but they will tend to recover completely with little or no permanent bonds being formed and regain their original shape soon after the load is removed.

7.4.2 Fracture

The main effect of particle fracture at this stage is to create clean fresh surfaces which assist bonding. Fragments will also fill up interparticle voids. The extent of the fracture depends mainly on the material being tabletted. As a result of fracture the external surface area of the particles increase even when the punch movement becomes small.

7.4.3 Plastic deformation

Probably the most important single mechanism which occurs during compaction is plastic deformation. This is especially true for metals[26] and ceramics. To illustrate this some results obtained by pelleting crushed crystals are presented.

The materials used were minerals with a range of hardness from gypsum

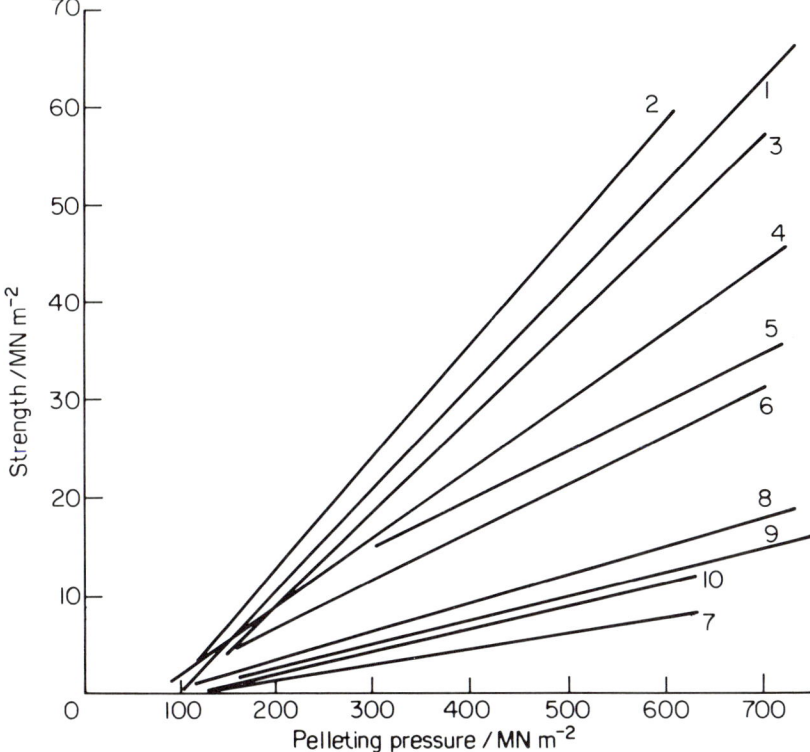

Figure 7.6. Effect of pelleting pressure on axial crushing strength of compacted calcite particles of different sizes

Curve	1	2	3	4	5
Size/μm	0—6	6—13	13—26	26—53	53—76
Curve	6	7	8	9	10
Size/μm	76—210	210—420	420—600	600—850	850—1000

(2 on Moh's scale) to fluorite (4 on Moh's scale). They were first ground and separated into size fractions by sieving, between 1000 and 120 μm, and by air separation below 120 μm. Each of ten size fractions was then compacted to make about 20 compacts, at pressures ranging from the minimum at which a compact could be ejected whole from the die to the maximum obtainable in the equipment. Compaction was performed in a cylindrical 8.5 mm diameter case-hardened steel die. This was closed at its lower end by a stationary punch.

As an aid to pellet ejection, and to reduce the pressure drop in the die during compaction, the die walls were lubricated with graphite by pre-forming and ejecting a graphite compact between each test. The required amount of powder was then placed in the die and compressed by moving the top punch downwards at a constant rate of 5.0 mm min^{-1}, using an Instron TT/cm Universal Testing Machine. A compact length of 11 mm was selected because below this value the axial compressive strength depended on the pellet length. Longer pellets were not used because the pressure drop due to frictional forces in the die would then have produced undesirable density distributions. Constant length was achieved by presetting the movement of the crosshead and the maximum load was recorded by use of the standard Instron 5000 kg load cell. The density and porosity

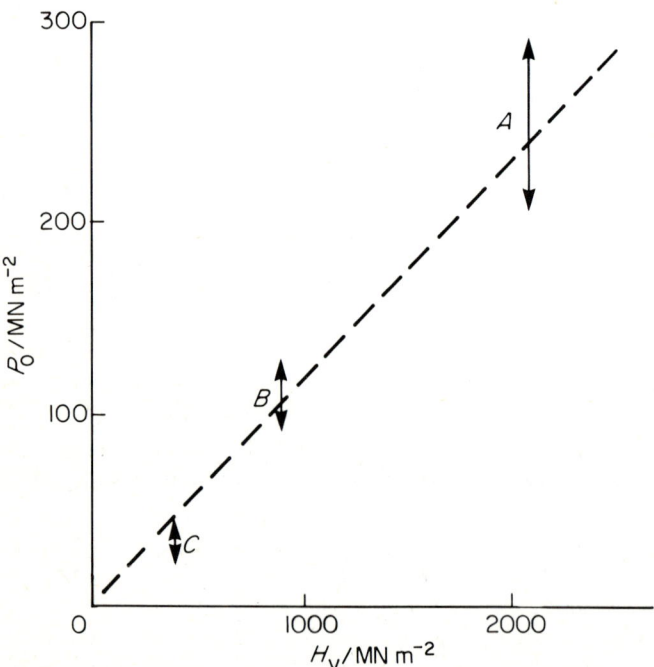

Figure 7.7.   *Pressure needed to give significant pellet strength as a function of Vickers hardness H_V for A, fluorite; B, calcite; C, gypsum*

of each compact were calculated from its weight and dimensions. Crushing tests were performed by axial compression of each compact between flat, polished steel platens at a rate of 5.0 mm min^{-1}. The results obtained showed the following features:

1. To obtain significant strength a characteristic pressure had to be exceeded (*Figure 7.6*). Above this pressure the strength increased approximately linearly with increasing pressure. At a fixed pressure the strength increased as the particle size decreased. To a first approximation the critical pressure was independent of particle size.
2. The critical pressure increased with the hardness of the mineral being pelleted as shown in *Figure 7.7*, which shows that the critical pressure is approximately proportional to the hardness of the material. Although the experiments were done on only three minerals, additional literature data on sodium chloride and ferrous powders support this conclusion.
3. The observed behaviour suggested that after the particle rearrangement and fracture which occurs during the initial stages of pelleting, continued compaction results in elastic deformation and plastic flow at asperities. Only plastic flow produces significant areas of permanent contact, and it is these which give the pellet its strength. The paper by Hardman and Lilley[6] discusses the relative importance of plastic deformation and fracture for sucrose, sodium chloride and coal.

7.5   Strength-producing mechanisms

On the whole in this subject there is a shortage of good, incisive, experimental results. Once areas of intimate interparticle contact have been formed, the second part of the bonding process is the formation of permanent attractive forces across these contact areas. Solid bridges between particles may be formed if sufficient heat is generated by friction at points of contact to cause local melting followed by solidification. Goetzel[27] considered that, with metal powders, insufficient heat was produced to cause melting, but with non-metallic powders of low thermal conductivity and low melting point this effect may contribute appreciably to the strength of a compact. The hardening of a binder can produce solid bridges of high strength between particles and when water or some other liquid is present in the mix, drying of the compact and crystallization of a dissolved substance may form solid connections between particles.

The presence of a liquid in a compact can cause an increase in strength owing to surface tension effects giving rise to forces much higher than the van der Waals forces acting between dry particles. The nature and magnitude of these forces and their effect on compact strength have been discussed by Newitt and Conway-Jones[28], Rumpf[29], Pietsch and Rumpf[30] and Cheng[31]. The strength of a compact increases with the amount of liquid present up to the point where there is sufficient liquid to fill the voids in the compact. Several writers have attempted to derive the expressions for the attractive forces between solid particles in terms of the van der Waals forces acting between molecules. London[32] discussed the attractive force between two atoms and Hamamaker[33], making the assumption of additivity of forces in the case of three or more atoms, obtained results which did not agree well with measurements of forces. Casimir[34] calculated the attractive force between metal plates and Lifschitz[35] developed a more general method for predicting the forces between macroscopic surfaces.

Commercially the effect of pressing speed on the properties of the compact is of great importance, and it is surprising that more attention has not been given to this problem. Long and Alderton[36] performed compaction tests on fine wax powder, using three pressing speeds. They found that the amount of air trapped increased, with increase in pressing speed. Even at lower speeds very little air escaped towards the end of the compaction process; at this stage the compaction curve shows that the only effect of increasing the pressure is to compress the entrapped air. Whitman[37] reported that in the compaction of dry soils the strain rate had little or no effect. Kunin and Yarchenko[38] compressed three materials, moulding powder, graphite and rocksalt, at different strain rates up to the same final pressure. They showed that the density of the compacts formed decreased with increasing strain rate. Wang and Davies[39] found, however, that on compacting an iron powder, less pressure was needed to achieve a given density at higher pressing speeds. The results of Whitman, Kunin and Yarchenko and Wang and Davies are contradictory, probably because they were investigating different ranges of pressure and strain rate. Smith[40] discusses the application of different binders to the production of ceramic ware.

7.6   Load removal and stress relaxation

The die still exerts a radial stress on the compact after the applied load is removed. The relationship of die wall pressure to average punch pressure during the compaction of several materials has been determined by Hiestand *et al.*[41]. For most compacted substances a cycle similar to that shown in *Figure 7.8* is obtained.

Long[42] pointed out that failure of the compact could be caused by the radial stresses within the die. Ejection forces are an appreciable fraction of the maximum force. They can be reduced by wall lubrication. As the pellet emerges from the die residual stresses can cause lamination. This is due to the stress concentration at the die exit when part of the pellet is still in the die and the remainder is out of it. In Section 7.1 the behaviour of some idealized viscoelastic materials was mentioned. Many compacts exhibit such time dependent stress-relaxation behaviour after die compression (*Figure 7.4*). If the relaxation time is less than the residence time of the compact in the die the ejection forces will be low. However, for many substances the relaxation times are very long. For example, some compacts show the effects of unrelaxed stresses for up to 100 hours after pelleting unless they are annealed. These stresses can be detected by measuring (*a*) the increase of

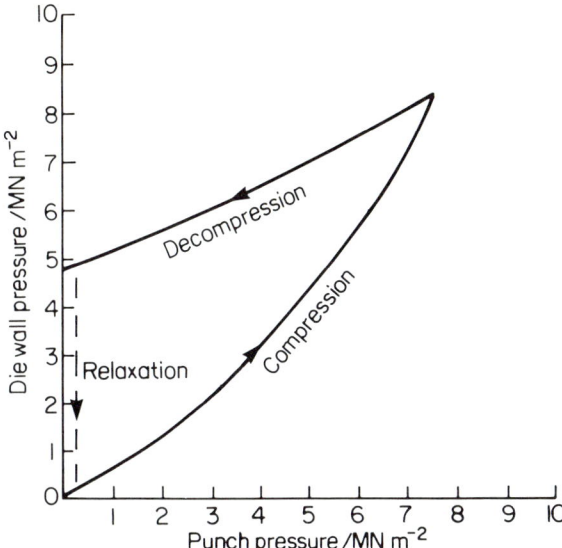

Figure 7.8. Axial and radial pressure cycle

crushing strength with annealing time, (*b*) the release of thermal energy upon immersion as a function of annealing time, and (*c*) the small but measureable increase of pellet length with time after ejection.

Data showing the decay of punch pressures with time after compaction have been given by Hiestand[41]. The decay of stresses after compaction generally exhibits a spectrum of times rather than a single relaxation time [cf. *Figure 7.4(c)*].

7.7 Material properties

So far emphasis has been placed on mechanical processes rather than material properties, although the appropriate physical behaviour has been mentioned at each compaction stage. An aspect of material properties which is common to most compacted materials is the effect of particle size on tablet properties. Research on sieved fractions has been done by many workers[8-14]. Although the stress system in a compact is generated in a fraction of a second, it can take hours to relax. The behaviour of different materials, e.g., ceramics and pharmaceuticals, is again alike and indicates the predominance of plastic behaviour in particle compaction.

It is generally accepted that the strength of granular materials, S, increases with decreasing particle or grain size, d, i.e. $S = S_0 d^{-n}$ where n is between about one third and one half. Whether this is due to an increase in the total interparticle surface area with decreasing particle size or to an effect of particle size during the strength measurement has not been fully explained.

It has recently been suggested by Ridgeway[43] that one reason why small particles promote higher strengths is because of the transition from brittle to plastic behaviour which can occur with decreasing specimen size[44–47]. The change from brittle to plastic behaviour as size decreases is however, relatively sudden (within about one decade of particle size) whereas the increase in strength with decreasing size is continuous over several decades of size.

7.8   Powder compaction equations

In view of the large number of different equations which have been suggested to link applied pressure to densification, they merit some discussion. Howakito and Lüdde[48] helpfully list 15 equations using a uniform notation (*see* Section 11.3). Such equations are of little use unless they enable a physical model to be critically evaluated. Most powders which are die-compacted are nearly spherical in shape; hence their uncompacted porosities are similar. As we have seen, changes in volume with increasing pressure are due to the action of several simultaneous mechanisms. The resultant changes in volume, however, are similar for different materials; the overall volume–pressure relations are simple functions and can be described by a variety of simple equations. It is therefore difficult to use compaction data to distinguish between equations or mechanisms. The usefulness of such equations would be increased if they were obtained over wider pressure ranges and were capable of predictive rather than interpolative use.

7.9   Tabletting defects

Compacting mechanisms can be viewed from an additional direction by considering the faults which may be produced during compaction. In this way the critical processes are highlighted. Defects are described in two parts: first, those visible immediately after pellet ejection and secondly, those which become evident during subsequent processing.

7.9.1   Immediately visible defects

7.9.1.1   Lamination [see Figure 7.9(a)]

The compact breaks into several thin discs perpendicular to its axis. They are generally believed to be caused during pellet removal. Perfect pellets may be produced under identical conditions in a split die from which the pellet is removed without axial ejection. If the same pellet is forced out of the die by an axial movement, it is found to be laminated. Long[42] has described their formation and minimization in some detail. Powder 'characteristics', density, die wall lubrication, die design and binder can each affect the onset of lamination.

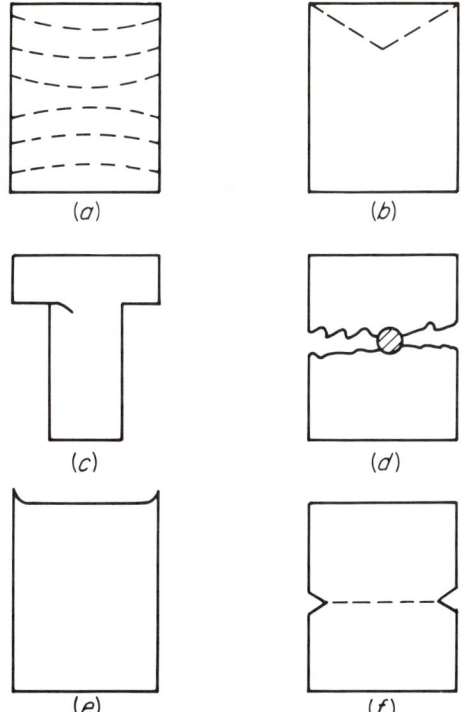

Figure 7.9. Compact defects: (a) *lamination* (b) *capping* (c) *localized cracks* (d) *spontaneous cracking* (e) *flash or skirting fault* (f) *weak equator*

7.9.1.2 Capping [*Figure 7.9(b)*]

This defect is sometimes associated with lamination but it usually refers to compacts from which only a single conical end-piece becomes detached. In pharmaceutical tablets with convex ends capping results in the removal of a curved end. This can be cured by improved lubrication since, as Rees and Shotton observe[23], its cause is excessive die wall friction during compaction. Nystrom *et al.*[49] discuss capping tendency in pharmaceutical tabletting. This defect is often evident when the pellet is still within the die. Sticking to the top punch can induce it. Rutter and Sucker have recently studied the variables that affect pharmaceutical tablet capping[50].

7.9.1.3 Localized cracks [*Figure 7.9(c)*]

Cracks form at a change of thickness due to non-uniform compression or recovery in adjacent parts. Those produced during compaction can often be detected by their shiny surfaces. On the other hand those cracks which are caused during the unloading part of the cycle generally have a matt surface. These faults are avoided by designing tool movement so that compaction is uniform in spite of differences in thickness.

7.9.1.4   *Spontaneous cracking after ejection* [*Figure 7.9(d)*]

Irregular fracture occurs usually perpendicular to the axis and abnormal particles are detected on the fracture surfaces. This is caused by a small amount of either very hard or highly elastic material in the feed which expands by a different amount from the matrix material. It is often due to the use of recycled material and is prevented by using more uniform feed. Similar cracks are formed at high compaction rates owing to trapped air[51].

7.9.1.5   *Flash or skirting fault* [*Figure 7.9(e)*]

With worn punches a small quantity of powder is compacted in the annular gap around the punch tip. The defect is cured by regrinding or fitting new punches.

7.9.1.6   *Die fouling*

If powder adheres to the punch face it becomes progressively harder and compacts will no longer have flat end-faces but instead will reflect the shape of the adhering layer.

7.9.1.7   *Weak equator* [*Figure 7.9(f)*]

A weak region near the centre line of the compact is caused by inadequate compaction as a result of excessive pressure drop within the powder. It is avoided by using more or better lubricant or decreasing the compact length.

7.9.2   Defects apparent after further processing

Instances where intrinsic properties are affected by imperfect compaction can be found in electrical ceramics. The following are examples.

1. If a high permittivity dielectric is 'soft pressed' or laminated, it will fail to vitrify completely and residual microporosity will result in quite drastic reduction in the level of permittivity obtained.
2. Similarly, slight residual porosity can result in drastic lowering of insulation resistance owing to moisture pick-up.
3. Non-uniform or soft pressing can prevent the uniform microstructure which is particularly important in products such as textile thread guides, pump sealing rings, and resistor rod formers on which a subsequent resistive film is deposited.

7.10   Conclusions

1. Although a wide variety of materials are die-compacted the mechanical steps involved are remarkably similar. They differ in degree rather than in kind.

2. Experimental results from measurements on one type of product have application to other materials. There is considerable scope for cross-fertilization between different industries. Quantitative correlations between stress and density distributions now seem imminent.
3. The past few years have seen significant progress in our understanding of compact mechanisms. This is due to two steps:
   (a) The application of three-dimensional stress analysis with Mohr failure envelopes to die compaction; and
   (b) Careful experimental measurements. Empirical relationships have been replaced by reliable experimental results.
4. Some topics which merit further attention are:
   (a) The material properties underlying strength production or bond formation.
   (b) More use of scanning electron microscopy to observed particle deformation.
   (c) The extent and importance of temperature rises.

References

1. KIRSOP, W., *Australian J. Pharm.* **45**, s73, Suppl. 19 (1964)
2. HECKEL R. W., *Trans. Am. Instn mech. Engrs*, **221**, 1001 (1961)
3. ONODA, G. Y. and HENCH, L. L.. 'Ceramic Processes before Firing', Wiley, New York (1978)
4. JAMES, P. J., *Powder Metall. Int.* **4**, 1 (1972)
5. TRAIN, D. and LEWIS, C. J., *Trans. Instn chem. Engrs*, **40**, 235 (1962)
6. HARDMAN, J. S. and LILLEY, B. A., *Proc. R. Soc. A* **333**, 183 (1973)
7. YARTON, D. and DAVIES, T. J., *Powder Metall.*, **11**(1), 115 (1963)
8. KOERNER, R. M., 'Conference on Powder Metallurgy', New York (1970)
9. SMITH, T. A., *Trans. Br. Ceram. Soc.*, **61**, 523 (1962)
10. HUFFINE. C. L. and BONILLA, C. F., *J. Am. Instn chem. Engrs* **8**, 490 (1963)
11. LEISER, D. B. and WHITTEMORE, O., *Bull. Am. Ceram. Soc.* **4**, 714 (1970)
12. VIJAYAN, S. and VENKATASWARLU, D., *J. Inst. Engng, India*, **49**, 58 (1969)
13. VARMA, Y. B. G. *et al.*, *J. chem. engng Data*, **13**, 498 (1968)
14. HEINS, H. *et al.*, *Pharm. Ind.* **3**, 155 (1969)
15. KOSTELNIK, M. C., KLOUT, F. H. and BEDDOW, J. K., *Int. J. Powder Met.* **4**(4), 19 (1968)
16. HIRCHHORN, J. S., *Powder Technol.* **4**, 1 (1970/71)
17. STRIJBOS, S., *Powder Technol.*, **18**, 209 (1977)
18. STRIJBOS, S., *Powder Technol.*, **18**, 187 (1977)
19. AKETA, Y., *et al.*, *Tech. Reports Osaka Univ.* **15**, 81 (1965)
20. TRAIN, D., *Trans. Instn chem. Engrs*, **35**, 258 (1957)
21. KUCZYNSCKI, G. C. and ZAPLAYTYNSKJ, I., *Trans Am. Instn mech. Engnrs*, **206**, 215 (1976)
22. HIRCHHORN, J. S., 'Introduction to Powder Metallurgy', American Institute of Powder Metallurgy, New York (1969)
23. REES, J. E. and SHOTTON, E., *J. Pharm. Pharmac.* 73 (1967)
24. BRIDGEWATER, J., *Trans. Instn chem. Engrs*, **56**, 157 (1978)
25. ONODA, G. Y., 'Ceramic Processes before Firing', Wiley, New York (1978)
26. HIRCHHORN, J. S., *Int. J. Powder Met.* **5**, 35 (1969)
27. GOETZEL, C. G., 'Treatise on Powder Metallurgy', Interscience, New York (1949)
28. NEWITT, D. M. and CONWAY-JONES, *Trans. Instn chem. Engrs*, **36**, 422 (1958)
29. RUMPF, H., 'International Conference on Agglomeration', Philadephia, p. 379 (1961)
30. PIETSCH, W. and RUMPF, H., *Chemie-Ingr-Tech.* 39, 885 (1967)

31. CHENG, D. C. H., *J. Adhes.* 2, 82 (1970)
32. LONDON, F., *Z. Phys* **63**, 245 (1930)
33. HAMAMAKER, H. C., *Recl. Trav. chim. Pays-Bas Belg.* **55**, 1015 (1936)
34. CASIMIR, H. B. G., *Proc. K. ned. Akad. Wet.* **51**, 793 (1948)
35. LIFSCHITZ, E. M., *Soviet Phys. JETP* **2**, 73 (1956)
36. LONG, W. M. and ALDERTON, J. R., *Powder Metall.* **3**(6), 52 (1960)
37. WHITMAN, R., '*Proceedings of Fourth International Conference on Soil Mechanics and Foundry Engineering*', Butterworths Scientific Publications, London, 1, 207 (1957)
38. KUNIN, N. F. and YARCHENKO, B. D., *Plast. Massy* **10**, 33 (1966)
39. WANG, S. and DAVIES, R., *Machinery*, **113**, 830 (1968)
40. SMITH, T. A., *Trans. Br. Ceram. Soc.*, **5**, 23 (1962)
41. HIESTAND, E. N. *et al.*, *J. pharm. Sci.* **66**, 511 (1977)
42. LONG, W. M., *Powder Metall.* **3**(6), 73 (1960)
43. RIDGEWAY, K., Lecture to The Chemical Society, December 1978
44. ROESLER, F. C., *Proc. phys. Soc. B* **69**, 981 (1956)
45. BENBOW, J. J., *Proc. phys. Soc. B* **78**, 970 (1961)
46. KENDALL, K. *Proc. R. Soc. A* **361**, 245 (1978)
47. PUTTICK, K. E., SHAHID, M. A. and HOSSEINI, M. M., *J. Phys. D* **12**, 195 (1979)
48. HOWAKITO, K. and LÜDDE, K. H., *Powder Technol.* **4**, 61 (1970/71)
49. NYSTROM, C., *et al.*, *Acta Pharm. Suecia* **15**, 226 (1978)
50. RUTTER, A. and SUCKER, H. B., *Pharm. Tech. Int.* 25 (1980)
51. LONG, W. M. and ALDERTON, J. R., *Powder Metall.*, **3**(6) 52 (1960)

CHAPTER 8

Fluidized bed granulation

A. W. Nienow

Department of Chemical Engineering, University of Birmingham, Birmingham

8.1 Basic fluidized bed concepts[1]

A fluidized bed is formed whenever an upward flowing fluid is passed through a bed of particulate material at a sufficiently high velocity. The minimum fluidization velocity U_{mf} is the velocity at which the upward force due to the pressure drop through the particulate material becomes equal to the force due to the bed weight acting downwards. An empirical relationship for U_{mf} (m s^{-1}) is:

$$U_{mf} = 7.9 \times 10^{-3} \, d_p^{1.82} \, \Delta\rho^{0.94} \, \mu^{-0.88}$$

when particle diameter, d_p, density difference $\Delta\rho$, and viscosity μ, are in SI units and $Re_{mf} \leqslant ca.\ 10$. Re_{mf} is the Reynolds number at minimum fluidization velocity.

With gas fluidized beds where the volumetric gas flow rate Q (velocity, v) is greater than that for minimum fluidization Q_{mf}, the excess gas Q_B generally flows through in the form of bubbles, i.e.

$$Q_B = Q - Q_{mf}$$

These bubbles transport upwards in their wake a volume flow of solids approximately equal to one third of the bubble flow. These two statements represent the two phase theory of fluidization in its most elementary form as applied to solids movement.

As a result of this movement of solids, the bed is very well mixed and an order of magnitude estimate of the mixing time, t_C, is[2]

$$t_C = H_{mf}/\{0.6(U - U_{mf})[1 - (U - U_{mf})/U_B]\}$$

where U_B is the bubble rise velocity and H_{mf} is the unfluidized bed height. The rapid mixing causes the bed to be remarkably isothermal (with equivalent conductivities many times that of silver for example) and also, because of particle movement over the surfaces, very high bed-surface heat transfer coefficients are obtained.

If a fluidized bed contains particles of different sizes or different densities, then vertical segregation may occur. The denser (or without a density difference, the larger) tend to segregate to the bottom. The material with this tendency has been called jetsam and, under certain conditions, it may become defluidized[3].

8.2   Definitions and applications[4]

In the production of a dry powder or granules from a solution, slurry or paste by heating in a fluidized bed, the dry product is the particulate phase of the fluidized bed and is in direct contact with the fluidizing gas, usually hot air or combustion products (*see Figure 8.1*).

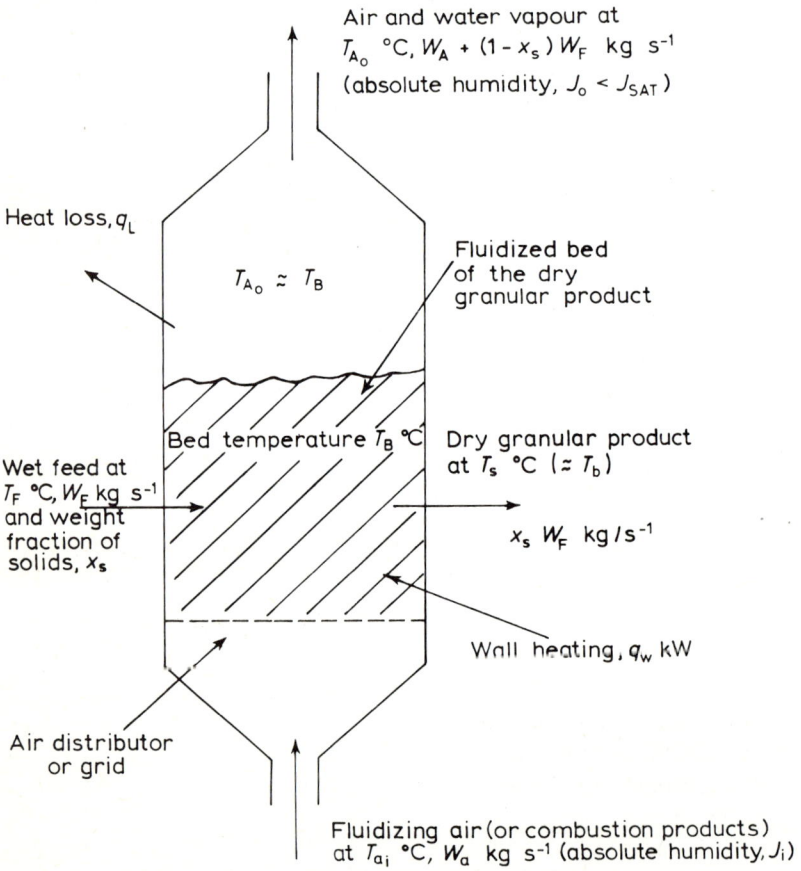

Figure 8.1.   Diagrammatic representation of a fluid bed granulator

It potentially competes with (*a*) spray driers (*b*) crystallizers (*c*) granulators (*d*) calcination reactors. It may be considered as an extension of a fluidized bed drier.

8.2.1 Possible advantages

1. Much smaller equipment (cf. voidage of spray drier and fluidized bed).
2. Enhanced thermal efficiency (cf. spray drier, and *see* Section 8.6.2).
3. Temperature stability and control (general properties of fluidized beds).
4. Elimination of other processing steps (cf. crystallization).
5. Sequence flexibility in batch granulation (cf. other granulators).
6. May produce special particle properties.

8.2.2 Possible disadvantages and problems

1. Continuous stirred tank reactor (CSTR) residence time distribution (cf. spray drier). This is capable of adjustment[1, 5].

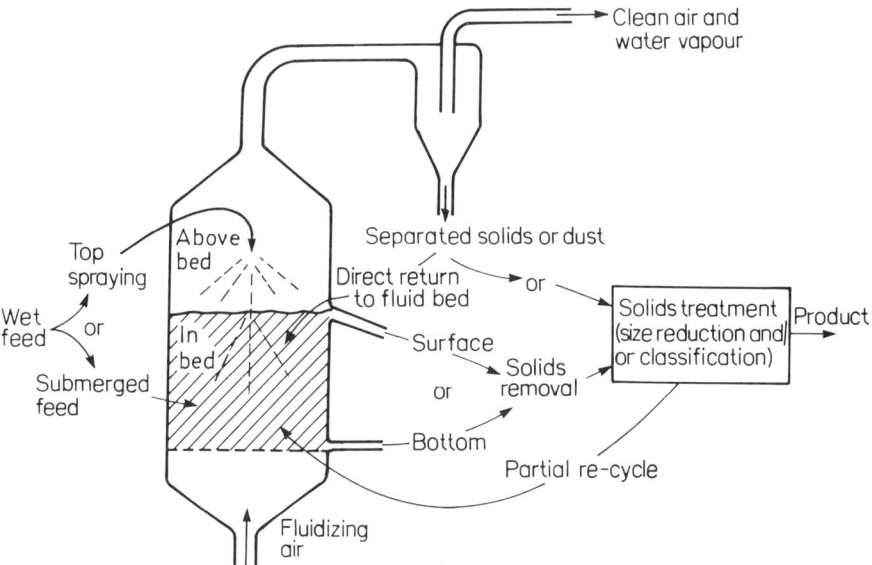

Figure 8.2. *Variations on the basic process; note that the population balance must be satisfied*

2. Will it work? The problems are:
   (*a*) Quenching and/or caking may occur.
   (*b*) The population balance must be satisfied.
   (*c*) The heat and mass balances must be satisfied.

8.3 Variations on the basic process: practical difficulties

8.3.1 Caking

Solid deposition on spray nozzles or on walls may occur. This may lead to nozzle blockage. These phenomena are very dependent on nozzle positioning.

8.3.2   Population balance

Sufficient fine particles must be formed: (*a*) at the spray nozzle[6,7]; (*b*) by jet grinding[8]; and (*c*) by fines separation from product and recycle[9]. Satisfying the population balance is mainly a problem for continuous operation.

8.4   Quenching

8.4.1   Wet quenching

This term is introduced to signify loss of fluidization due to excessive liquid causing oversized and uncontrolled particle agglomeration either over the whole bed or in a localized region[4]. Segregation of wet jetsam clumps generally occurs owing to size and density effects.

8.4.2   Dry quenching

Recent work[10] suggests that an equally important cause of defluidization arises from the formation of large dry agglomerates following the drying out and stabilization of localized wet quenching. Segregation due to size may even occur before quenching.

8.4.3   Other processes in which quenching may occur

Examples are coking[11], sludge incineration[12], and tailings combustion[13].

8.5   Mass and moisture balance

$$W_F + W_A = W_A + W_F(1 - x_s) + x_s W_F$$

$$\{\text{Air} + \text{vapour}\} \quad \{\text{Product}\}$$

where x_s is the initial weight fraction of 'dry' solids in the feed, where 'dry' means that degree of dryness which is acceptable in the product.

$$J_0 = \frac{W_F(1 - x_s) + J_i W_A}{W_A}$$

If

$$T_{A0} < T_{\text{boiling point}}$$

where $T_{\text{boiling point}}$ refers to the temperature at which the liquid phase boils; then to prevent wet quenching,

$$J_0 < J_{SAT}$$

where J_{SAT} is the saturation humidity.

8.6   Heat balance

Datum: ambient temperature

Heat input: $W_F c_F T_F + W_A T_{A_i} c_A + q_w$

Heat output: $W_A c_A T_{A_0} + (1 - x_s)(c_v T_{A_0} + \lambda)W_F + x_s W_F c_s T_s + q_L$

where c_F, c_A, c_v and c_s refer to the specific heats of feed, air, vapour and solid product, respectively, and λ is the latent heat of evaporation.

Provided the bed is well-fluidized, i.e., $U - U_{mf}$ is sufficiently high to give good solids circulation, then

$$T_{A_0} \approx T_s \approx T_B \neq f(\text{position in the bed})$$

8.6.1   Comparison with a spray drier

The points to be made are:

1.   In a spray drier, $q_w = 0$ since all heat comes in with the air.
2.   Spray drying in the co-current mode also gives $T_s \approx T_{A_0}$ but $T_s = f(\text{position})$
3.   In the counter-current mode, $T_s \approx T_{A_i}$

8.6.2   Thermal efficiency, η_H

Neglecting heat losses, i.e., assuming a well-insulated bed, then $q_L = 0$. If η_H is defined as

$$\eta_H = \frac{\text{Heat in} - \text{Heat wasted}}{\text{Heat in}}$$

and we neglect the small terms

1. $W_F c_F T_F$
2. $x_s W_F c_s T_s$ since x_s is usually small
3. $(1 - x_s)c_v T_{A_0} W_F$ since $W_A \approx 10(1 - x_s)W_F$

then

$$\eta_H = \frac{W_A c_A T_{A_i} + q_w - W_A c_A T_{A_0}}{W_A c_A T_{A_i} + q_w} = \frac{W_A c_A(T_{A_i} - T_B) + q_w}{W_A c_A T_{A_i} + q_w}$$

Therefore, a significant heat input through the wall, q_w, dramatically increase the efficiency. Typical heat transfer coefficients (wall–bed; > ca. 200 W m^{-2} K^{-1}) should make this possible[14]. However, caking possibilities are enhanced and, in practice, heat input through walls has not been established.

If $q_w = 0$, efficiency can still be enhanced by using a high value of T_{A_i}.

8.7 Particle growth mechanisms: dynamic equilibrium[15]

8.7.1 Types of particle

Most fluidized bed granulation processes produce only one of three possible types of particle.

1. Layered or onion-ring granules approximately spherical and often large (1 to 2 mm).
2. Multiparticle agglomerates.
3. Porous particles (especially if accompanied by reaction).

8.7.2 Dynamic equilibrium

The solvent and solute are acting together to cause particle growth either by depositing solid on the outside of other particles (onion-ring) or by binding particles together (agglomerates).

Opposing growth by agglomeration is the particle movement throughout the bed. This movement is itself proportional to the excess gas velocity $(U - U_{mf})$ giving rise to bubbles and wake transfer. As the particles move, they collide with each other and these collisions may lead to breakage of agglomerate bonds. The forces involved in break-up due to the collisions are inertial and therefore proportional to the fourth power of the particle (or agglomerate) size.

8.7.3 The 'spray' region[16]

X-ray viewing has shown that a jet does not exist. However, thermocouple probes clearly show a low temperature region close to the nozzle where evaporation is taking place (*see Figure 8.3*).

In this region agglomeration occurs, initially owing to liquid bridges (*see Figure 8.4*). Liquid bridges are partially or completely converted into solid bridges provided the heat and mass balance requirements are met. The spray zone is continuously fed with solids as they move through the bed, into the zone and out again.

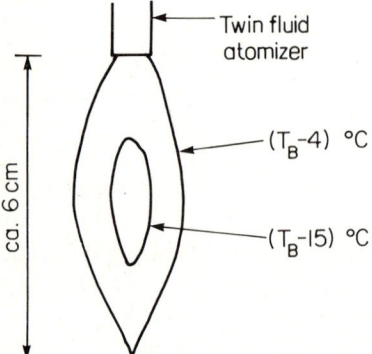

Twin fluid atomizer

(T_B-4) °C

(T_B-15) °C

ca. 6 cm

Figure 8.3. The nozzle region

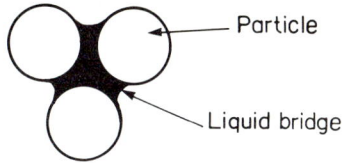

Figure 8.4. *Idealized agglomerate*

8.7.4 The isothermal remainder of the bed

Solids are circulated out of the spray zone into the remainder of the bed which acts as a heat sink. Whether the agglomerates leaving the spray zone cause dry quenching, controlled agglomerate growth or 'onion-ring' growth to occur, depends on the relative magnitude of the binding forces and the break-up forces.

8.7.5 Excessive binding forces: dry quenching

This situation develops if:

1. Poor atomization is present.
2. A high rate of binder addition is used especially in conjunction with a high binder bond strength.
3. The solvent or solution is of high viscosity before or during drying.

8.7.6 Controlled agglomeration

If the above parameters set out in Section 8.7.5 are counterbalanced by a sufficiently high excess gas velocity to cause some particle break-up, then controlled agglomeration occurs (*see Figure 8.5*). This type of growth often leads to a maximum stable agglomerate since inertial break-up forces are approximately proportional to the fourth power of the diameter.

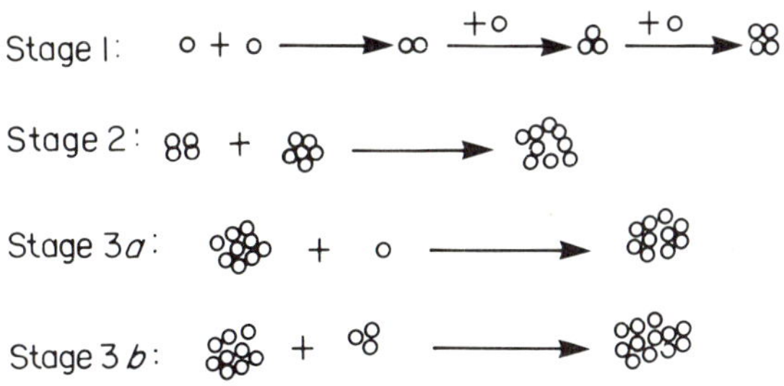

Stage 1: o + o ⟶ ∞ $\xrightarrow{+o}$ ⅋ $\xrightarrow{+o}$ ⅋

Stage 2: 88 + ⅋ ⟶ ⅋

Stage 3*a*: ⅋ + o ⟶ ⅋

Stage 3*b*: ⅋ + ∞ ⟶ ⅋

Figure 8.5. *Stages in agglomerate growth*

8.7.7 'Onion-ring' growth

This type of growth occurs if either large break-up forces are present owing to high excess gas velocities and/or large bed particles; or with low binding forces because of a very weak binder or one in very low concentration.

The stage 1 agglomerates shown in *Figure 8.5* break up in their time in the bed away the spray zone. Growth occurs owing to the collection of broken solid bridges on the surface leading to a sputtered, matt appearance.

8.7.8   Porous particles[17]

If porous particles are present or form, then the wet initial agglomerates may not develop and the whole bed acts as an evaporator or reactor.

8.8   Growth models and rates

8.8.1   'Onion-ring' growth

The rate of growth if an 'onion-ring' mechanism is found is given by the relationship:

$$4a^3 + 6a^2 d_p + 3d_p^2 a = (M_b \rho_s d_p^3)/2\rho_b M$$

where M_b is the mass of binder of density ρ_b which is uniformly distributed in a layer thickness, a, on a mass M, of particles of size d_p. The growth rate is inherently low[10], typically of the order of 10—50 μm h^{-1}.

8.8.2   'Agglomeration' growth

This is best modelled by a modification of Sherrington's model, originally developed for moist agglomerates produced in drum agglomerators[10,18]. For the same material, growth rates are much faster, typically[10] 100—500 μm h^{-1}.

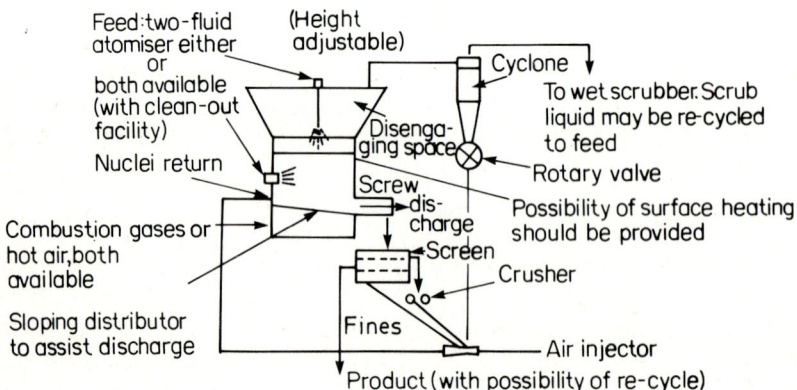

Figure 8.6.   Flexible pilot plant for continuous operation

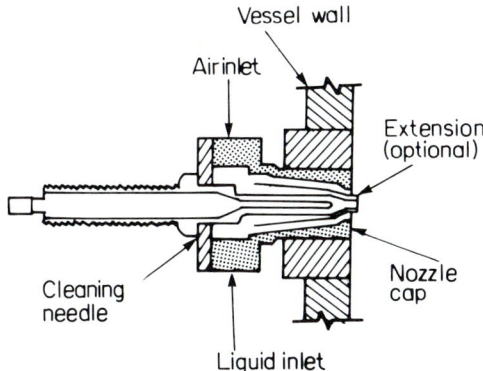

Figure 8.7. A typical two-fluid atomizer

8.9 Batch *versus* continuous operation

8.9.1 Batch

Batch operation is the most common way of operating in pharmaceutical applications. The binder solute is not the same as the bed material and the agglomeration mechanism is the most desirable. Many steps may be completed in the same equipment.

8.9.2 Continuous

For true continuous operation, 'onion-ring' growth is easier to control because it is further away from the quenching region. The product is free flowing, spherical and strong.

8.10 Pilot plant testing

This is essential because of the difficulties of quenching and caking, of ensuring that the population balance can be met and to ascertain the type of granule formed. Satisfactory instrumentation and controls have not yet been established for detecting wet or dry quenching or just a fall in the quality of fluidization. *Figures 8.6* and *8.7* indicate a possible pilot plant and twin fluid atomizer.

8.11 The use of inert 'nuclei'

In this type of equipment, large hollow plastic particles (e.g., isodimensional cylinders up to 2.5 cm diameter) are fluidized. The solute is deposited on them and subsequently breaks away as a fine powder or tiny flakes (to be collected by cyclones). It is particularly popular for slurries which would themselves be difficult to handle and dry, e.g., organic dyes or chicken slurry. The gas velocities must be high (up to 3 to 4 m s^{-1}).

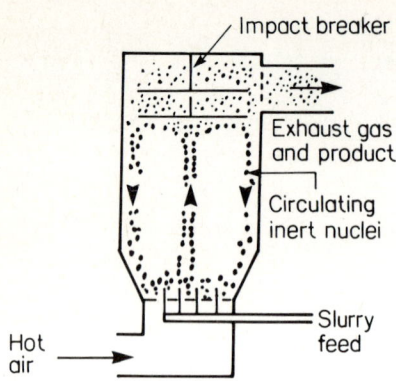

Impact breaker

Exhaust gas and product

Circulating inert nuclei

Slurry feed

Hot air

Figure 8.8.   Diagrammatic representation of Calmic drier[19]

Acknowledgements

This chapter is a shortened version of a similar chapter published in *Fluidisation*, 2nd Edn, Academic Press, 1983. Permission has been given for the use of *Figures 8.1, 8.2* and *8.6*.

References

1. KUNII, D. and LEVENSPIEL, O., *'Fluidization Engineering'*, Wiley, New York (1969)
2. ROWE, P. N., *Chem. Engng Sci.* **28**, 979 (1973)
3. NIENOW, A. W. and CHIBA, T., *'Mixing of Particulate Solids'*, pp. F1—F19, Institute of Chemical Engineers, Rugby (1981)
4. NIENOW, A. W. and ROWE, P. N., unpublished work for Separation Processes Service, AERE, Harwell (1976)
5. BERAN, Z. and LUTCHKA, J., *Chem. Engr. Lond.* 678 (1975)
6. LEGLER, B. M., *Chem. Engng Prog.* **63**(2), 75 (1967)
7. PHILOON, W. C., *et al.*, *Chem. Engng Prog.* **56**(4), 106 (1960)
8. BJORKLUND, W. J. and OFFUTT, G. I., *A.I.Ch.E. Symp. Ser.* **69**(128), 123 (1973)
9. JONKE, A. A., *et al.*, *Nucl. Sci. Engng* **2**, 303 (1957)
10. SMITH, P. G., Ph.D. Thesis, University of London (1980)
11. DUNLOP, D. D., *et al.*, *Chem. Engng Prog.* **54**(8), 39 (1958)
12. WALL, C. J., *et al.*, *Chem. Engng* 14th April, 77 (1975)
13. COOKE, M. J. and HODGKINSON, N., Institute of Fuel Symposium Series, No. 1, pp. c2-1 (1975)
14. BOTTERILL, J. S. M., *'Fluid-Bed Heat Transfer'*, Academic Press, New York (1975)
15. SMITH, P. G., and NIENOW, A. W., *'Particle Technology'*, pp. D/2/K/1—D/2/K/14, Institution of Chemical Engineers, Rugby (1981)
16. SMITH, P. G. and NIENOW, A. W., *Chem. Engng Sci.* **37**, 950 (1982)
17. SMITH, P. G. and NIENOW, A. W., Sixth Institution of Chemical Engineers Residential Meeting, London (1979)
18. SHERRINGTON, P. J., *Chem. Engr. Lond.* 201 (1968)
19. Anonymous, *Chem. Engng* 19th March, 78 (1973)

CHAPTER 9

Compact characterization

N. G. Stanley-Wood
Schools of Studies in Chemical Engineering and Powder Technology, University of Bradford, Bradford

9.1 Strength of materials: fundamentals

In the compaction of particulate solids in a die the fabrication technique can be considered to be first the feed of the particulate solid into the die, secondly the compaction of the powder mass into a coherent body and thirdly the ejection of the compact from the die. In all three stages (the flow, deformation and strength of the assembly of particles) the application of the principles of powder or soil mechanics can be used to evaluate the stresses and strains within a compacting and/or compacted powder. Before elastic/plastic models that represent the mechanical behaviour of solids can be described in relation to the stress required to overcome cohesion, internal friction, and the yield criterion of materials it is essential that various parameters and terms used in the characterization of the strength of materials be defined.

In the study of strength of materials and the behaviour of solids during deformation there are varying characteristics which can be observed in the relationship and response of materials when subjected to stresses. In engineering terms when a material is subjected to a stress (load/cross-sectional area) and the strain (extension/original length) is observed there are usually three separate phenomena which can describe the mechanical characteristics and the behaviour of the material.

9.1.1 Elastic strain

The simplest and most widely used concept is when a material undergoes a linear stress–strain relationship. The elastic characteristics of the material such as Young's modulus and Poisson's ratio can be evaluated. With a perfectly elastic body the elastic strain induced by the application of a load can be recovered by unloading the material. The dimensional extension, usually length, is due to the stretching of the atomic bonds and changes in the internal energy of the loaded and unloaded specimen.

9.1.2   Plastic strain

In the case of ductile materials these materials are capable of undergoing plastic strain (non-recoverable strain on unloading) when loaded beyond the elastic limit. The process occurs mainly because of the sliding of planes of atoms over each other and is probably the most important phenomenon in the understanding of the behaviour of solids when subjected to stress. The plastic deformation of materials occurs non-homogenously by means of lattice faults (dislocations) within the crystal structure of materials. Observation of the stress–strain relationship determines the onset of yielding and plastic flow of a solid.

9.1.3   Fracture

The terminology of fracture is confusing because there are various methods of classification either in terms of the kind and mode of fracture or in the engineering terminology in which fracture can be either ductile or brittle, depending on the amount of plastic flow present.

Ductile fracture usually means breakage or the separation of crystal planes and reduction in size which proceeds slowly when external loads are applied and which involve considerable plastic flow of the material and/or cleavage plane. Brittle fracture means that the crack/breakage and separation of materials spreads rapidly through the material gaining energy for crack propagation from the elastic energy of the surrounding material[1].

9.1.4   Stress

The stress on a solid can be defined in two ways; either the engineering stress which is the force applied divided by the original area before application of force, or the true stress which is the force applied divided by the instantaneous area over which the force is acting. Throughout this discussion true stress will be used when the term stress is indicated.

If a force is normal to a plane area over which it is acting then a normal stress is produced. If the force is parallel to the plane a shear stress is produced. A force acting at an arbitrary angle can be resolved into normal and shear components of stress (*Figure 9.1*). Note that only forces can be resolved and not stresses. In the study of both elasticity and plasticity usually only the deformation of a body is important. In general, however, after application of a load the total displacement is made up of three parts.

1.   Body displacement/translation.
2.   Body rotation.
3.   Body distortion.

To prevent displacement/translation, force must occur as equal and opposite pairs acting on parallel planes and to prevent rotation, shear forces

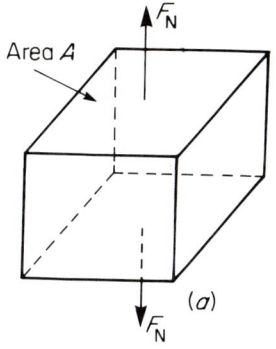

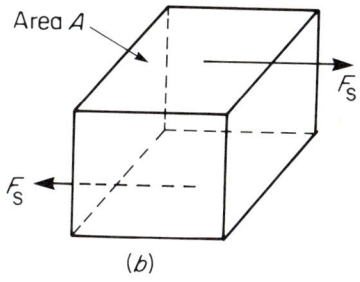

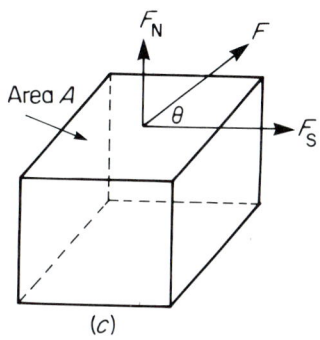

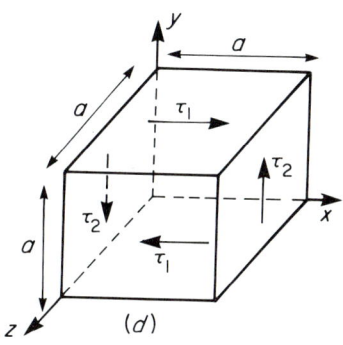

Figure 9.1. Normal and shear stresses
(a) *Normal stress* $\sigma = F_N/A$
(b) *Shear stress* $\tau = F_S/A$
(c) *Resolving force F:*

$$\sigma = \frac{F\sin\theta}{A} \qquad\qquad \tau = \frac{F\cos\theta}{A}$$

(d) *Complementary shear stresses:* $\tau_1 = \tau_2$

on perpendicular planes must act in pairs in opposite directions. *Figure 9.2* shows a small cube within a solid body subjected to external forces. The stress on each face of the cube is resolved into components which are parallel to the axes of the cube $(X_1, X_2$ and $X_3)$. Each component is described as σ_{ij}, where the first index (i) indicates the face on which stress acts and the second index (j) the direction. The index of the face (i) is that of the axis normal to it, the index of the direction (j) is that of the axis parallel to it. Thus σ_{31} acts on the cube face normal to the X_3 axis, in the X_1 direction. Only three faces are considered; the other three faces are equal and opposite. There are therefore nine components, and these are arranged to give the stress tensor

$$\begin{pmatrix} \sigma_{11} & \sigma_{12} & \sigma_{13} \\ \sigma_{21} & \sigma_{22} & \sigma_{23} \\ \sigma_{31} & \sigma_{32} & \sigma_{33} \end{pmatrix}$$

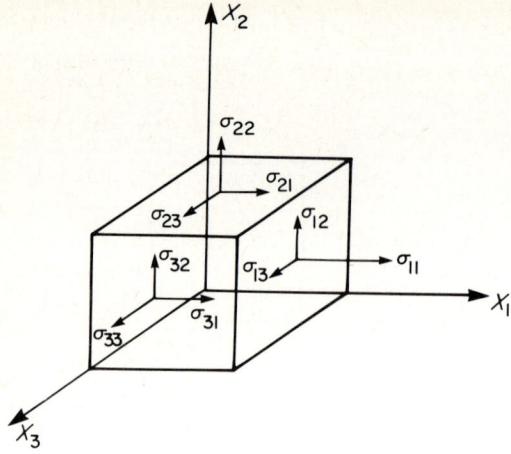

Figure 9.2.   Indexing of stresses: σ_{ij} indicates that the stress acts on the i-face (normal to the i-axis) in a direction parallel to the j-axis; σ_{ii} is therefore a normal stress and $\sigma_{ij}(i \neq j)$ is a sheer stress

The diagonal terms (σ_{11} σ_{22} σ_{33}) are all normal stresses. (A positive normal stress is called a tensile stress and a negative one is a compressive stress.) When dealing with compressive stresses in later sections the compressive stresses will be taken as positive. This sign convention is however the exact opposite to the literature on the Theory of Elasticity (Timoshenko and Goodier) and Plasticity (Nadai, Prager).

It is always possible to find three mutually orthogonal *principal* cleavage planes which will have zero shear stress components. The directions of the normals to these planes are the *principal stress components* of the principal stress directions or axes.

If the orientation of the cube axes is changed, the stress tensor will change. It can be shown that in some orientation the shearing stresses can always be reduced to zero, whereupon the stress tensor reduces to

$$\begin{pmatrix} \sigma_1 & 0 & 0 \\ 0 & \sigma_2 & 0 \\ 0 & 0 & \sigma_3 \end{pmatrix}$$

The faces of the cube in this orientation are called the *principal (stress) planes* and the corresponding normal stresses are called the *principal stresses, σ_1 σ_2* and *σ_3*. These are defined such that $\sigma_1 > \sigma_2 > \sigma_3$. The principal planes are often obvious from the geometry of the system and this is one of the advantages of describing the system in terms of the principal stresses. There are still six variables to define the stress system: three for the principal stresses and three for the orientation of the principal axes. In an anisotropic material the direction of the axes in relation to the material property axes may be important. If $\sigma_1 = \sigma_2 = \sigma_3 = \sigma$ the system is said to be under *hydrostatic* stress, and $\sigma = -p$ where p is the hydrostatic pressure on the body.

9.1.5   Hydrostatic (spherical) and deviatoric stress

The principal stress tensor may be separated into two parts:

$$
\begin{pmatrix} \sigma_1 & 0 & 0 \\ 0 & \sigma_2 & 0 \\ 0 & 0 & \sigma_3 \end{pmatrix}
$$

$$
= \begin{pmatrix} \tfrac{1}{3}(\sigma_1 + \sigma_2 + \sigma_3) & 0 & 0 \\ 0 & \tfrac{1}{3}(\sigma_1 + \sigma_2 + \sigma_3) & 0 \\ 0 & 0 & \tfrac{1}{3}(\sigma_1 + \sigma_2 + \sigma_3) \end{pmatrix}
$$

$$
+ \begin{pmatrix} \sigma_1 - \tfrac{1}{3}(\sigma_1 + \sigma_2 + \sigma_3) & 0 & 0 \\ 0 & \sigma_2 - \tfrac{1}{3}(\sigma_1 + \sigma_2 + \sigma_3) & 0 \\ 0 & 0 & \sigma_3 - \tfrac{1}{3}(\sigma_1 + \sigma_2 + \sigma_3) \end{pmatrix}
$$

The first part is seen to represent a hydrostatic stress

$$\sigma_h = \tfrac{1}{3}(\sigma_1 + \sigma_2 + \sigma_3) \equiv p \tag{9.1}$$

The second term is called the *deviatoric* stress (or reduced stress)

$$\sigma_{di} = \sigma_1 - \sigma_h \equiv q \tag{9.2}$$

The hydrostatic component causes a change of volume (dilation) but no change of shape (distortion). This is because there are no shear stresses on planes of any orientation (as can be shown analytically or deduced from reasons of symmetry). Every plane is identical and therefore a principal plane. It has been found that hydrostatic stress does not cause plastic flow even at very high stress levels. The deviatoric stress components, on the other hand, produce no dilation since their sum is always zero, but do produce shear stresses on other planes with consequent distortion and plastic flow.

9.1.6   Strain

Strain is a measure of the deformation of a body. There are, as for stress, two types of strain: normal strain and shear strain. Normal strain occurs in straightforward stretching, and is defined as the fractional change in length on deformation. Linear strain is the total change in length divided by the original length,

$$e = \int_{l_0}^{l_1} \frac{dl}{l_0} = \frac{l_1 - l_0}{l_0} \tag{9.3}$$

Natural, or logarithmic, strain is the incremental change in length divided by the instantaneous length, $d\varepsilon = dl/l$, and the total natural strain resulting from a change in length from l_0 to l_1 is given by:

$$\varepsilon = \int_{l_0}^{l_1} \frac{dl}{l} = \ln \frac{l_1}{l_0} \tag{9.4}$$

Linear strain is larger than natural strain for the same extension in tension; the reverse is true in compression. For the small range of strains encountered in elastic deformation the difference between the two is negligible (as for nominal and true stress) and linear strain is generally used, as being mathematically simpler (*Figure 9.3*). Shear strain occurs when parallel planes are moved relative to one another keeping the separation distance constant, such as occurs in the distortion of a rectangular block into a parallelepided (*Figure 9.4*).

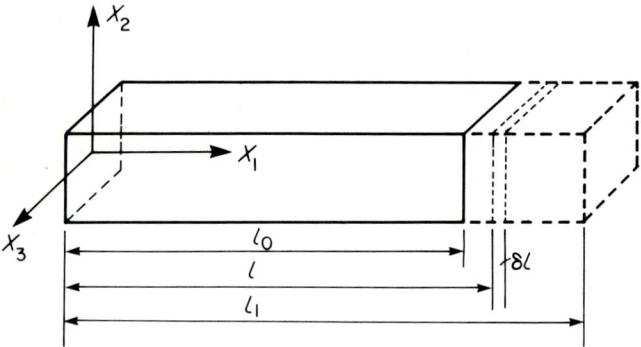

Figure 9.3.   Uniaxial normal strain

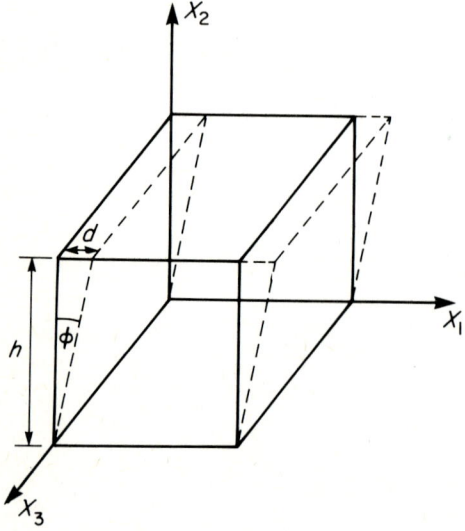

Figure 9.4.   Shear stress and strain

9.1.7   Principal stress space

The principal stresses $(\sigma_1 \; \sigma_2 \; \sigma_3)$ on a point in a continuum can be used as cartesian coordinates to define any point in a three-dimensional space or *principal stress space.*

Principal stress space is used to represent the theories of yield strength of plastic materials. Large changes of hydrostatic or spherical pressure/stress have no effect on the deviatoric stress component so there is no change in either the shape and distortion of the material body or the yielding strength. The yield functions and representation of the plastic behaviour of metals can be described by either Tresca in the form of a regular hexagon or by von Mises in which the hydrostatic diagonal is the locus of a circle. All states of stress within the enclosed surface allow material to be in a stable equilibrium. Outside these hexagonal or circular cylinders is a state of yielding.

9.1.8   Elasticity

The mathematical theory of elasticity is based upon two laws. The first is Hooke's law, which states that the stress in an elastic body is directly proportional to the strain, that is,

$$\sigma = \text{constant} \times \varepsilon \tag{9.5}$$

The constant is called the elastic modulus E. The second is the principle of superposition. This states that if a stress σ_A produces a strain ε_A and a stress σ_B produces a strain ε_B then application of $(\sigma_A + \sigma_B)$ together will produce a strain equal to $(\varepsilon_A + \varepsilon_B)$. This holds provided that $(\sigma_A + \sigma_B)$ does not exceed the limit of proportionality.

Three elastic moduli are recognized. Young's modulus E relates tensile stress to tensile strain; the shear or rigidity modulus G relates shear stress to shear strain; and the bulk modulus K relates hydrostatic stress to dilational strain. The other constant characteristic of an elastic body is Poisson's ratio v, which relates longitudinal to lateral strain in simple tension. The value of v is 0.25—0.33 for most metals, and about 0.1 for solids with highly directional bonds, such as strongly covalent crystals. For there to be no change in volume on deformation, v should equal 0.5.

Various relations can be shown to exist between the moduli:

$$K = \frac{E}{3(1 - 2v)}$$

$$G = \frac{E}{2(1 + v)} \tag{9.6}$$

$$E = \frac{3G}{1 + G/3K}$$

Hence only two of the four constants need to be known for any material.

9.1.9   Criterion of yielding for plastic materials

A ductile material yields in a simple tensile test at some stress level, say Y. This is the only principal stress present, the other two being zero. A yield criterion is required which will indicate when the same material will yield under combined stresses, that is, when there is more than one principal stress.

Experiments have shown that the application of a pure hydrostatic stress system will not produce yielding. This is not unexpected since a hydrostatic stress produces only dilation whilst it is known that plastic deformation involves no volumetric change. Thus any yielding criterion must be independent of the hydrostatic component of a stress system, or, in other words, yielding occurs when a function of the deviatoric stress components reaches a critical value, which depends on the particular material and its history. It should be noted that the concept of a yield criterion is not restricted only to the initial loading but also applies during strain hardening.

When the principal stresses are plotted along coordinate axes a point in the coordinate space thus defined represents a particular stress system, that is, one with principal stresses given by the coordinates of the point. For each material some points will represent stress systems which do not cause yielding, and some represent systems which cause the material to yield. A surface, called the yield surface, can be constructed so as to separate these two sets of points; points on this surface are those which just satisfy the yield criterion. Since a line making equal angles with the axes, that is $\sigma_1 = \sigma_2 = \sigma_3$, represents hydrostatic stress only, the yield surface will be centred on this line.

Various shapes for the yield surface are possible, representing different criteria of yielding. For any isotropic material it must have threefold symmetry about the hydrostatic axis, and, if the yield stress in tension is equal to that in compression, the symmetry will be sixfold. This is assumed in the theory of plasticity.

The simplest yield criterion is Tresca's maximum shear theory of 1864, which states that yielding occurs when the maximum shear stress exceeds a critical value. This can be expressed mathematically, when $\sigma_1 > \sigma_2 > \sigma_3$ as

$$F = \sigma_1 - \sigma_3 - 2k = 0 \qquad (9.7)$$

where k is the critical value or maximum shear stress for yield to occur and F is the yield fraction.

An alternative yield function, that of von Mises, can be expressed in a similar manner as

$$F = (\sigma_2 - \sigma_3)^2 + (\sigma_3 - \sigma_1)^2 + (\sigma_1 - \sigma_2)^2 - 2Y^2 = 0 \qquad (9.8)$$

where Y is the von Mises yield stress obtained in axial tension. The von Mises yield surface is a circle with the line $\sigma_1 = \sigma_2 = \sigma_3$ as its axis (*Figures 9.5(b)* and *9.6*).

The Tresca yield surface is a hexagonal prism with the line $\sigma_1 = \sigma_2 = \sigma_3$

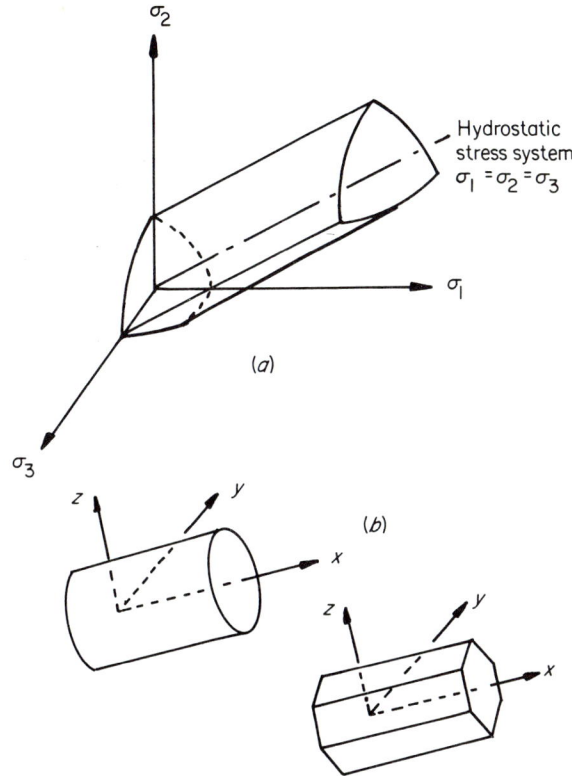

Figure 9.5. Yield surfaces in stress space

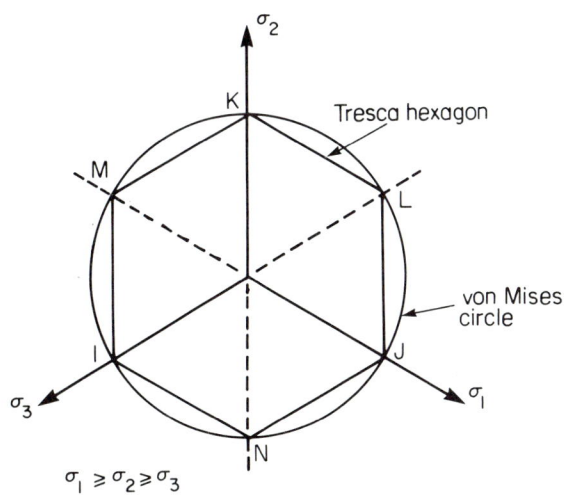

Figure 9.6. Yield loci of Tresca and von Mises

as its axis. Its projection on a plane normal to this line is a hexagon, as shown in *Figure 9.6* and also in *Figure 9.5*.

9.1.10   Plasticity

Once the stress system satisfies the yield criterion, the laws of elasticity are no longer sufficient and a plasticity theory is required to relate the stresses and strains in the plastic regime. Both elastic (recoverable) and plastic (non-recoverable) strains will be present although the former can sometimes be neglected. The value of the initial yield stress, or the yield criterion, is a function of the material composition and its history. In addition, the plastic deformation is irreversible, that is, it is not dependent on the instantaneous value of the stress system—for example, the plastic strains do not become zero when the stresses are removed.

In order to develop a mathematical theory of plasticity, various simplifying assumptions have to be made. In many applications the plastic strains are very large compared with the elastic strains, which can be neglected. The material model is then called *rigid-plastic*. Another simplification which is often made is to ignore the strain hardening and assume the flow continues at constant stress. The material is then said to be perfectly (or ideally) plastic. This is reasonable, except that the value of the yield stress is depen-

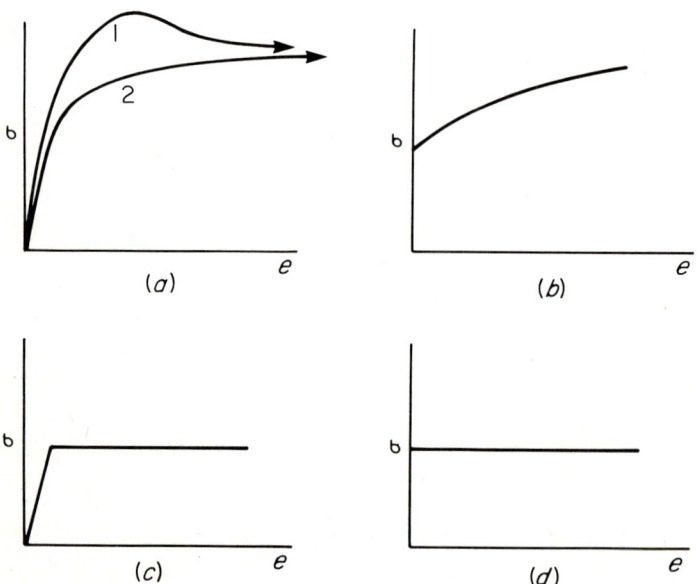

Figure 9.7.   *Material models used in continuum plasticity theory.*
(a)   *Real material*
(b)   *Rigid-plastic model*
(c)   *Elastic-ideally plastic model*
(d)   *Rigid-ideally plastic model*

dent on the strain rate which may vary across the deforming body. Finally a combination of both approximations can be used, giving a material called *rigid-ideally plastic*. The various material models are illustrated in *Figure 9.7*.

The salient feature of the theory of stress and strain in relationship to plasticity can be illustrated by using a two dimensional state of stress.

Assuming that the normal stresses σ_1 and σ_2 are principal stresses and that the third principal stress is zero, the state of stress may be represented by a point (stress point) on the plot of σ_1 against σ_2. The curve which encloses the elastic stress points, that is, the curve containing stress points which can be reached without the occurrence of plastic deformation, is termed the yield curve. In ordinary material, such as a work-hardening material, the stress point must move outside the yield curve in order for plastic deformation to occur. The yield point must be exceeded. A new yield curve is thus established which may or may not resemble the old yield curve. Stress changes inside the new curve will again only cause elastic phenomena. Strain coordinates ε_1 and ε_2 can be superimposed on the stress coordinates σ_1 and σ_2 to permit plotting of stress and strain on the same diagram. The total strain due to the change in loading of a specimen consists of an elastic component and a plastic component. Taking only the plastic component ε^p which is normal to the yield curve and directed outwards, Drucker[2] has introduced the concept of stability in which the product of the stress increment σ_{12} and the associated plastic strain increment ε_{12} must be positive or zero. Mathematically

$$\sigma > \varepsilon^p \geqslant 0$$

Plastic materials are thus stable in the sense that they only yield for stress increments σ that satisfy the above equation, otherwise the material remains elastic but is work-hardened.

9.2 Soil mechanics stress–strain curves for granular materials

9.2.1 Coulomb yield criterion

In soil mechanics and allied subjects the observed stress–strain curve of granular material is generally that shown in *Figure 9.7(a)*. The upper curve (1) with a distinct peak or failure is typical of dense or overconsolidated materials while the lower curve (2) is typical of less dense or normally consolidated material which eventually undergoes no volume change with increase in shear stress. Failure of solids does not imply fracture but the value where the shear stress reaches a maximum value. The early work on soil mechanics and consolidated clays, developed from Coulomb[3] and Rankine[4], treated soils as perfectly elastic solids, while Sokolovski[5] and De Josselin de Jong[6] used the theory of perfectly plastic soils. There is however still a wide divergence between observed and assumed behaviour of granular materials[7]. Schwartz and Weinstein[8], by treating a compacting powder mass

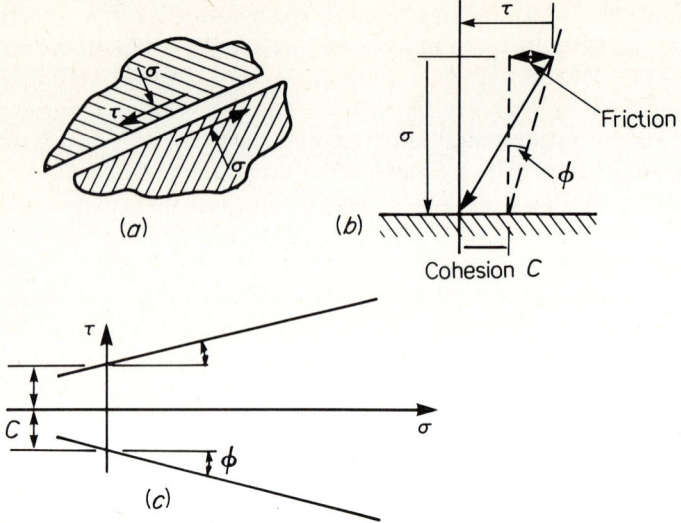

Figure 9.8.   Coulomb's failure criterion

as an aggregate of individual particles interacting with each other through the mechanism of interparticle friction, used the yield criterion of Coulomb (*Figure 9.8*) to evaluate the stresses on compacts. This was a departure from the deformation of metal using the yield criterion of Tresca or of von Mises. The assumption of a frictional model as a means of predicting compacting loads and stress distributions in a powder suggested, however, the use of the concepts of soil plasticity and the Coulomb yield criterion. A similar approach was used by Jenike and Shield for bulk solids[9].

The Coulomb yield criterion states that failure occurs when the shear stress on any plane in a material reaches a critical value which is linearly dependent on the normal stress on that plane. It is this dependence on normal stress that distinguishes the Coulomb yield criterion from the Tresca or maximum shear criterion commonly used for metals.

Coulomb in his work on the rupture surfaces and the mechanisms of sliding contact of soils suggested that the parameters of cohesion and friction of materials must be overcome during slippage along a ruptured surface. In the early theories, peak stresses which caused the rupture of soil required materials to be treated as perfect elastic solids and peak shear values were used in the stress–strain curves. The Coulomb equation, implied in Coulomb's work, was to assume that the stress component σ normal and τ tangential to a rupture plane at failure could be expressed mathematically as

$$\tau_f = C + \sigma_f \tan \phi \tag{9.10}$$

where τ_f and σ_f are the shear and normal compaction stresses at failure and C is the cohesion due to bond forces of either microscopic or macroscopic

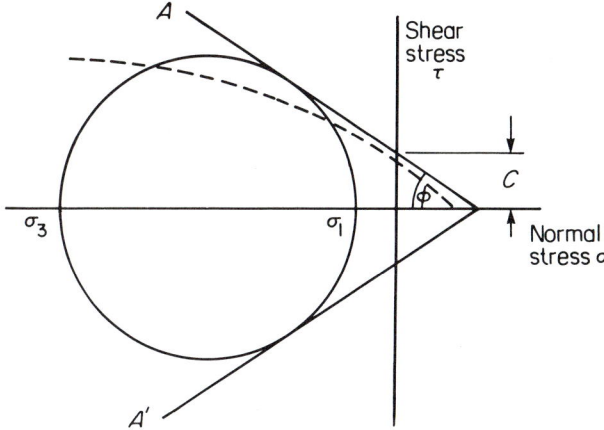

Figure 9.9. The Mohr–Coulomb yield criterion. The Mohr–Coulomb yield locus is represented by the cone AA', and the broken line indicates the general shape of an experimentally determined yield or failure locus

origin between particles and ϕ is the angle of internal friction; the tangent of which is the slope of a line encompassing the failure stress state plots on a Mohr diagram.

Equation 9.10 plotted in *Figure 9.9* on the Mohr diagram coordinates of σ and τ in principal stress space where the yield or failure surface is an irregular hexagonal pyramid with its axis equally inclined to the axes of the space[10]. For increasing normal stress the shear stress at failure increases and all successive states produce a failure envelope.

In metals, which exhibit no deformation under hydrostatic stress, the yield surface is thought to extend indefinitely along the space axis where $\sigma_1 = \sigma_2 = \sigma_3$. For the failure of a compressible mass of granular material considerable flow will occur on hydrostatic loading and the failure surface is required to intersect its axis[8].

9.2.2   Closure of Mohr–Coulomb failure cone

Drucker, Gibson and Henkel[2] suggested the closure of the yield or failure surfaces with a base cutting the axis of equal stress. Such a closure produces a combination of interacting yield or failure surfaces.

1. The pyramid faces at which flow will continue at a constant state of stress satisfying the Coulomb yield criterion.
2. A work hardening surface, the pyramid base, at which flow results in compaction and the strengthening of the material.

Schwartz and Weinstein, compacting a uranium dioxide powder

(ammonium diuranate) of particle sizes less than 10 micrometres in a 0.395 inch diameter die over a pressure range of 0—1500 p.s.i. showed a verification of the Coulomb equation that shear stress increased with increase in normal stress; the values of cohesion C and angle of internal friction ϕ being 700 p.s.i. and 28.4 degrees respectively. Thus although the Coulomb yield criterion was originally proposed for the behaviour of loose granular material its concept can be extended to the failure of a coherent powder compact.

In the process of compaction of powders at higher pressures than those used in soil mechanics, compaction involves a decrease in volume. Use of the Mohr–Coulomb yield criterion or equation seems therefore an inconsistency with reality if used to predict failure and/or slip of compacted powdered or granular material.

Drucker and his associates[2] qualitatively closed the Mohr–Coulomb failure cone with a spherical base when considering the consolidation process of work-hardening plastic materials. Although it is convenient to consider the linear Mohr–Coulomb equation as yield curves for a perfectly plastic material it is necessary at times to consider the stress points on the closing end cap for some materials. Plastic deformation on loading may involve volume decrease or increase depending on prior loading history. No simple yield or failure surface can explain all observable phenomena. The closing spherical end caps are probably not the only shape and other convex yield or failure surfaces should be tried[2]. Schwartz and Weinstein[8] used the Coulomb yield criterion closed with a pyramidal base to describe the compaction of granular material while Jenike and Shield used a flat base[9].

Suh[11] criticized the use of the Mohr–Coulomb yield criterion because it predicted that the yield locus was a straight line with constant slope which was inconsistent with experimental data[12,13]. Suh thus proposed a modified Mohr–Coulomb equation which gave equation 9.11 for the condition at which there was no volume change on failure:

$$\bar{J}_1 = k - \frac{n\tau\beta}{2\phi}\left(\sin\frac{\pi}{2}\frac{\theta}{\phi}\right)\left(\cos\frac{\pi}{2}\frac{\theta}{\phi}\right)^{n-1}\left(\frac{fJ_2^{\frac{1}{2}}}{M}\right)^{1/2} \tag{9.11}$$

where $\bar{J}_1$ is $\frac{1}{3}J_1$ where J_1 is the first invariant of stresses and equivalent to $\sigma_1 + \sigma_2 + \sigma_3$. $J_2^{\frac{1}{2}}$ is the second invariant of stress and is a function of

$$[(\sigma_1 - \sigma_2)^2 + (\sigma_2 - \sigma_3)^2 + (\sigma_3 - \sigma_1)^2]$$

k, n and β are positive constants dependent on material properties, $f = \frac{2}{3}$, ϕ is the friction angle of the material;

$$\tan\phi = \frac{(fJ_2^{\frac{1}{2}})^{1/2}}{\bar{J}_1 + k}$$

and

$$M = [(1 - \bar{J}_1 + k)^2 + fJ_2^{\frac{1}{2}}]$$

The space of the yield or failure surface is determined by the constants

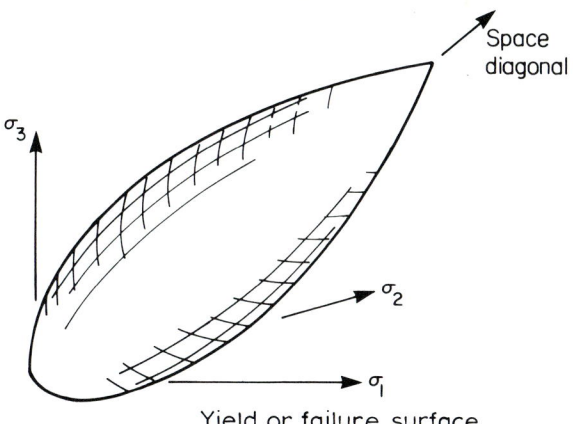

Figure 9.10.   Principal stress space

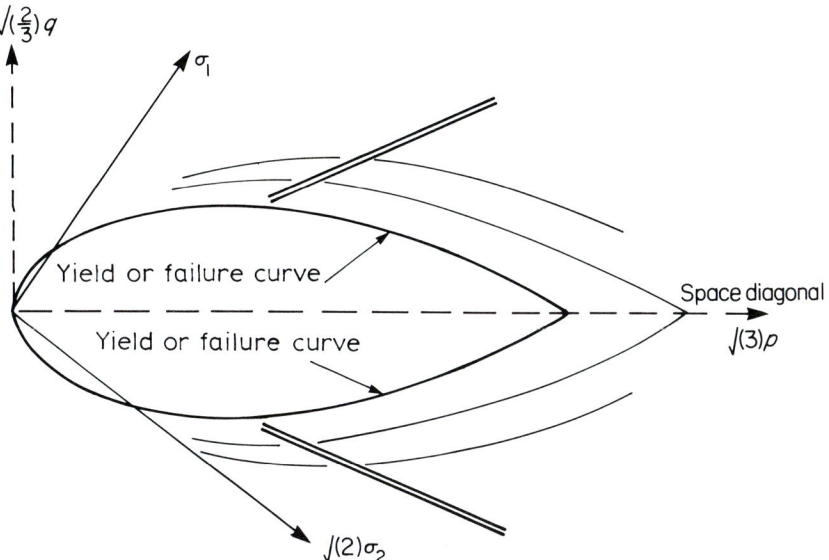

Figure 9.11.   Reorientated principal stresses

k, n, p and ϕ (*see Figure 9.12*). If $\bar{J}_1$ is smaller than the value calculated from equation 9.11 the material will undergo compaction and if $\bar{J}_1$ is larger the material will expand upon yielding or failure.

9.2.3   Granta gravel and Cam clay

Schofield and Wroth[14] criticized the use of Coulomb's failure criterion because its use led to erroneous predictions of high rates of change of volume during shear distortion. Using the Drucker concept of stability Schofield and Wroth developed the conceptual model of an ideal rigid–

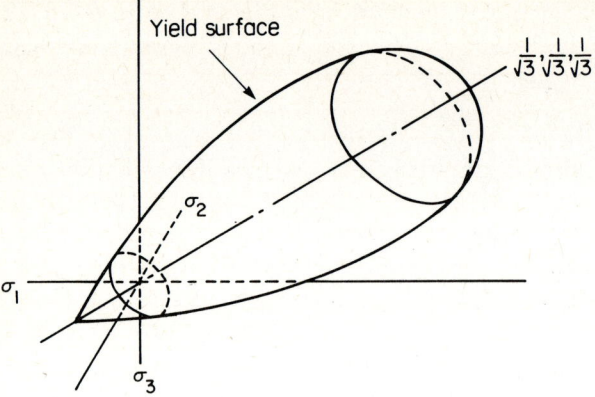

Figure 9.12.   *The proposed yield surface (after Suh[11])*

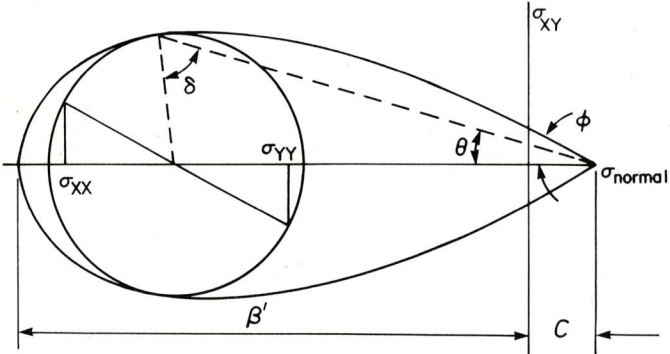

Figure 9.13.   *A yield locus for a plane strain case with the Mohr circle*

plastic continuum which they called Granta gravel and a model of an ideal elastic–plastic continuum which they called Cam clay to formulate a simple critical state model. Both model names derived from the river around Cambridge.

The theoretical Granta gravel and Cam clay models have a yield or failure surface as shown in *Figures 9.10* and *9.11* which, when orientated on the axial test plane and sectioned to include the space diagonal and axis of longitudinal stress, give the pear shaped curve of *Figure 9.11*. The pointed tip on the space diagonal occurs at relatively high pressure and the flank, parallel to the space diagonal, at a lower but not zero pressure.

The space diagonal axis in *Figures 9.10* and *9.11* has the units $\sqrt{(3)}p$ where p, the hydrostatic or spherical pressure $= (\sigma_A + 2\sigma_R)/3$ and the perpendicular axis has units $\sqrt{(2/3)}q$ where q the axial deviator stress $= \sigma_A - \sigma_R$ (equations 9.1 and 9.2; $\sigma_A \equiv \sigma_1$, $\sigma_R \equiv \sigma_3$). The axes p and q will be used later without the multiplying factors $\sqrt{3}$ and $\sqrt{(2/3)}$. σ_A and σ_R are the longitudinal and radial stress respectively. The values $\sqrt{3}$ and $\sqrt{(2/3)}$ can be obtained from a geometrical study of the space diagonal arranged equally from all three principal stresses and an axis at right angles to the space diagonal respectively (*Figures 9.12* and *9.13*).

9.3   Volume reduction in unidimensional consolidation

In soil mechanics the time-dependent process of volume reduction when a soil specimen is compressed is termed consolidation and usually is measured in a consolidometer (oedometer) or an axial compression cell (also known as a triaxial cell). When a material is consolidated observation of the piston and evaluation of the voids ratio with variation in vertical stress will show the resulting equilibrium when plotted as in *Figure 9.14*.

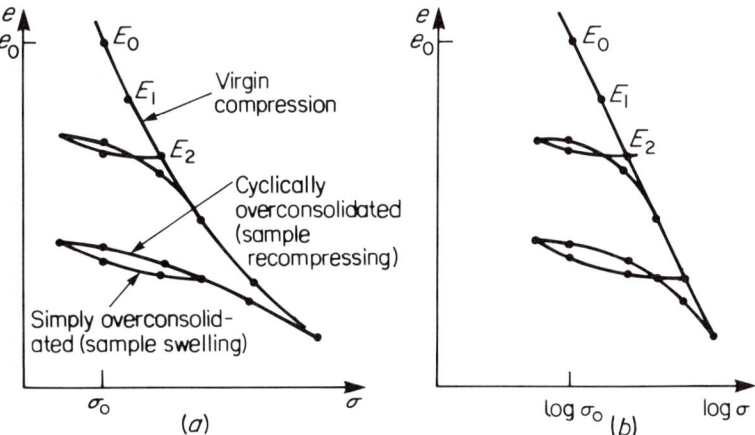

Figure 9.14.   *Terzaghi equilibrium states*

In *Figure 9.14(a)* the vertical stress σ on an arithmetic scale shows the results of a sequence of loading and unloading. The high stress loops occur when the specimen is allowed to recover and swell and then be reconsolidated. The outermost curve connecting E_0, E_1 and E_2 is known as the virgin compression line. *Figure 9.14(b)* shows the same results plotted with σ on a logarithmic scale, the virgin compression line becoming a straight line. Terzaghi[15] deduced equation 9.12 to fit the above line.

$$e = e_0 - C_c \log_{10} \left(\frac{\sigma}{\sigma_0} \right) \tag{9.12}$$

where C_c is the compression index and e the void ratio.

The Terzaghi expression can also be expressed in terms of natural logarithms for the virgin compression line

$$v = v_0 - \frac{\lambda \ln p}{p_0} \tag{9.13}$$

where v and v_0 are specific volumes at experimental and datum levels respectively and p and p_0 hydrodynamic (spherical) pressure at experimental and datum levels respectively. λ is the compression constant for an irreversible process. For a reversible process such as the swelling and recompression curves of *Figure 9.14*, equation 9.13 can be represented by

$$v = v_0 - k \ln \left(\frac{p}{p_0} \right) \tag{9.14}$$

9.3.1   Hvorslev surface

Studies of the strength of clays in Terzaghi's laboratory succeeded in forming together the concept of stress and Coulomb's concept of failure. Although it was known that the specific volume of material changes under changing stress and it was also known that yield strengths of materials vary with specific volume, Coulomb's equation (9.10) takes no account of this change. Hvorslev used shear boxes of Terzaghi design (*Figure 9.15*) to observe the rise and fall of a porous stone in the shear boxes which indicated dilation or compression of material during the shearing process and before powder failure.

The initial data of the water content w *versus* stress σ for the virgin compression, swelling and recompression curves appears confusing when compared with the result at failure. If however for every water content on the dashed curve (failure) an equivalent compression pressure σ_e is obtained

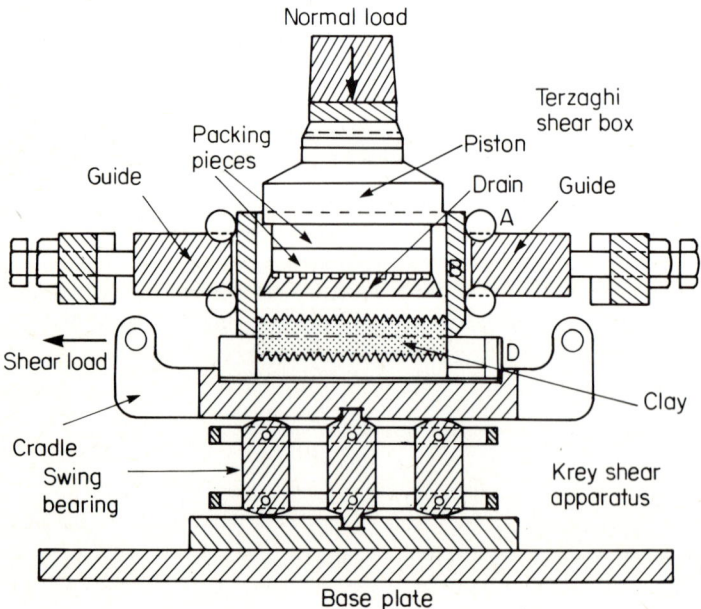

Figure 9.15.   Shear apparatus used by Hvorslev

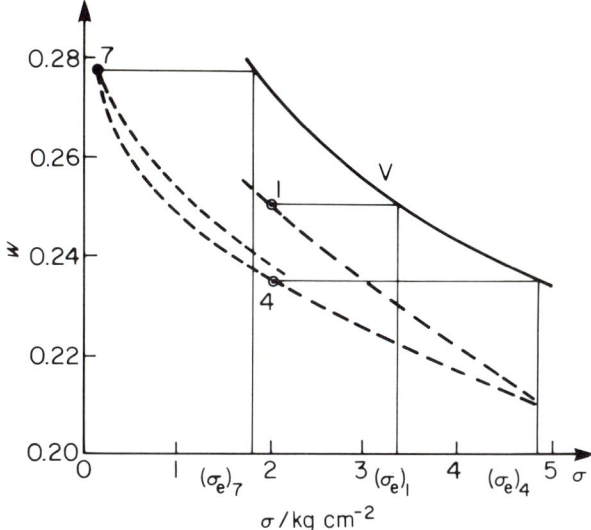

Figure 9.16.   Derivation of equivalent pressure (after Hvorslev). V is the virgin compression curve

from the solid curve (virgin compression, *Figure 9.16*) Coulomb's equation can be modified to become:

$$\frac{\tau}{\sigma_e} = k_0 + \frac{\sigma}{\sigma_e} \tan \phi \qquad (9.15)$$

All data then fall on one straight line (*Figure 9.17*). This then gave rise to the Hvorslev–Coulomb surface in normal stress, shear stress and volume space (*see Figure 9.19*).

Figure 9.18 shows that in range II, $0.05 < (\sigma/\sigma_e) < 0.6$, failure occurs while in range I, $0.6 < (\sigma/\sigma_e)$ no failure is observed. This observation, which was not appreciated until much later, indicates, at *c*, the near vicinity of the critical state[14].

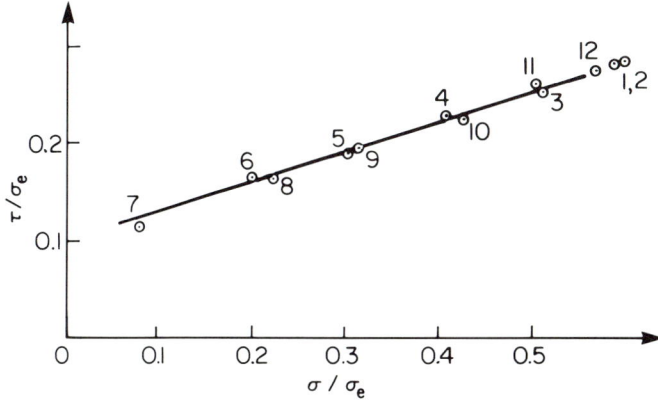

Figure 9.17.   Data of failure (Wiener Tegel V)(after Hvorslev)

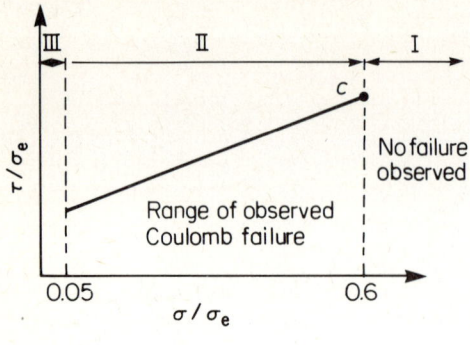

Figure 9.18.   Range of observed data of failure

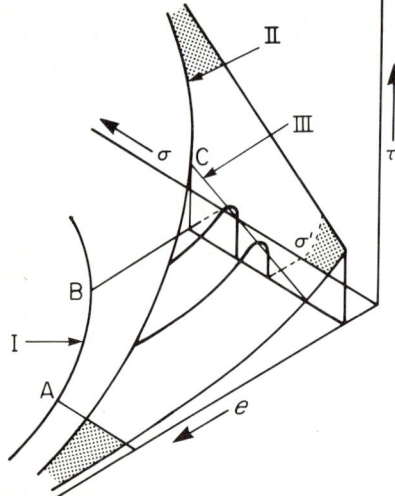

Figure 9.19.   The Hvorslev failure surface. I, normal consolidation line; II, critical void ratio; III, yield or failure locus

Hvorslev[16] was the first to derive a relationship between the shear stress (τ) at failure to the normal stress (σ) and the void ratio (e) of a particulate solid. The condition of failure is therefore defined by a surface known as the *Hvorslev surface* in the three dimensional system whose coordinates are (σ, τ and e). The Hvorslev surface has been determined for some soils and has the form shown in *Figure 9.19*.

9.3.2   Critical state

It was suggested by Schofield and Wroth[14] that in a soil shear test a critical state was reached in which unlimited shear strain could be applied to the soil without further change. The points determined by these τ, σ and e values when plotted in three dimensional space lay on a unique *critical state line*. The path on the failure surface had an end point at which, when reached, any further strain took place without change in the shear stress or void ratio under a given normal stress (σ', *Figure 9.19*).

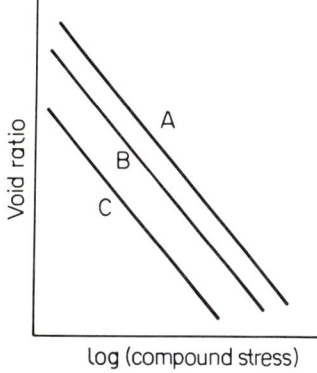

Figure 9.20.   Linearized projections in the void ratio–compound normal stress plane; A, compaction line; B, critical state line; C, ultimate tensile strength line

All such points lying on this single *critical state line* (CSL) or *critical void ratio* (CVR) are shown in *Figure 9.19*. The critical state line when projected on to the void ratio *versus* log normal stress plane (*Figure 9.20*) is found experimentally to be

1. Parallel to the normal consolidation line;
2. To have a constant ratio of τ to σ for any point on it.

Consequently the idealized soil can be assumed to satisfy the requirements of a CSL by two equations

$$e = K - \lambda \ln \sigma \quad \text{(or} \log \sigma) \tag{9.16}$$

$$\tau = M\sigma \tag{9.17}$$

where K is the void ratio at unit normal stress.

Roscoe also showed that when projected on the τ, σ plane the critical state line gives a straight line passing through the origin. The results from Jenike shear tests give however no information about the voidage of the specimen although now a modification is available to show changes in the depth of the specimen[17].

A typical Hvorslev surface is shown in *Figure 9.21*. In a shear test the values of τ and σ follow paths on the failure surface going up to the critical state line. Projection of these path lines on to the τ *versus* σ plane gives the familiar family of Jenike yield or failure loci (*see also* Chapter 5). Three lines can be drawn on the void ratio–normal stress plane (*e versus* σ). These are

1. The consolidation line which represents the changes in void ratio, or specific volume with consolidation pressure in the absence of shear.
2. The line representing the variation of the ultimate tensile strength of the material with void ratio or specific volume.
3. The projection of the critical state line.

The above three lines are shown in *Figure 9.21*. Wroth[7] suggested that if the void ratio or specific volume on consolidation was plotted against the

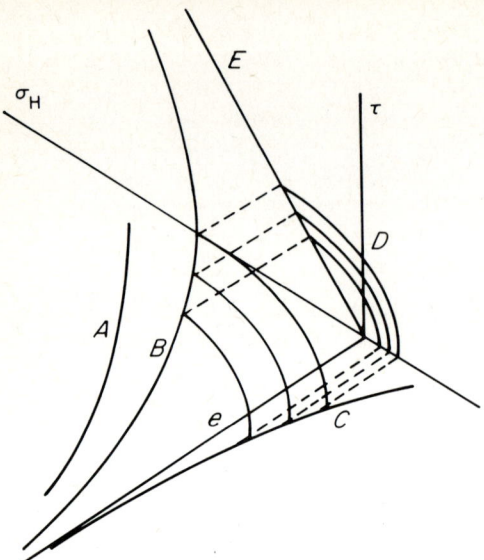

Figure 9.21.   The Hvorslev failure surface; A,
consolidation; B, *critical state;* C, *tensile
strength;* D, *yield or failure locus;* E, *projection
of critical state line*

logarithm of the stress the consolidation line and the critical state line
would be straight parallel lines (*Figure 9.20*). For the consolidation line to
be straight on logarithmic scales the relationship must be of the form of
equation 9.16.

9.3.3   Application of the principles of soil plasticity to the compaction of particulate solids

Although a large amount of work has been done since the initial work of
Shapiro and Cooper and Eaton[18] in the development of a pressure–
porosity relationship and an understanding of the compaction and bonding
mechanism of powders, little work on the forces involved in simple closed-
die compression of particulate material was done before Schwartz and
Weinstein[8]. They proposed a model using the methods of soil plasticity as a
means of predicting the compaction loads and stress distribution in powder
compacts. It is known that the deformation in metals and the yield criterion
can be evaluated from the models of Tresca or von Mises. By treating the
compacting mass as an aggregate of individual particles interacting with
each other Schwartz and Weinstein suggested the use of the Coulomb yield
criterion to explain the punch loads, die wall friction and stress distributions within powder compacts. The pressures involved in the compaction
of powders are much higher than those used either in soil mechanics to
predict failure of soils or in experimentation to predict the flow of solid

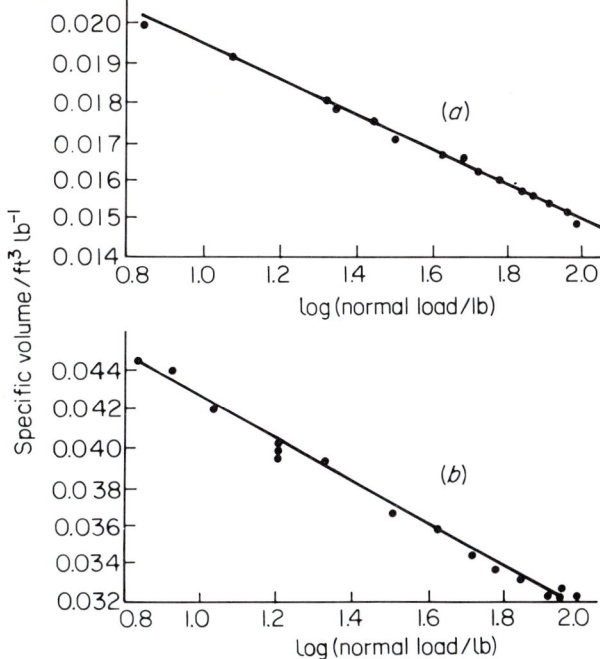

Figure 9.22.   Results of consolidation tests for (a) *titanium dioxide and* (b) *penicillin*

from a hopper. There seems therefore an inconsistency with reality to use the Mohr–Coulomb yield criterion to predict failure and/or slip of materials in compaction.

Results from both low and high stresses and the evaluation of powder failure and wall yield loci has, however, been rewarding in the understanding of the mechanism and stress distribution of compacted materials.

Williams and Birks[17] investigated the relationship expressed by equation 9.16 with a series of experiments on titanium dioxide and penicillin. The bulk density of each powder was measured in a Jenike cell under various consolidating loads without shear. The results shown in *Figure 9.22* are for specific volume (which is a linear function of the voids ratio) plotted against the logarithm of the consolidating pressure. In each case the points lie closely on a straight line showing the relationship

$$v = K_2 - m \log \sigma_c \tag{9.18}$$

where v is the specific volume and is related to the void ratio e by

$$v = \frac{1}{\rho_s} (1 + e)$$

where ρ_s is the powder density and σ_c is the applied normal consolidating stress. K_2 and m are constants for a given material and equation 9.18 is of the same form as equation 9.16 derived by Schofield and Wroth[14]. The

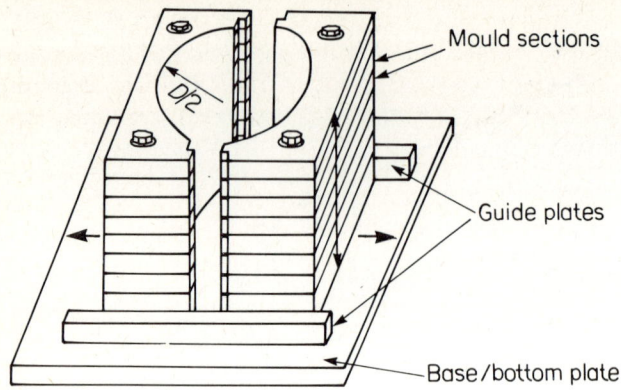

Figure 9.23.   Schematic diagram of split mould used by Williams, Birks and Bhattacharya

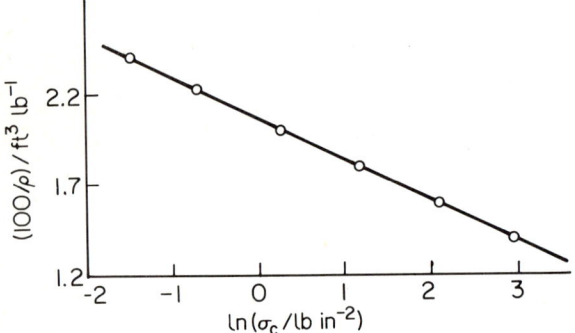

Figure 9.24.   Compaction characteristics of titanium dioxide

maximum normal load applied to titanium dioxide and penicillin (*Figure 9.22*) was 100 pounds. Williams, Birks and Bhattacharya[19], to obtain failure properties at stresses higher than those usually obtained in a Jenike shear cell, designed a direct compaction and failure measurement instrument with a split mould (*Figure 9.23*). They showed that equation 9.18 found from a simple powder consolidated in a shallow smooth walled cylindrical container was verified and gave when the specific volume or reciprocal of the bulk density was plotted against the natural logarithm of the applied consolidating stress a straight line (*Figure 9.24*) with a value of

$$v = \frac{1}{\rho_b} = 0.0206 - 0.0023 \ln \sigma_c$$

where ρ_b is in lb ft^{-3} or v in ft^3 lb^{-1} and σ_c in p.s.i. ($= $ lb in^{-2}).

They also showed that if a specimen of given height is prepared by compacting in a number of increments, N, the unconfined yield strength of such a compact is uniquely determined by the mean bulk density or specific volume independent of the number of increments by which it was formed. *Figure 9.24* shows values of the reciprocal bulk density plotted against the logarithm of either the normal stress or, since unconfined yield strength with titanium dioxide was directly proportional to normal stress, the unconfined yield strength.

9.4 Compaction of powders

9.4.1 Low stress compaction

Birks and Muzaffar[20], in a preliminary investigation of the mechanisms of volume reduction at low stress, determined with the use of a modified

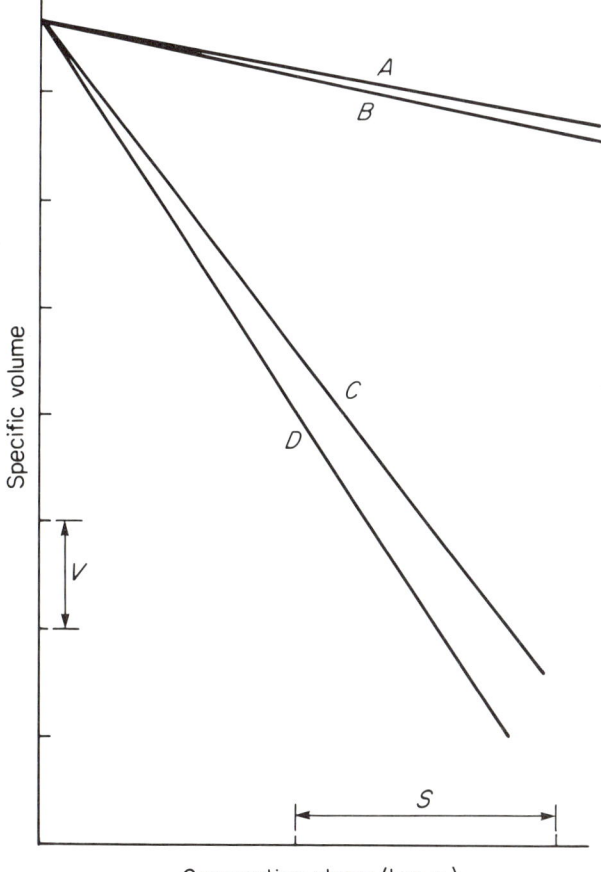

Compacting stress (log σ_c)

Figure 9.25. Compaction curves of 'simple' powders of Type 1; A, Britmag; B, calcined alumina; C, titanium dioxide; D, Calofort U. In Figures 9.25, 9.26, 9.27 and 9.28 the distance V represents a specific volume of 10^{-3} ft^3 lb^{-1} and the distance S represents $0.5 \log_{10}$ (number of cycles)

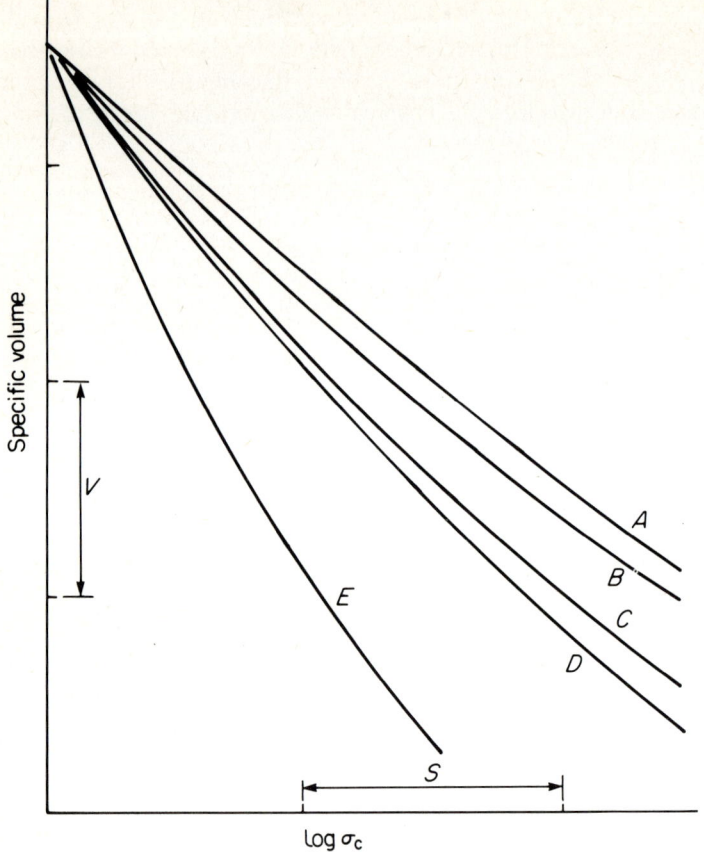

*Figure 9.26.   Powders with a concave upwards compaction
curve of Type 2; A, B, C, D, various grades of Portland
cement; E, titanium dioxide at high stress*

standard soil mechanics oedometer working in the range 80—1450 lb, the
compaction or volume reduction of ten different particulate solids. From
their observations the behaviour of the powders on compaction can be
classified into four different phenomena by the shape of the compaction
curve.

1. Straight line curve: titanium dioxide, Britmag, alumina and Calofort U
    (*Figure 9.25*).
2. Concave curve to X axis: Aerosil, molybdenum disulphide and spher-
    ical plastic powder (*Figure 9.26*).
3. Convex curve to X axis: titanium dioxide at high stress and Portland
    cement (*Figure 9.27*).
4. Variable curves: flowers of sulphur and milk powder (*Figure 9.28*).

The shape of the curves obtained were interpreted as being due either to

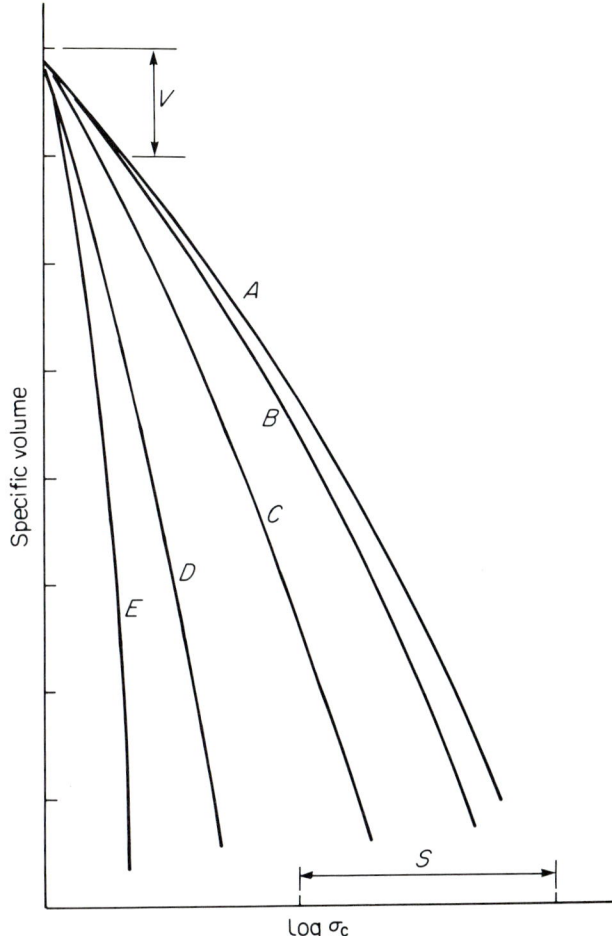

Figure 9.27.   Powders with a convex upwards compaction curve of Type 3; A, fine spherical plastic powder; B, unsieved titanium dioxide at low stress; C, molybdenum disulphide; D and E, two grades of Aerosil silica

the relief of stress by particle rearrangement within the powder or, if this mechanism of stress relief was no longer possible without interference from surrounding particles, to deformation before any subsequent rearrangement. Deformation of the particle assembly can occur by

1.   Plastic deformation of particle asperities.
2.   Elastic deformation of the interior of the particle.
3.   Plastic deformation of the whole particle.

The method of deformation depends on the material under compaction. For large particles plastic deformation of particle asperities predominates

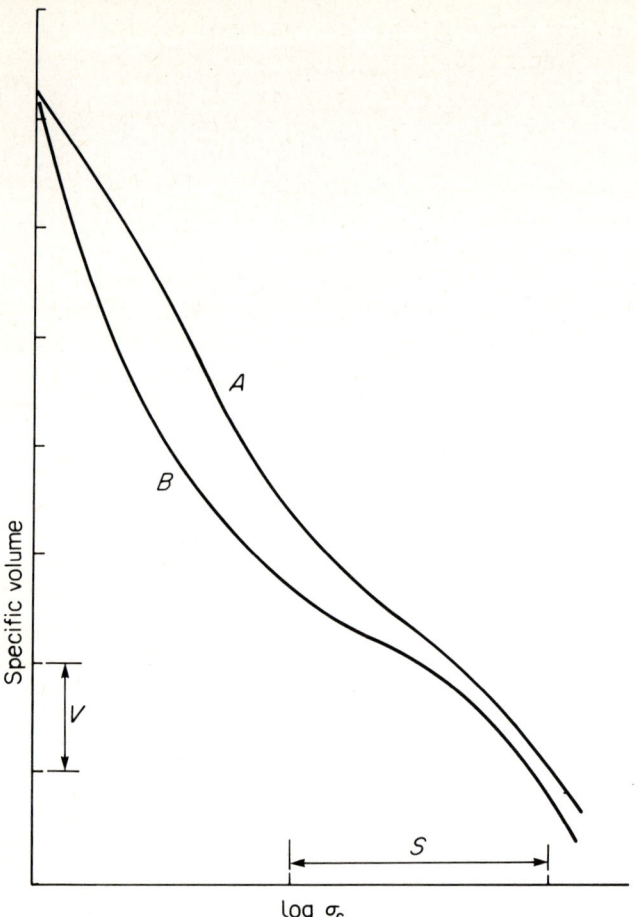

*Figure 9.28.   Powders with a variable compaction curve of Type 4; A,
flowers of sulphur; B, milk powder*

while for finer particles plastic deformation of the whole particle occurs.
The last mechanism of deformation modifies the mechanism of compaction.

9.4.1.1   Type 1

Birks and Muzaffar[20] stated that when a Type 1 curve was observed, which
could be mathematically expressed as

$$v = K_2 - m \log \sigma_c \tag{9.18}$$

the volume reduction was solely by particle rearrangement. This was deduced from the fact that TiO_2 particles are hard and at the low initial
stresses used the titanium dioxide particles were not subjected to plastic
deformation stresses. Fragmentation was also disregarded owing to the

small particle size and low stress; thus particle rearrangement was the only possible mechanism for volume reduction.

9.4.1.2   Type 2

The shape of this volume reduction *versus* compaction stress relationship shows a curve concave to the stress axis which indicates that the opportunity for particle rearrangement has become restricted and instead of the stress being accommodated by an increase in particle–particle contact the extra stresses are transferred to the existing particle contacts. Transfer of the extra increasing stress to an established particle–particle contact may however result in the contact breakdown. The overall particle contacts then will decrease with increase in normal stress and the compaction volume reduction curve will be shaped as in *Figure 9.26*. The character of whether particle contact is achieved by plastic or elastic deformation has not been examined.

9.4.1.3   Type 3

The shape of this volume reduction *versus* compaction stress relationship shows a curve which is convex to the stress axis. This convex curve occurs with a particle structure which is easily broken or deformable. This shape of curve has been observed by Duffield and Grootenhuis[21] when spherical copper particles were compressed to a gross deformable plastic stage just before complete powder to metal solidification. Type 3 phenomena can thus be associated with a compaction mechanism which relies on the fragmentation and deformation of particles.

9.4.1.4   Type 4

The shape of this volume reduction *versus* compaction stress relationship shows a curve with both a concave and convex attitude to the stress axis. Type 4 curves occur with materials which have low melting points and a rough particle structure and consolidate in a series of sudden collapses. The explanation offered for these sudden changes or collapses was coupled with a high strain rate and friction melting at the particle contacts. Melting reduced friction and allowed a considerable amount of deformation before the contacts solidified.

9.4.2   Medium stress compaction

When Schwartz and Weinstein[8] compacted uranium dioxide powder in the range 0 to 1500 lb in^{-2} they found that the shear stress, τ, and normal stress, σ, relationship (*Figure 9.29*) was linear and that the powder yield criterion from Coulomb was satisfied and tenable during compaction for every point in the compact. The material was assumed to fail at every

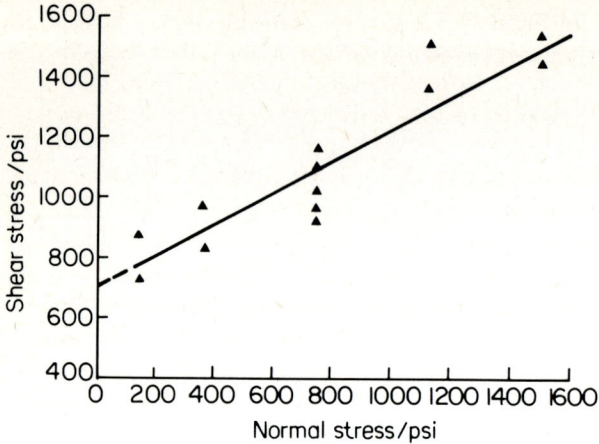

Figure 9.29.   Shear strength of pressed pellets as a function of normal loading: $c = 700$ *p.s.i.,* $\phi = 28.4$ *degrees*

point of contact to produce the observed densification and behaved as a rigid plastic material. In recent years however several authors[13,22,23] have found that the failure behaviour of metal powders compacted at medium pressures (stress) deviates from that of a rigid-plastic powder. The powder yield or failure locus properties can be such that:

1. The stress state represented by a Mohr stress circle below the powder yield locus can still cause certain plastic deformation. This is in oppo-

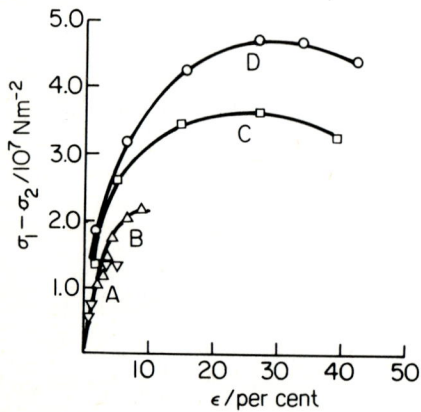

Figure 9.30.   Plot of the principal stress difference $\sigma_1 - \sigma_2$ *against the axial strain* ε *during triaxial testing at various confining pressures* σ_2

Curve	A	B	C	D
$\sigma_2/10^7\,\mathrm{N\,m^{-2}}$	0.25	0.5	1.25	2.0

sition to the assumption that if the stress state, as represented by a Mohr circle, was below the failure locus deformation did not occur.

2. Irreversible plastic deformation only occurred if the Mohr circle touched the powder yield or failure locus.

For ferric oxide powder of particle size 0.03—0.05 micrometre compacted in 20 mm diameter punch and die equipment, Strijbos *et al*[23] showed that the powder yield locus could be calculated from knowledge of the relationship between the principal stress, σ_2, at pressures of 0.25, 0.5, 1.25 and 2.0×10^7 N m^{-2} and axial strain (*Figures 9.30* and *9.31*). Comparison of the powder yield locus with the wall yield locus showed that the wall yield locus lay below the powder yield locus and compacted powder compacts cannot be considered as a rigid-plastic material because the powder yield locus was not linear. This was also seen from the work of Suh[11] and Jenike *et al*[12]. Fukumari and Okada[24] compressed crystalline potassium chloride of particle size 208—350 μm in a 20.45 mm diameter die at a constant rate of upper punch displacement of 0.46 mm s^{-1} and also showed that the powder yield locus was not linear and was above the wall yield

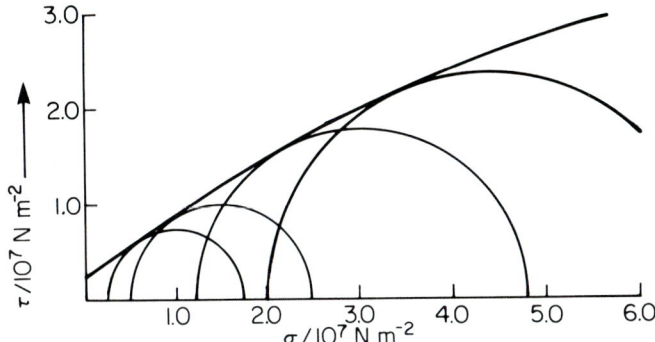

Figure 9.31. Powder yield locus constructed with the help of failure stresses following from Figure 9.30

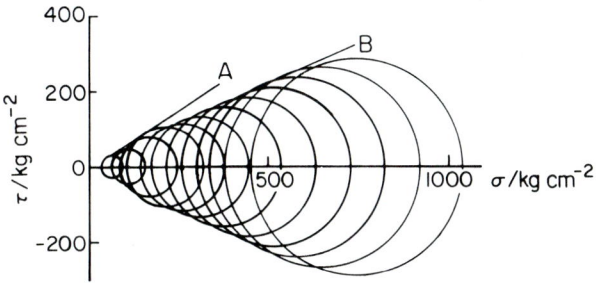

Figure 9.32. Mohr circles for potassium chloride powder; for A and B see text

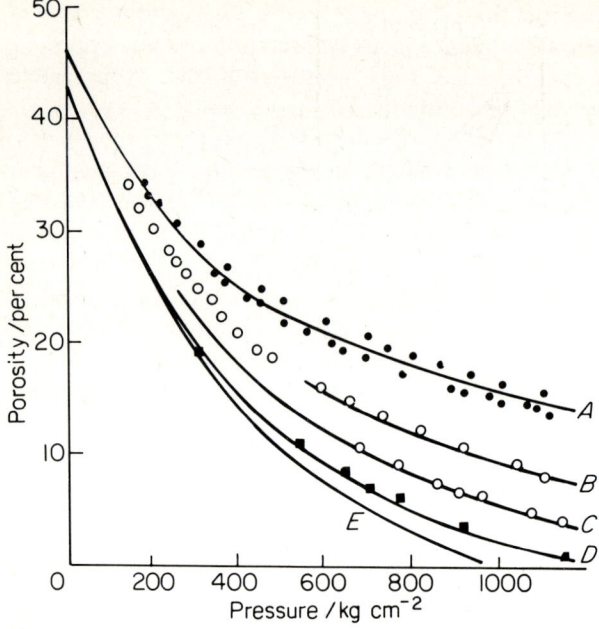

Figure 9.33.   Pressure–porosity relationships for potassium chloride powder

Curve	A	B	C	D	E
Weight of powder/g	23.5	15.67	7.83	3.92	0 (*extrapolation*)

locus. The powder yield or failure locus could however be approximated to two linear relationships (*Figure 9.32*):

$$\tau = 0.653\sigma \text{ (curve } A)$$

$$\tau = 0.378\sigma + 33.8 \text{ (curve } B)$$

where τ and σ are in kg cm^{-2}. They also determined the pressure–porosity relationship of potassium chloride with different die fills as well as the wall yield locus (*Figure 9.33*).

9.4.3   High stress compaction

9.4.3.1   *Rendulic and Hvorslev failure or yield surfaces*

Roscoe[25], after developing a shear cell to measure accurately the volume changes which occurred with consolidating soils under different shear and normal stress conditions, found that by using the Hvorslev relationship between σ and e, a unique three dimensional surface could be defined, termed the Hvorslev surface (*Figure 9.34*, surface *ABCDE*). Progressive failure of a granular sample of soil produced loading paths in three dimen-

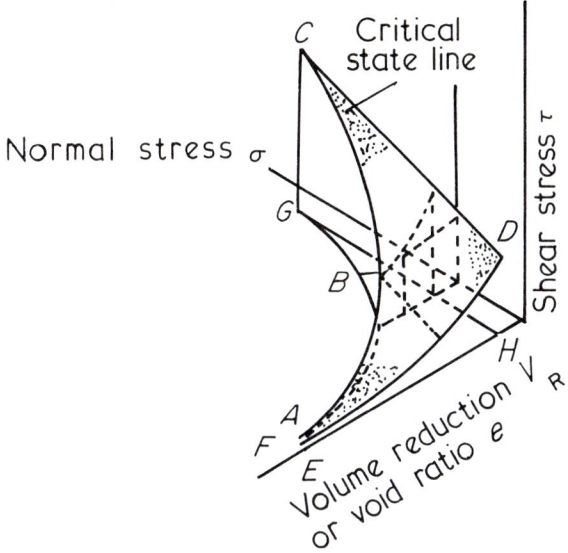

Figure 9.34. The Hvorslev failure surface. ABCDE *defines the Hvorslev surface and* FGH *its projection on the normal stress–void plane.*

sional space, the final portion of these loading paths falling on this unique surface and terminating at a unique *critical state line* (CSL).

Roscoe, Schofield and Wroth[26] used the concept of the Hvorslev surface to determine the degree of shear necessary to cause the flow or failure of clays and granular material consolidated to a known value in triaxial tests. They stated that the critical void ratio could be defined in two alternative ways. One was concerned with the change in volume of the consolidated material when the effective normal stress remained constant (the drained test) and the second was the change in effective stress when the void ratio remained constant (the undrained test). Experimental evidence showed that in each type of triaxial shear test the loading paths obtained with clays and granular material had a coincidental surface as their envelope in σ, τ and e space. The surface defined was the yield surface and the space it contained the yield domain. The family of all possible loading paths must thus lie either within the yield domain or on the yield surface. To check the CSL hypothesis Roscoe, Schofield and Wroth sheared, in the Roscoe shear apparatus, cohesionless 1 mm diameter steel balls which had no danger of individual particle breakage when subjected to a maximum normal stress of 100 lb in^{-2} (690 kN m^{-2}) at different normal stresses. Loosely packed steel balls (normally consolidated) underwent a volume contraction along the surface (Rendulic surface: *Figure 9.35*) until a terminal point was reached on the CSL whilst densely packed steel balls (overconsolidated) underwent a volume expansion along the Hvorslev surface until the critical volume line

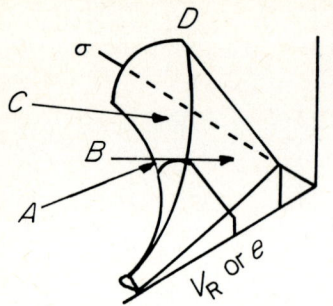

Figure 9.35.   *Rendulic and Hvorslev surfaces and domains:* A, *normal consolidation line;* B, *Hvorslev surface;* C, *Rendulic surface;* D, *critical state line*

(CVL) was reached. Glass beads and sand gave similar results although the degree of scatter, which was attributed to particle breakage, was greater.

Calladine[27, 28], in explaining the mechanical properties of plastic and elastic compressible clays and the effect of shear on normal and readily consolidated clays from a microstructural and interparticle viewpoint, quoted the work of Rendulic[29], who had performed tests with increasing deviatoric stress on isotropically normally consolidated material. The Rendulic contours in principal stress space showed that the stress paths taken were members of a family for both drained and undrained tests. Combination of the Rendulic and Hvorslev surfaces (*Figure 9.35*) have been used by many workers in the field of soil mechanics to produce a boundary surface which supports the hypothesis of Roscoe that the critical state line divides the three dimensional representation of σ, τ and e space into either a surface for normally consolidated (loosely packed and eventual volume reduction with shear) material or a surface for overconsolidated (densely packed and volume expansion with shear) material.

Wroth and Basset[30] also stated that the CSL was of fundamental importance because it divided samples into two distinct categories. Normally consolidated samples with initial stress states on the far side of the CSL from the origin (which they termed the wet side) experienced a reduction in volume when sheared while overconsolidated samples on the near side of the CSL to the origin (which were termed the dry side) showed an increase in volume when sheared.

The shear tests on consolidated cohesive soils are concerned with interparticle forces and the bonds produced contribute to the mechanical behaviour of soils. The interparticle bonds produced during soil shear tests are usually achieved with little particle fracture owing to the low initial consolidation pressures used. Billam[31] studied the effect of higher soil consolidation stresses, in the range 3.0×10^{-3} to 10 MPa, together with the fracture of particles, on the stress–strain relationships of chalk, crushed anthracite, limestone and Ham River sand. He found that the principal stress ratios at failure, the ratios of σ_3 and σ_1 when an assembly of particles under shear began to fail or flow, were dependent upon the surface texture, particle fracture strength and internal particle porosity.

Work done at the University of Bradford[32] on the compaction of plastic

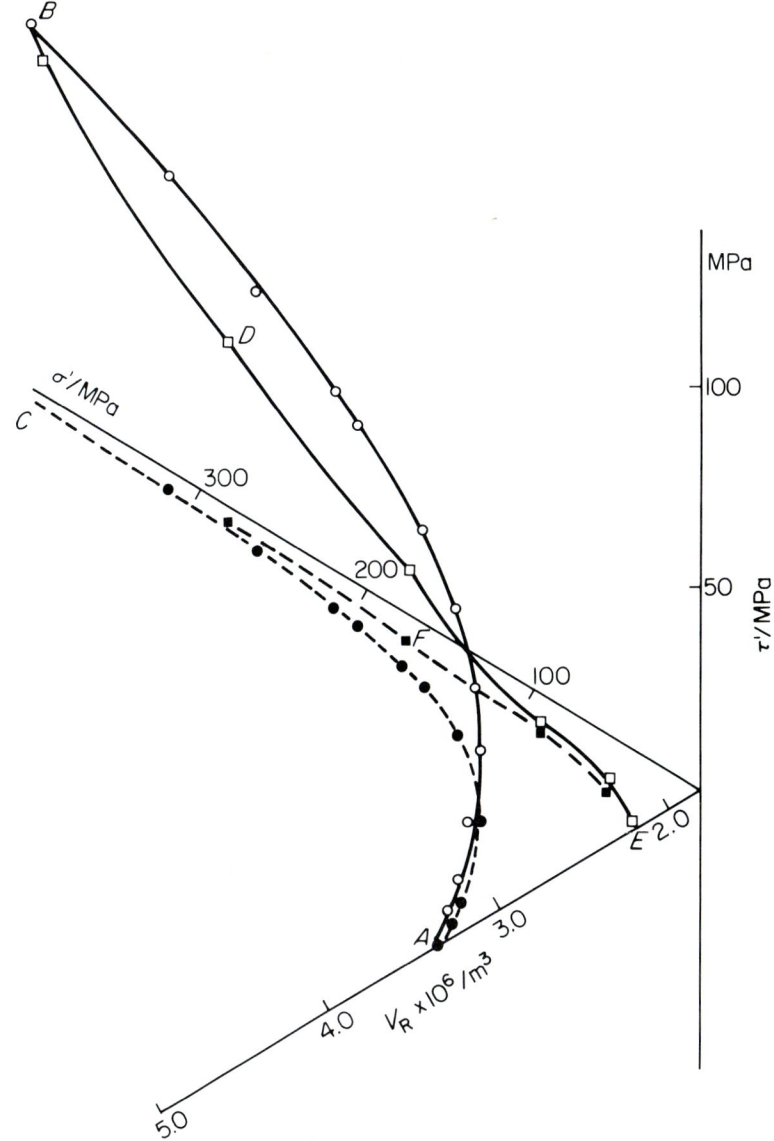

Figure 9.36. Three dimensional space relationships for polystyrene compaction and relaxation pathways

○   σ', τ', V_R *Compaction curve*
□   σ', τ', V_R *Relaxation curve*
●   σ', V_R *Compaction projection*
■   σ', V_R *Relaxation projection*

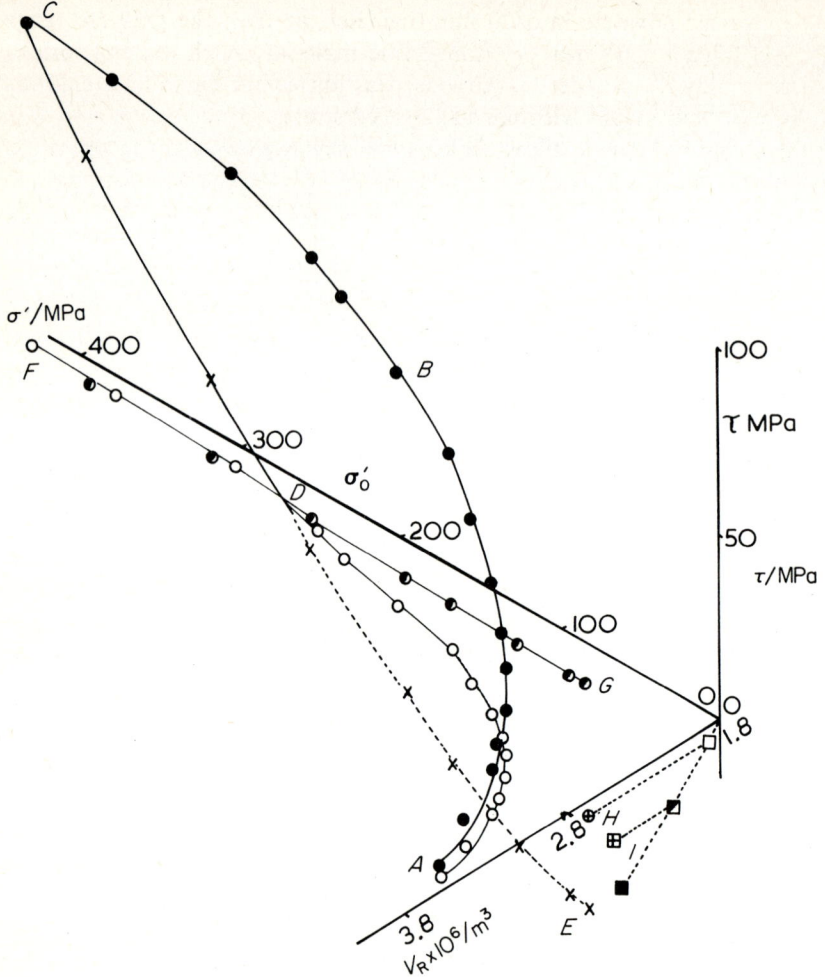

Figure 9.37. Three dimensional space relationships for sodium chloride (3.27 kN s^{-1})
compaction and relaxation pathways

○	σ', V_R	Compaction = 248 MPa
●	σ', V_R, τ'	Compaction = 248 MPa
◐	σ', V_R	Relaxation = 248 MPa
×	σ', V_R, τ'	Relaxation = 248 MPa
□	σ', τ'	Relaxation = 62 MPa
◪	σ', τ'	Relaxation = 155 MPa
■	σ', τ'	Relaxation = 248 MPa
⊞	σ', τ', V_R	Relaxation endpoint = 155 MPa
⊕	σ', τ', V_R	Relaxation endpoint = 62 MPa

and inorganic powders in a 20 mm diameter die over the pressure range 32—248 MPa has shown that for plastic material which did not form a coherent body the powder failure locus was linear, but for sodium chloride the failure locus varied with normal applied stress.

Figure 9.36 shows in three dimensional representation the compaction and relaxation paths when non-compactable polystyrene was subjected to uniaxial compaction. Curve AB is the τ', σ' and V_R space path for the compaction process which occurs in the Rendulic yield domain while curve AC is the two dimensional projection on to the $\sigma'-V_R$ plane. The relaxation curve in τ', σ' and V_R space (BDE) may still be within the Rendulic domain as the critical state line could not be determined with the uniaxial compaction apparatus. The relaxation curve in the σ' *versus* V_R plane is line CFE which for non-compactable material does not show negative values of shear stress.

Figure 9.37 shows the compaction and relaxation paths obtained from sodium chloride compacted at 248 MPa at a loading rate of 3.27 kN s^{-1}. Curve ABC is similar to that seen in *Figure 9.36*, curve AB. The relaxation curve CDE (*Figure 9.37*) begins to show a negative shear stress at point D which corresponds to σ'_0 in the τ' *versus* σ' plane. The end point of the relaxation curve for 248 MPa occurs at point E. Points H and I are the end points of the relaxation curves for sodium chloride compacted at 155 and 62 MPa respectively. Projection of the endpoints E, I and H on to the σ' *versus* negative τ' axis shows a linear relationship which may be regarded as the compact relaxation locus.

References

1. GRIFFITH, A. A., *Phil. Trans. R. Soc. A* **221**, 163 (1920)
2. DRUCKER, D. C., GIBSON, R. E. and HENKEL, D. J., *Trans. Am. Soc. civ. Engrs* **122**, 338 (1975)
3. COULOMB, C. A., *Mém. Acad. r. Sci. Paris* (1976); from ref. 7
4. RANKINE, W. J. M., *Proc. R. Soc. A* **147**, 9 (1857)
5. SOKOLOVSKI, V. V., '*Statics of Soil Media*', Butterworths, London (1960)
6. De JOSSELIN de JONG, G. *See* ref. 7
7. WROTH, C. P. and BASSETT, R. H., *Geotechnique* **15**, 32 (1965)
8. SCHWARTZ, E. G. and WEINSTEIN, A. S., *J. Am. Ceram. Soc.* **48**, 346 (1965)
9. JENIKE, A. W. and SHIELD, R. T., *J. appl. Mech B* **81**, 599 (1959)
10. SHIELD, R. T., *J. Mech. Phys. Solids* **4**, 10 (1955)
11. SUH, N. P., *Int. J. Powder Met.* **5**, 69 (1969)
12. JENIKE, A. W., ELSEY, P. J. and WOOLEY, R. H., *Proc. Am. Soc. Test. Mater.* **60** (1960)
13. SCHWARTZ, E. G. and HOLLAND, A. R., *Int. J. Powder Met.* **5**, 79 (1969)
14. SCHOFIELD, A. and WROTH, P., '*Critical State Soil Mechanics*', McGraw-Hill, New York (1968)
15. TERZAGHI, K. and PECK, R. M., '*Soil Mechanics in Engineering Practice*', Wiley, New York (1961)
16. HVORSLEV, M. J., '*Ingeniorvidenskabelige Skrifter A*', No. 45 (1937)
17. WILLIAMS, J. C. and BIRKS, A. H., *Rheol. Acta*, **4**(3), 170 (1965)
18. COOPER, A. R. and EATON, C. F., *J. Am. Ceram. Soc.* **45**, 97 (1962)
19. WILLIAMS, J. C., BIRKS, A. H. and BHATTACHARYA, D., *Powder Technol.* **4**, 328 (1970/71)

20. BIRKS, A. H. and MUZAFFAR, S. A., '*Symposium on Powder Flow and Storage*', University of Bradford (1971)
21. DUFFIELD, A. and GROOTENHUIS, P. '*Symposium on Powder Metallurgy*', Special Report No. 58, Iron and Steel Institute, London (1954)
22. KOERNER, R. M. and McCABE, W. M. '*Proceedings of 1972 Powder Metallurgy Conference*', pp. 225–241
23. STRIJBOS, S., RANKIN, P. J., KLEINWASSINK, R. J., BANNINK, J. and OUDEMANS, G. J., *Powder Technol.* **18**, 187 (1977)
24. FUKUMORI, Y. and OKADA, J., *Chem. pharm. Bull.* **25**, 1610 (1977)
25. ROSCOE, K. H., '*Proceedings of the Third International Conference on Soil Mechanics*', **1**, 186 (1953)
26. ROSCOE, K. H., SCHOFIELD, A. N. and WROTH, C. P., *Geotechnique*, **8**, 22 (1958)
27. CALLADINE, C. R., *Geotechnique*, **13**, 250 (1963)
28. CALLADINE, C. R., *Geotechnique*, **21**, 391 (1971)
29. RENDULIC, L., *Bauningeneur*, **18**, 459 (1937)
30. WROTH, C. P. and BASSETT, R. H., *Geotechnique*, **15**, 32 (1965)
31. BILLAM, J., '*Proceedings of Roscoe Memorial Symposium*', Cambridge University, March (1971)
32. STANLEY-WOOD, N. G. and ABDELKARIM, A. M., *Powder Metallurgy Intern.*, **14**, 135 (1982)

CHAPTER 10

Instrumentation of tablet machines

H. S. Thacker
Manesty Machines Limited, Speke, Liverpool

10.1 Introduction

Tablet making machinery is used in many industries to produce a wide variety of compacts with an even wider range of applications. Many of the compacts produced on tablet machines are far removed from what is normally understood by the word 'tablet'. The types of machine used also vary widely, from simple hand operated machines varying, in the case of the pharmaceutical industry, to sophisticated high speed machines capable of producing over 750 000 tablets per hour and, on the powder metal side, to high energy presses capable of exerting forces in excess of 10 MN (1000 tonnes force). The instrumentation used on these presses can vary almost as widely as the presses themselves. It is only intended to cover here the development and current use of instrumentation in pharmaceutical compaction processes. It is, however, interesting that the instrumentation now used in the pharmaceutical industry appears to have been developed from fundamental research carried out by powder metallurgists.

The purpose of most of the early work was to investigate the compaction process itself rather than to control the machines. It was often carried out with idealized materials on equipment which differed considerably from the normal compaction equipment used by the pharmaceutical industry. Nevertheless, it resulted in basic relationships being devised, many of which are still accepted today.

Later investigators sought to apply these findings to the conditions existing in commercial tablet production and this work has divided itself into two distinct fields: (*a*) the investigation of the compaction process under production conditions and (*b*) the control of the tablet presses. The methods used to obtain the information about the compaction process is often common to both fields.

10.2 Instrumentation of single acting machines

10.2.1 Measurement of punch forces

Some of the earliest work on measuring the forces involved in compressing a tablet was carried out by Brake[1] and later by Higuchi and his coworkers[2]

The method used was to fit strain gauges to the frame of the machine in order to measure the small elastic deformations which occur in the machine when the tablet is compressed.

Bonding the gauges to the machine frame had the obvious disadvantage that the gauges were remote from the site where the forces were being applied and may not therefore give a true representation of the compaction forces at the punch face. Shotton *et al.*[3,4] fitted strain gauges to the upper punch and lower punch holder in order to make the measurements as near the active site as possible and thus eliminate any error resulting from remote readings (*Figure 10.1*).

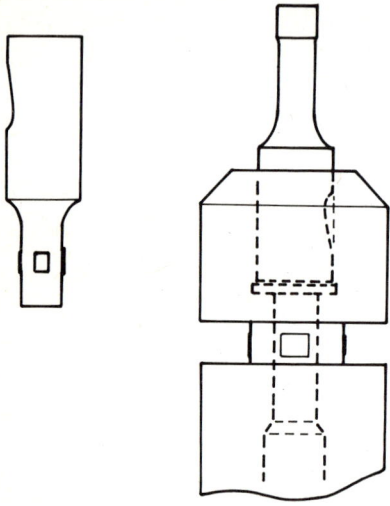

Figure 10.1.   *Strain gauges on upper punch and lower punch holder on an* F3 *machine*

10.2.2   Die wall forces

The main purpose of the work by Higuchi *et al.* had been to investigate the compaction process for a variety of pharmaceutical materials and to investigate such aspects as the effect of lubricants. An important parameter in this study was the effect of friction and shear at the die wall–tablet interface and two of Higuchi's coworkers investigated this area. Nelson[5] measured die wall forces by using a conventional die with a rectangular hole perpendicular to the die in which a third punch was inserted. The load on the third punch was detected by means of a Baldwin load cell. Windheuser *et al.*[6] removed a segment of the die wall to increase sensitivity and render the stress forces more parallel at the area where strain gauges were fitted (*Figure 10.2*).

10.2.3   Calibration

Calibration of the strain-gauged punches was relatively easy as a known load could be applied to the punches and the response measured. The

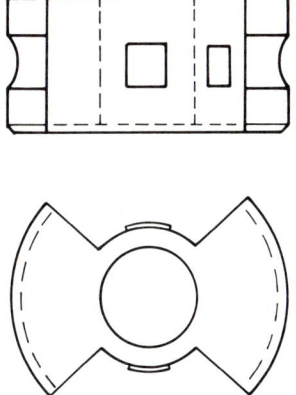

Figure 10.2. Die fitted with strain gauges

calibration of the die was slightly more difficult but this was achieved by filling the die with oil or grease and thereby hydraulically transmitting the force from the punches to the die wall[5,6]. This method presents some difficulty in sealing the die and it has been shown[6-9] that a rubber plug placed between the punches can be used as this will behave elastically under compression and transmit the force to the die wall.

10.2.4 Punch position

It has also been found desirable to measure punch position during the compression cycle as a knowledge of punch position enables specific points in the cycle to be determined. This can be achieved by the use of a Linear Variable Differential Transformer (LVDT) (*Figure 10.3*). The moveable core of the transformer can be connected to the punch so that the punch movement unbalances an AC bridge circuit. The output of the bridge is attenuated to produce a DC voltage which is directly proportional to the punch

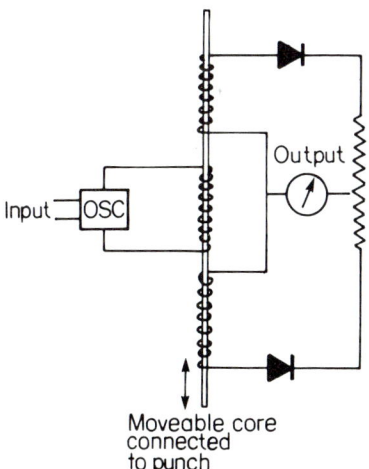

Figure 10.3. Typical circuit for a linear variable differential transformer

displacement. Their resolution is theoretically infinite but is limited in practice by the quality of the mechanical link. They are unaffected by temperature and can be used at frequencies up to 500 Hz.

10.3   Instrumentation of rotary tablet machines

10.3.1   Punch instrumentation

The instrumentation of single acting machines led to a much greater understanding of the compaction process. Many machines, instrumented in ways similar to those described, are in current use and provide much useful information on the compaction properties of both experimental formulations and production batches. However, the single acting machine cannot reproduce all the conditions which exist on a rotary tablet machine and consequently the problems of instrumenting rotary machines were investigated. The main difficulty in instrumenting the punches on a rotary machine, is retrieving the signal from the moving punch. Two methods present themselves, radio telemetry or slip rings.

In the early 1960s when this work was started the slip ring systems available were not suitable for use in the dusty environment of a tablet machine and the electrical noise generated by the dust on the rings interfered with the signal being measured. Since that time slip ring design has improved considerably and one British pharmaceutical company has a high speed tablet machine, extensively instrumented and using slip rings to retrieve the signals.

The early work was however carried out by use of radio telemetry. Shotton[10] and Fuehrer[11] both reported on instrumented rotary machines using this method. At that time semiconductors were still in their infancy and the radio transmitters were quite large. They were housed in the punch guide holes adjacent to the punch being monitored. Consequently, it was not possible to use the machine with a full set of tooling. It has been reported by several subsequent workers in this field that the machine could not be run at full speed because of the radio link. This was not true as there is no mechanical reason why a machine cannot be run at full speed irrespective of the number of stations of tooling fitted. The possible discrepancy between the results obtained under these conditions and what would actually occur in normal tabletting lay in the fact that preceeding or succeeding stations to the one being monitored were omitted. This situation undoubtedly affects die filling and in turn the compaction conditions, although the difference in signals obtained by this method and those that would have been obtained had a full set of tooling been used are probably minimal. It did however trigger a vast amount of work dedicated to measuring the punch forces from each successive station in turn.

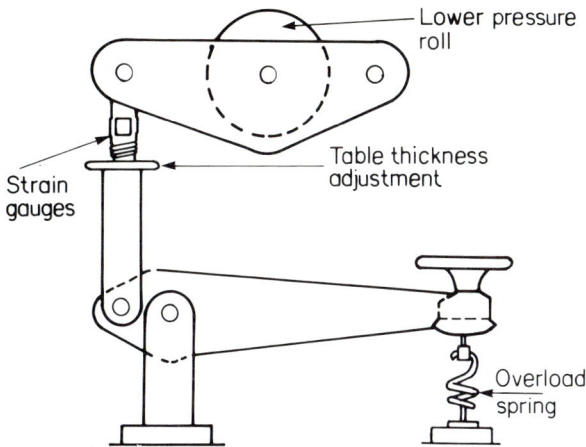

Figure 10.4.   Strain gauges on the compression release mechanism

10.3.2   Machine instrumentation

Knoechel *et al.*[12] reverted to the early ideal of instrumenting various parts of machines remote from the punch, particularly the compression release mechanism (*Figure 10.4*). They also instrumented the ejection cam although the signals received from the cam were of doubtful value because they were a function of the ejection force being applied by several punches which were on the cam at the same time.

Deer *et al.*[13] endeavoured to bring the machine frame instrumentation as close to the punch as possible by instrumenting a modified pressure roll. While this offered an excellent research tool it was not sufficiently robust for use under production conditions.

10.3.3   Comparison of punch and machine methods

Wray[14] combined the methods used by Shotton and Knoechel by instrumenting both the punches and the machine frame. He showed that signals from the machine frame did give a true representation of the compression forces. He also improved the monitoring of the ejection force by dividing the ejection cam so that it registered the ejection force from only one punch at a time.

10.4   Force measuring systems

10.4.1   Strain gauges

The strain gauges used in most of the early development work on tablet press instrumentation consisted of a flat winding of high resistance wire mounted on a paper backing. This was bonded to the punch or some other part of the machine which was being stressed during the compaction of the tablet. The gauge was then deformed simultaneously with the member to which it was bonded. The deformation of the gauge caused a change in the dimensions of the wire which resulted in a change in its electrical resistance. The change in resistance was measured by making the gauge one of the arms of a Wheatstone bridge. The output of the bridge was monitored either by an oscilloscope or recorded using an ultraviolet recorder.

Temperature compensation was achieved by bonding a matched reference gauge to an unstressed piece of metal with the same heat capacity as one being stressed and sited so that it was subjected to the same variations in temperature. The reference gauge was then used in the opposite arm of the bridge. Additional gauges were normally used to compensate for bending or other stresses which may have been produced by the compaction force or other factors.

Modern gauges are usually made by etching foil to form the resistance element. These have the advantage that the flat surface of the foil is much easier to bond to the metal surface than the round wire. Gauges are produced in a wide variety of shapes and sizes. The shape, size and total resistance of the gauge should be selected to suit the particular application. Various series of gauges are available with temperature coefficients corresponding to particular metals so that by selecting the correct gauge for the metal being stressed, temperature compensation between the gauge and the metal is automatic.

Semiconductor strain gauges are also available and these can be 50—100 times more sensitive than the wire or foil gauges. They do not appear to have been used to any great extent in tablet machine instrumentation and the reasons given vary; but they appear to be difficult to bond accurately owing to their small size, and they are more temperature sensitive.

It is also good practice whenever possible to use a full bridge, positioned and wired so that the only stress measured is the one in which you are interested.

In early work it was common practice to use AC energization of the bridge as DC amplifiers were inherently unstable and caused drift problems. However, the rapid advances in electronics have enabled extremely stable DC amplifiers to be produced and it is now possible to use DC measurement.

10.4.2 Piezoelectric force transducers

An alternative system of measuring the compaction forces, by means of piezoelectric force transducers[15], was investigated by Marshall[16] and Ridgway[17,18]. The transducers contain a quartz crystal which, when subjected to an external force, develops an electrical charge proportional to the force applied. The charge can be converted to a DC voltage by means of a charge amplifier and measured on an oscilloscope or ultraviolet recorder, in a similar manner to strain gauges. This system has a greater sensitivity than strain gauges, does not require bonding to the machine, and is less temperature sensitive. They have been successfully used on single acting machines by mounting them in the upper and lower punch holders (*Figure 10.5*). Lindberg[19] developed a system similar to that of Windheuser[6] for measuring radial forces at the die wall but using a piezoelectric force transducer as the measuring element.

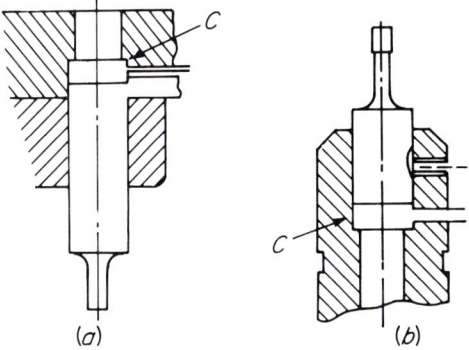

Figure 10.5. Enlarged detail of piezoelectric crystal thrust unit on (a) upper and (b) lower punch. The position of the crystal unit is indicated by C

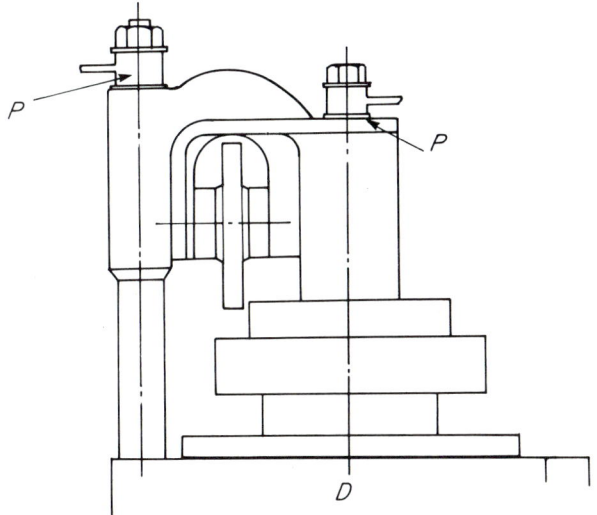

Figure 10.6. Piezoelectric transducers on single rotary machines; the positions of piezo-crystal thrust washers are indicated by P, and D is the die table

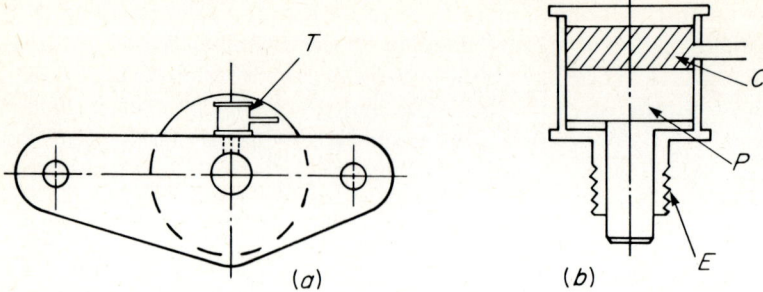

Figure 10.7.   (a) *Piezotransducer fitted on lower pressure roll;* (b) *enlarged detail of crystal thrust unit;* T, *crystal thrust unit;* C, *crystal;* P, *piston;* E, *screwed end fitting*

These force transducers have an advantage over strain gauges for the instrumentation of rotary machines because they can be located with a minimum amount of structural alteration to the machine (*Figures 10.6 and 10.7*). They can also be used for measuring ejection forces by fitting them to the ejection cam mounting bolts or, if individual punch ejection forces are required, by detaching a section of the cam and fitting one or more transducers below it.

The disadvantage of using these force transducers lies in the fact that they will not give a stable indication of a static load, but they are ideally suited to measuring dynamic loads. It was therefore assumed that they would be suitable for measuring what are apparently dynamic forces involved in compacting tablets. Most of the initial work with piezoelectric force transducers was carried out on simple low speed machines on which the dies are widely spaced. Under these conditions they performed satisfactorily. Even when used on high speed machines with close die spacing they can still perform satisfactorily for materials with low compression ratios, i.e.,

$$\text{C.R.} = \frac{\text{volume of material before compression}}{\text{volume of the compressed tablet}}$$

However, conditions can exist on high speed tablet machines where the combination of die spacing and compression ratio of the material results in a situation where the dynamic forces contain a semistatic component. Under these circumstances the piezoelectric force transducer is not suitable for measuring punch forces. The semistatic component of the force results from the fact that two or more tablets are being compressed by one pressure roll at the same time. When this occurs the pressure does not return to zero between each tablet and the machine is subjected to a static force equal to the lowest level to which the compaction force falls before it starts to rise again for the next compression. It is this static component which the charge amplifier gradually loses, giving the impression that the mean force required to make the tablets is gradually falling.

10.4.3   Photoelasticity

The measurement of the radial forces on a die in a rotary tablet press presents particular problems owing to the close spacing of the dies, on high speed machines, leaving very little space to fit instrumentation. The only report in the literature of die wall forces on a rotary machine is by Ridgway[8]. He used the technique of measuring the fringe patterns developed in a poly(methyl methacrylate) die using high speed photography. This method has the advantage that point values can be obtained, as opposed to the measurement of radial forces over a larger area, as would be obtained from strain gauges. It is difficult to assess whether the greater yield of the poly(methyl methacrylate) die, when compared with a steel die, had any influence on the results.

10.5   Uses of instrumentation

10.5.1   The study of the compaction process

10.5.1.1   Force displacement curves

The most extensive use of instrumentation has undoubtedly been for the study of the compaction process itself and by far the largest part of this work has been carried out on single-acting presses. Reports on the study of variables such as compact porosity, disintegration time, dissolution time, and crushing strength, are all to be found in the literature.

The study of the effect of lubricants appears to have been of particular importance probably because of the fact that the quantity of lubricant used affects other tablet parameters such as crushing strength and disintegration time as well as the ejection force. In the early work on force and lubricants a relationship 'R' was used[20]. R was the ratio of the maximum upper punch force to the maximum lower punch force. This was later shown to be affected by the L/D ratio for the compact and the nature of the material being lubricated. This was superseded by measuring force–displacement (FD) curves and calculating the work done in compression which gives a better measurement of the effect of a lubricant.

In order to evaluate the work done in compression by means of an FD-curve, it is necessary to modify the tablet machine to enable a second compression cycle to be made on the compact before it is ejected from the die. An FD curve for a typical pharmaceutical material is shown in *Figure 10.8*. The lightly shaded area is a measure of the work done by the upper punch during compression and the more heavily shaded area represents the work of expansion of the compact which has been transferred back to the upper punch during decompression. The latter area is dependent upon the degree of elastic deformation which takes place. However, it is impossible to be certain that the total elastic recovery is measured as it may not all be transferred back to the upper punch. It is therefore necessary to compress

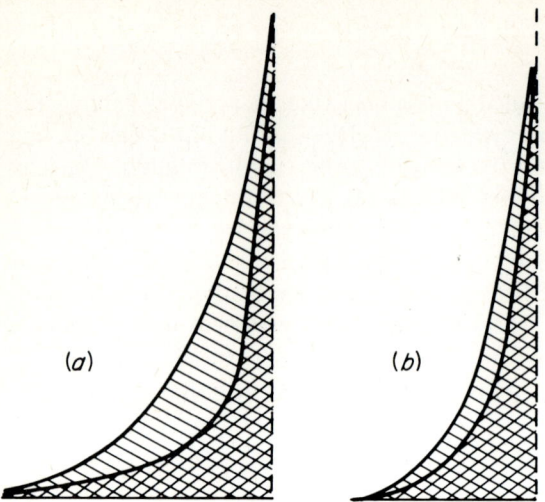

Figure 10.8.   Typical force–displacement curves for
(a) *the first and* (b) *the second compression*

the material a second time. If further elastic recovery has occurred after the
upper punch lost contact with the compact in the first compression, then
the work done in the second compression will exceed that recovered in the
first compression. The total amount of work done in compressing the compact is therefore the total work done in the first compression less that in the
second compression.

10.5.1.2   Pressure cycles

Perhaps some of the most interesting work which has come from studies
made on instrumented tablet machines is that of Leigh *et al.*[7] who plotted
the axial force against the radial force to give complete pressure cycles for a
variety of materials. This work was based on that of Long[21], the powder
metallurgists again showing the way in this field.

The general pressure cycle obtained is as shown in *Figure 10.9. OA*
represents elastic deformation. *AB* represents the stage after the elastic limit
is exceeded and plastic deformation and/or brittle fracture occurs. At *B* the
axial load is released and *BC* represents elastic recovery and when P_R is
greater than P_A by an amount which exceeds the elastic limit recovery
continues to a similar point to that occurring in *AB*. The line *OD* represents
the residual radial pressure in the compact.

The slope of *OA* is given by $v/(1 - v)$, where v is the Poisson ratio, therefore

$$P_R = P_A v/(1 - v)$$

For a fluid $v = 0.5$ and the slope of *OA* would be unity. Leigh found that
the behaviour of soft materials such as aspirin tended to approach that of a

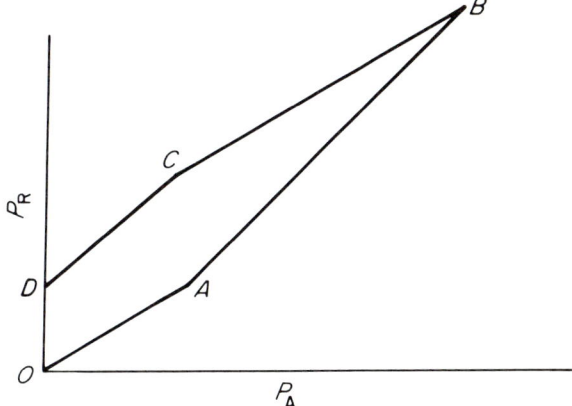

Figure 10.9.   A typical compression cycle

fluid whereas hard materials with a high modulus of elasticity such as sucrose gave lower slopes.

AB is a measure of the compressibility of the material. If it yields readily in plastic deformation and the yield stress *S* is constant then

$$P_R = P_A - 2S$$

and the slope of the line is unity. Sucrose crystals and sodium chloride crystals give slopes in the region 0.7—0.8 for this part of the cycle. In both cases the area of contact of the constituent particles increases considerably. The amount by which the slope is less than unity indicates the extent to which brittle fracture is occurring and a more complicated type of yield takes place described as a 'Mohr body'. Sucrose granules and Acetaminophen granules with slopes in the region of 0.3—0.4 are examples of this type of behaviour (*see also* Section 11.2.3).

The ratio of the lengths *OA* and *AB* give an indication of the amount of elastic as opposed to plastic deformation or brittle fracture which is taking place.

The decompression can be explained in a similar way but the most interesting factor is the degree of hysteresis and the magnitude of the residual stress *OD*. High values for *OD* such as are obtained for sodium chloride imply a strong compact although considerable force may be needed to eject it from the die. A lower value of *OD* as obtained with Acetaminophen indicates that the compact has failed or the deformation has almost completely reversed.

The area enclosed by the hysteresis loop is a measure of the unrecoverable mechanical energy. For aspirin this is small owing to the deformation being largely elastic, whereas for sodium chloride, where plastic deformation and brittle fracture are greater, the area is larger.

Acetaminophen is known to laminate when compressed alone but the lamination can be overcome by the addition of poly(vinylpyrrolidone) (PVP).

Leigh examined the pressure cycles for Acetaminophen with and without PVP. His results show a marked change in the slope of *CD* and the residual stress *OD* when 3 per cent was used.

From this work and that of others it has been shown that a good indication of the likely tabletting properties of a mixture or granulation can be obtained by using an instrumented press.

10.5.2   Tablet machine control

10.5.2.1   Development of a control system

Knoechel used his instrumented machine on production work and provided the machine operator with an oscilloscope which displayed the punch forces. The variation in die fill from one die to the next, i.e., the difference in tablet weights, causes compression forces to vary. By observing the average height of the peaks of the trace the operator found that he could tell if the mean tablet weight had increased or decreased from the original setting. It therefore assisted him in controlling the machine and resulted in improved product weight consistency. He could also tell from the variation in the heights of the individual peaks the variation of the individual tablet weights, from the mean. As this is usually governed more by the quality of the granules and the mechanical condition of the tablet machine, there was little he could do to control it but it did afford a batch-to-batch comparison of granule quality.

It was obviously a very short step from the operator controlling the machine by observing the oscilloscope to automatic machine control and this was patented by Knoechel *et al.*[22].

This system of machine control has been criticized by Deer and others[23,24] on the grounds that the relationship between tablet weight and compressional force does not hold good on a rotary machine because of such factors as variation in punch lengths, non-uniformity of die bores and asymmetry of pressure rolls. This criticism is valid if it is attempted to use the system to monitor or control individual tablet weights. Control of individual tablet weights by means of measuring the force used to compact them is impossible because once the tablet is compressed and the force measured it is too late to adjust the weight.

However, the factors cited by Deer are not infinitely variable. The rate of wear on punches and dies is very small, also punches do not move from one die to another and eccentricity of pressure rolls is normally extremely small; therefore, the error in the value of the compressional force is virtually constant for each particular station on a rotary tablet machine. Consequently the mean compressional force is a function of mean tablet weight and, as it is only possible to control the mean tablet weight and not the individual tablet weights, the system can and does work satisfactorily. The criticism does however serve to emphasize that if this type of tablet weight control system is to be used the accuracy of control which will be obtained will be a function of the accuracy of the tooling used and the mechanical condition of the machine.

10.5.2.2   *Commercial instrumentation and control of tablet machines*

So far all the instrumentation discussed has been concerned with measuring punch forces but the variable to be controlled is the tablet weight. The earliest commercially available tablet machine control system was introduced by Hoffliger and Karg[25]. This consisted of a unit which sampled the tablets from a group of machines in turn; weighed the individual tablets; gave a print-out of the tablet weights; calculated the mean weight and, if this differed from the required weight, initiated action to adjust the weight control mechanism of the machine in order to restore the weight to the required value. The main disadvantage of this type of unit lay in the fact that the sampling frequency was low and considerable quantities of tablets could be produced between checks, although this was no different from the situation when the operator was controlling the machine.

The need was for a system which could check every tablet and the next development came from work by Schupbach *et al.*[26] It was marketed as the Gretag T.P.G. 400. This consisted of a system which measured the compressional force for each tablet by means of strain gauges mounted on the tie bar. The signals were compared with preset limits and any individual tablets which gave signals outside the limits were rejected. This system failed owing to the fact that the forces measured for individual tablets on a rotary machine do not directly relate to the tablet weights, as already explained.

Two systems manufactured under licence from the Knoechel patents are available. The Thomas Tablet Sentinel (T.T.S.)[27] is available in North America and the Manesty Sentinel[28] in the rest of the world. The T.T.S. measures signals from successive stations and compares the average signal with preset limits. It adjusts the fill mechanism on the tablet machine to maintain the signal within the limits. The operation of the Manesty Sentinel is slightly different in that instead of averaging the signals, it operates with limits set extremely close to the required mean value and adjustments are made according to the proportion of the signals exceeding each limit. This has the advantage of holding the mean tablet weight as close as possible to the required value rather than between set limits.

Both these units have been extended to enable tablets giving signals outside secondary limits to be rejected and if the frequency of rejects exceeds certain limits the press is stopped. Other features included are tablet counts; facilities to batch preset quantities of tablets and indication or print out of the station number from which a signal exceeding the secondary (reject) level originates.

A modern version of the Hoffliger and Karg check weigher has been introduced by C. I. Electronics[29] known as the Watchdog Tablet Press

Control System. This has an advantage over the Hoffliger and Karg system in that a balance head is associated with every tablet outlet chute giving a very much higher frequency of weight measurement.

The output signals from one or more balance leads can be fed to a small computer which can provide and record statistical analysis of the tablet weights from each batch of tablets produced. The system is equally suited for use with capsule filling machines.

References

1. BRAKE, E. F., M.S. Thesis, Purdue University, West Lafayette, Indiana (1951)
2. HIGUCHI, T., NELSON, E. and BUSSE, L. W., *J. Am. pharm. Ass. Sci. Ed.* **43**, 344 (1954)
3. SHOTTON, E. and GANDERTON, D., *J. Pharm. Pharmac.* **12**, Suppl. 87т (1960)
4. SHOTTON, E. and GANDERTON, D., *J. Pharm. Pharmac.* **13**, 144т (1961)
5. NELSON, E., NAQVI, S. M., BUSSE, L. W. and HIGUCHI, T., *J. Am. pharm Ass. Sci. Ed.* **44**, 494 (1955)
6. WINDHEUSER, J. J., MISR, J., ERIKSEN, S. P. and HIGUCHI, T., *J. pharm. Sci.* **52**, 767 (1963)
7. LEIGH, S., CARLESS, J. E. and BURT, B. W., *J. pharm. Sci.* **56**, 888 (1967)
8. RIDGWAY, K., *J. Pharm. Pharmac.* **18**, 176s (1966)
9. SIXSMITH, D., Ph.D. Thesis, University of Bradford (1969)
10. SHOTTON, E., DEER, J. J. and GANDERTON, D., *J. Pharm. Pharmac.* **15**, 106т (1963)
11. FÜHRER, C., *Pharm. Ind.* **26**, 674 (1963)
12. KNOECHEL, E. L., SPERRY, C. C., ROSS, H. E. and LINTNER, C. J., *J. Pharm. Sci.* **56**, 109 (1967)
13. DEER, J. J., RIDGWAY, K., ROSSER, P. H. and SHOTTON, E., *J. Pharm. Pharmac.* **20**, 162s (1968)
14. WRAY, P. E., VINCENT, J. G., MOLLER, S. W. and JACKSON, G. J., Paper presented to American Philosophical Association Meeting, Dallas, Texas (1966)
15. KISTLER, A. G., Zurich, Switzerland
16. MARSHALL, K., Ph.D. Thesis, University of Bradford (1970)
17. RIDGWAY, K., DEER, J. J., LAZAROU, C. and FINLAY, P. L., *J. Pharm. Pharmac.* **23**, Suppl. 214s (1971)
18. RIDGWAY, K., DEER, J. J., FINLAY, P. L. and LAZAROU, C., *J. Pharm. Pharmac.* **24**, 265 (1972)
19. LINDBERG, N. O., *Acta Pharm. Suecia* **9**, 135 (1972)
20. STRICKLAND, W. A., NELSON, E., BUSSE, L. W. and HIGUCHI, T., *J. Am. pharm. Ass. Sci. Ed.* **45**, 51 (1956)
21. LONG, W. M., *Powder Metall.* **6**, 73 (1960)
22. U.S. Pat. 3 255 716; Br. Pat. 1 152 061
23. DEER, J. J., *Chemist Drugg.* 660 (1975)
24. LING, W. C., *J. pharm. Sci.* **62**, 2007 (1973)
25. Hoffliger and Karg, Waiblingen/Rems, W. Germany
26. SCHUPBACH, O., KESTENHOLTZ, F. and FURTWANGLER, R., *Pharm. Ind.* **30**, 743 (1968)
27. Thompson Engineering Inc., Hoffmann Estates, Chicago, Ill., U.S.A.
28. Manesty Machines Limited, Speke, Liverpool
29. C. I. Electronics, Churchfields, Salisbury, Wiltshire

CHAPTER 11

Compaction of ceramics

H. M. MacLeod

United Kingdom Atomic Energy Authority, Windscale Nuclear Power Development Laboratories, Seascale, Cumbria

11.1 Introduction and scope

The densification of powders is an operation common to a wide variety of disciplines and several compaction techniques have been developed to suit particular powder properties or product requirements. In every case force or energy is applied to a particulate system such that porosity is reduced and a densified product is obtained. Vibrocompaction employs gravitational acceleration, slip-casting employs surface tension forces, and for ferrous metals magnetic compaction forces have been employed. The most widely recognized compaction technique, however, is that of compressing the powder between loaded punches in a die of appropriate cross section.

The interdisciplinary nature of die-compaction is emphasized in the literature. It is rare to find a review of ceramic powder compaction that does not include references to metallurgical or pharmaceutical journals and a glance at the references at the end of this chapter will confirm this. Differences which arise between disciplines can be wholly ascribed to differences in powder properties and process requirements.

In common with metal powder compacts and unlike pharmaceutical tablets, the ceramic compact obtained from the die is very rarely the finished product. Usually furnace treatments follow at temperatures up to 800 °C to remove residual binders and lubricants and at temperatures up to 2000 °C to remove residual porosity and produce a dense homogeneous product. Because of this, powder properties often cannot be optimized to suit the compaction process and later process requirements may impose restraints on the choice of binder and granulation procedure adopted.

The purpose of this chapter is to survey the process of die-compaction in relation to ceramic powders and to review the material and process variables which affect its application and control.

11.2 Pressure transmission through powders

11.2.1 External force measurements

Unlike a fluid which transmits changes in pressure at a point uniformly throughout its mass, a powder subjected to an external force exhibits a characteristic resistance to relative movement between particles. When a loose powder bed of infinite extent is subjected to a static pressure the transmitted pressure decreases uniformly as distance from the pressure source increases[1]. If the powder is confined in a die, however, the pressure is no longer transmitted uniformly but conforms to a stress pattern imposed by die wall restraint.

The principal source of stress rearrangement in die compaction is die wall friction which is dependent on powder properties, state of compaction and interfacial conditions between powder and wall. As a result of an externally applied force F_A several reaction forces are induced in a powder

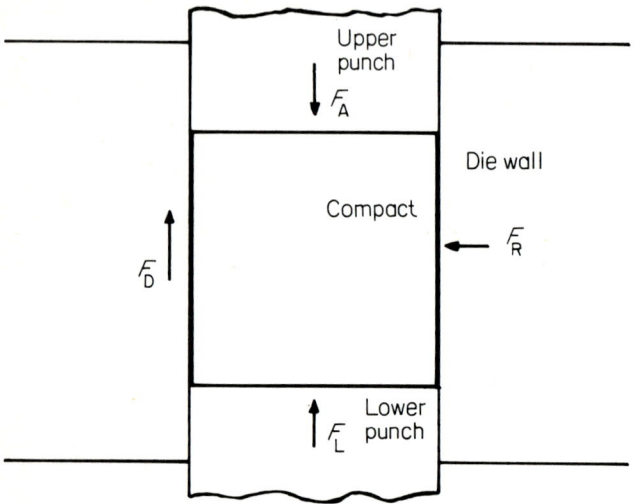

Figure 11.1.   Forces operating on a powder under compression

mass confined in a die. *Figure 11.1* shows schematically the forces induced in a cylindrical compact being compressed from one end. If the material being compressed were a perfect fluid, force would be transmitted iso-statically and

$$F_A = F_L = F_R$$

In a particulate system, however, friction is generated at the material–die wall interface as a result of induced radial force F_R. Radial force F_R is a function of the Poisson ratio, v, of the material[2] and is related to force F_A by the expression

$$F_R = \frac{v}{1-v} \cdot F_A$$

A proportion of the applied force is transmitted to the die body and appears as die wall reaction force, F_D. The remainder of the applied force is transmitted through the powder to the lower punch where it is measured as transmitted force, F_L. The force balance may be expressed as

$$F_A = F_L + F_D$$

If F_D is considered as losses due to friction and F_L as force available for densification[3], then F_D incorporates losses due both to die wall friction, f_w and interparticle friction, f_i. Interparticle friction would be dominant at an early particle rearrangement stage of densification and die wall friction for the rest of the process. Each component may be expressed in terms of radial force, F_R. Where μ_w is the coefficient of die wall friction and μ_i is the coefficient of interparticle friction:

$$F_D = (f_w + f_i)$$

$$= F_R \mu_w + F_R \mu_i$$

$$= F_R(\mu_w + \mu_i)$$

and the force balance may be rewritten

$$F_A = F_L + F_R(\mu_w + \mu_i)$$

Several authors used a static force balance technique to investigate frictional forces during compaction. Duwez and Zwell[4] plotted the pressure ratio P_L/P_A against compact H/D ratio and obtained a dimensionless curve of the form

$$\ln\left(\frac{P_A}{P_L}\right) = K_1\left(\frac{H}{D}\right)$$

Sheinharz and coworkers[5] confirmed that the ratio F_L/F_A was a constant for a powder depending on compact mass, die geometry and lubrication conditions. Later workers[6] used the ratio F_L/F_A, termed the *R*-value, as a measure of lubricant efficiency.

The extent of die wall and interparticle frictional forces can be estimated, using these simple relationships, from external measurements but information obtained by a static balance of forces in this way represents values averaged over the compact. Since F_R and μ_w vary over the compact length, i.e., distance from the applied force, the information gained is of limited value. In particular, since forces are resolved in terms of their axial and radial components, no information is gained about stress distribution and correlation of force measurements with compact density structure is impossible.

11.2.2   Studies of internal stress distribution

The basis for much of the work elucidating stress distributions in powder compacts was observation of the progressive rearrangement of powder in the die as applied force was increased. The methods adopted ranged from the forming of identifiable layers of powder in the die by a tamping technique before compaction[7] to the incorporation of materials opaque to X-rays whose deformation or measurement during compaction could be continuously followed by radiography[8,9]. Train[10] examined in detail the distribution of pressure within a compact of magnesium carbonate by placing manganin wire resistance gauges at uniformly spaced stations in the compact before compaction. Pressure reactions measured at a range of applied pressures corresponded closely to density levels determined by subsequent accurate sectioning of the compacts on a lathe (*Figure 11.2*). More recently beta-autoradiography has been used to determine density distri-

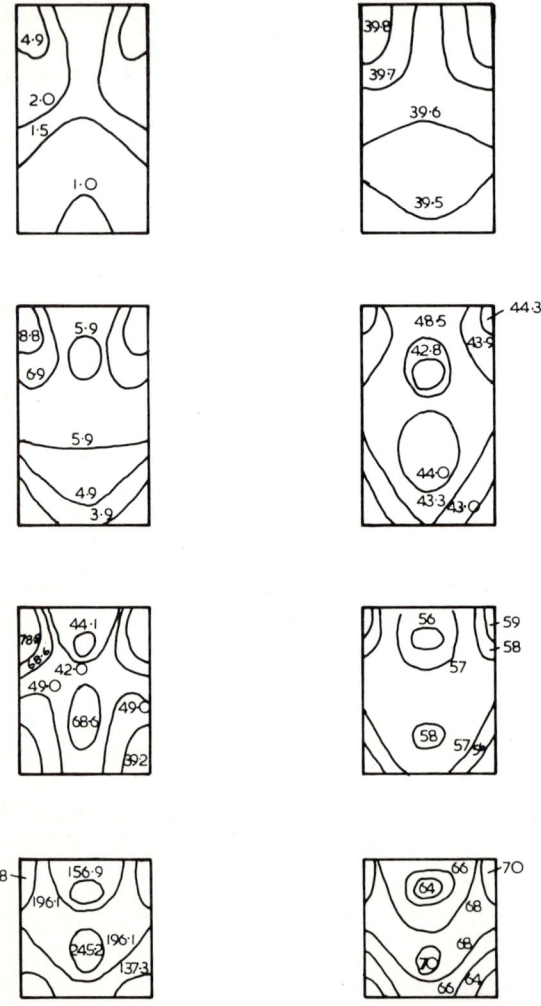

Figure 11.2. Relationship between observed pressure distribution (on left) and relative density distribution (on right) in compacts. From top to bottom, P_A is 2.79, 8.83, 65.8, and 200.0 MN m^{-2} respectively (after Train[10])

butions in uranium dioxide compacts produced under varying conditions of feed size, L/D ratio, applied force and lubrication condition[11]. Improved resolution of density distribution patterns has been obtained by the application of computer analysis to gamma and X-radiographs of 1—2 mm thick slices of lightly sintered electroceramic specimens, enabling over 2400 point density measurements to be made over the sample surface[12]. All measurements revealed that for single ended compaction the density was at a maximum at the outer circumference of the compact adjacent to the moving punch and at a minimum at the outer circumference adjacent to the stationary punch. More significantly, they showed that, while mean density decreased with increasing distance from the moving punch, a dense core of material existed on the compact axis near the stationary punch which was in many cases denser than material directly between it and the moving punch.

Seelig and Wulf[8] attributed the phenomenon of non-uniform density distribution to die wall friction and demonstrated that lubrication of the die wall was effective in reducing it. Train[10] supported their suggestion and proposed that capping and lamination faults in compacts were caused by non-uniform stress patterns generated. He proposed a simple explanation for the observed patterns of density distribution based on the development of high density wedges of material owing to high shearing forces at the die wall adjacent to the moving punch. *Figure 11.3* represents the forces developed in the compact. At equilibrium conditions the axial force is supported by the punch in direction a and by the powder mass in direction b. The radial force component is supported by the die wall in direction c and by the powder mass in direction d. The resultant force acting on the powder mass is represented by k and because of axial symmetry in the compact the pressure front may be considered as a conical surface with focal point B. Area B will show enhanced pressure and density values compared with its

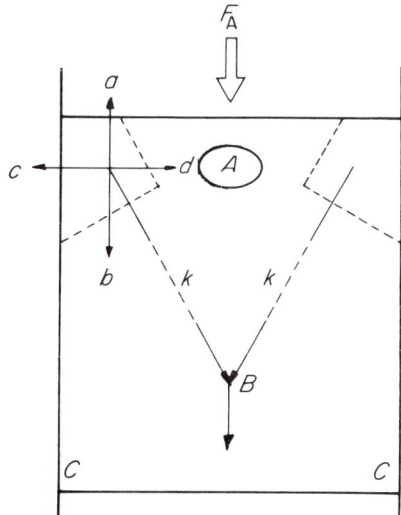

Figure 11.3. Development of pressure pattern (after Train[10])

surroundings because of effective pressure concentration. The wedge shaped area adjacent to the moving punch, bounded by dotted lines, will be subject to high shearing forces and will be highly densified. The central area *A* on the axis will be subjected to negligible shearing forces and will be partially protected from axial normal pressures by the vaulting effect of the high density wedges. Area *C* adjacent to the stationary punch will undergo no

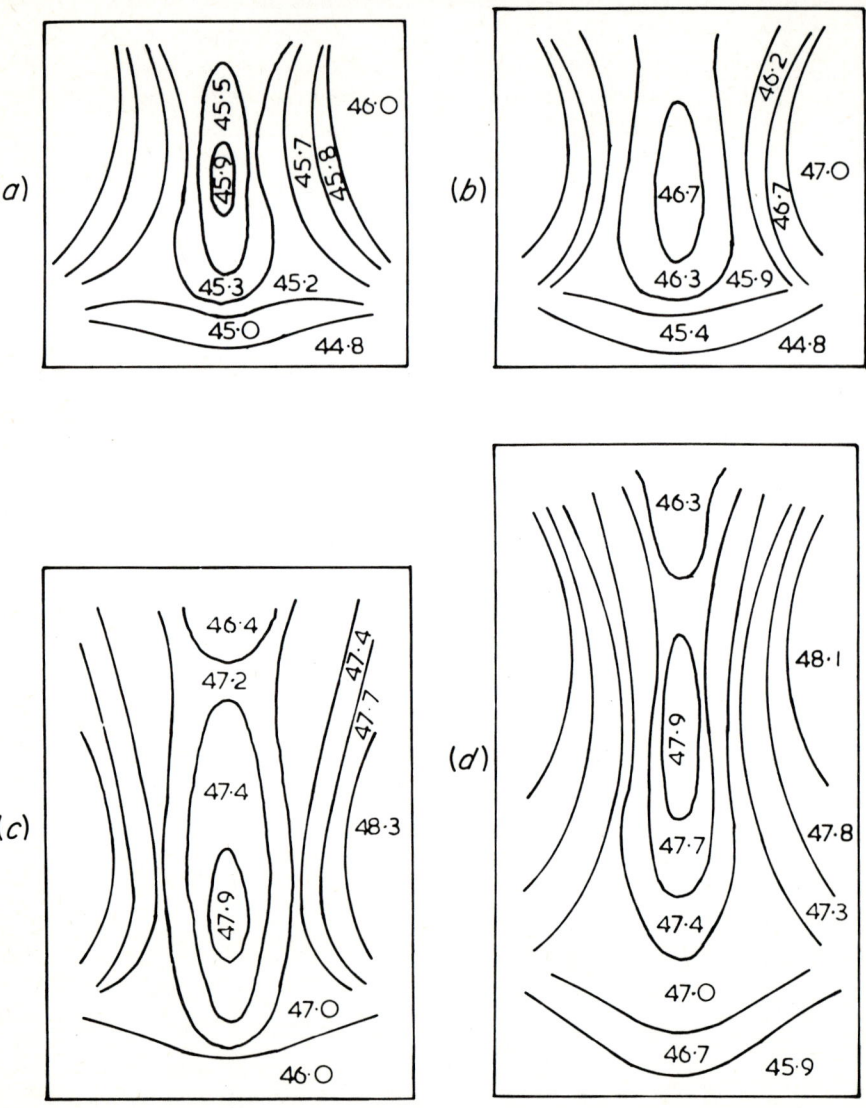

Figure 11.4.   *Density distribution patterns for uranium dioxide compacts of varying H/D ratio. A die wall lubricant was employed*

Pattern	(a)	(b)	(c)	(d)
P_A/MN m^{-2}	79.38	75.83	79.38	79.38
H/D	0.910	0.963	1.397	1.746

movement relative to the die wall so shearing forces will be absent. Density will be low since consolidation will depend on transmitted axial forces only.

The density patterns observed in UO_2 compacts by MacLeod and Marshall[11] were more complex than those reported by Train for magnesium carbonate (*Figures 11.4—11.7*). They identified three distinct density regions comprising an inner axial region and an outer peripheral

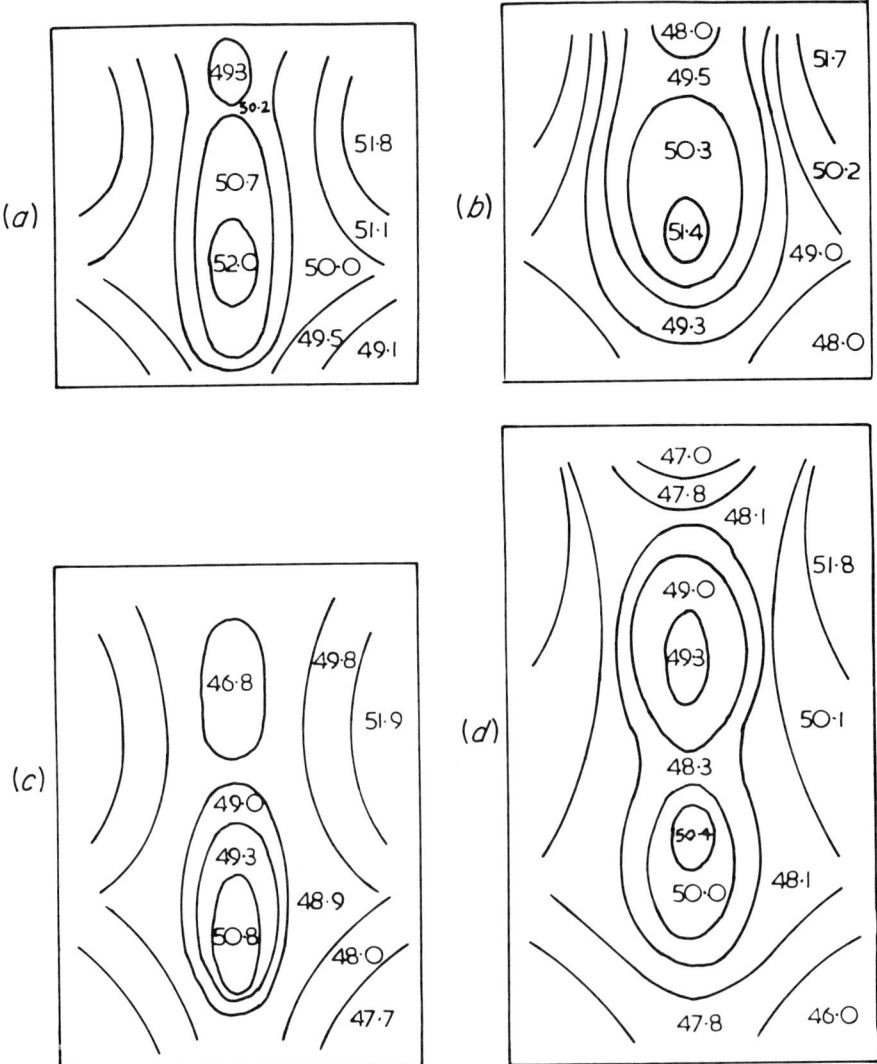

Figure 11.5.   Density distribution patterns for uranium dioxide compacts of varying H/D ratio, without any lubricant

Pattern	(a)	(b)	(c)	(d)
$P_A/MN\ m^{-2}$	160.77	157.37	160.77	157.37
H/D	0.974	0.998	1.341	1.695

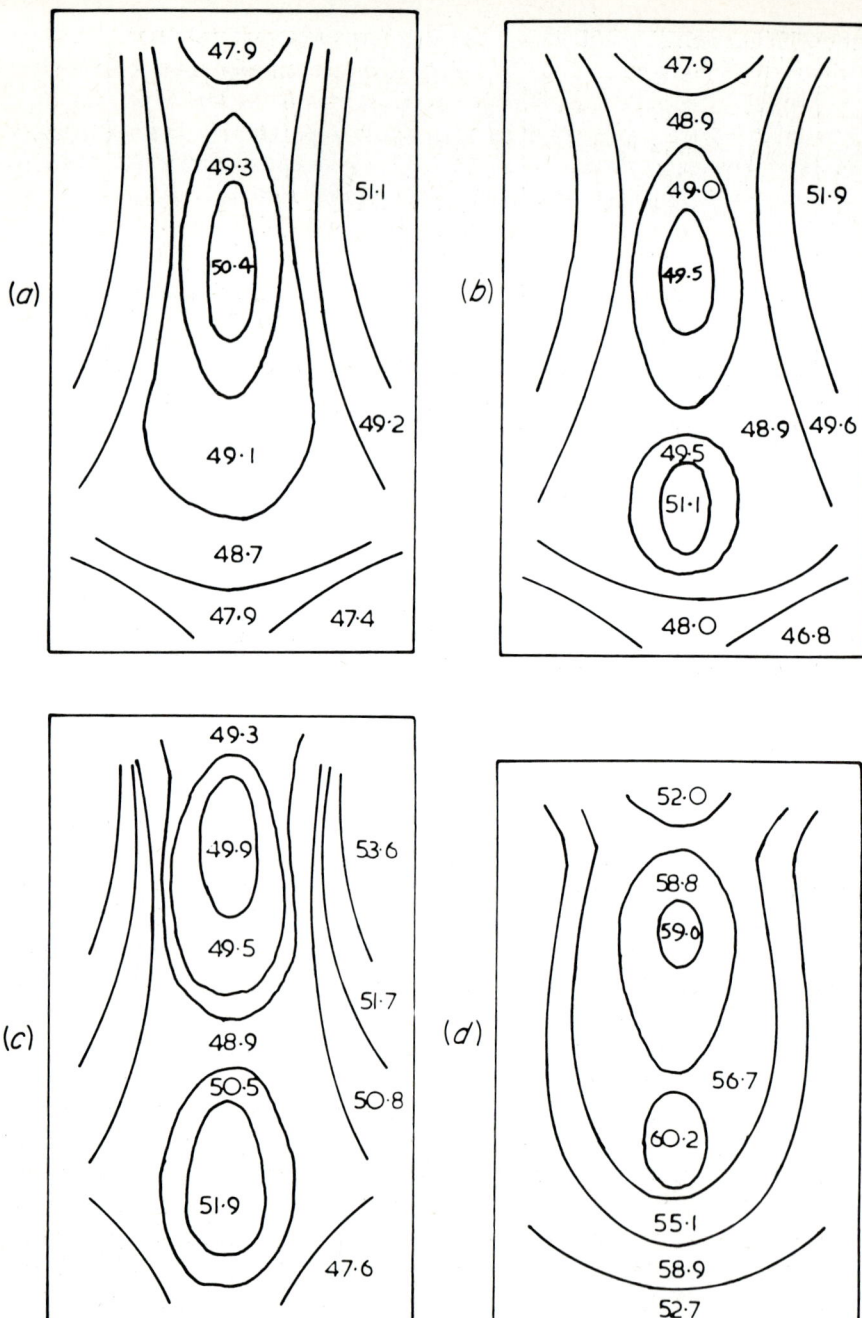

Figure 11.6. Density distribution patterns for uranium dioxide compacts prepared at a range of compaction pressures. A die wall lubricant was employed

Pattern	(a)	(b)	(c)	(d)
$P_A/MN\ m^{-2}$	76.91	157.37	237.68	500
H/D	1.688	1.675	1.623	1.489

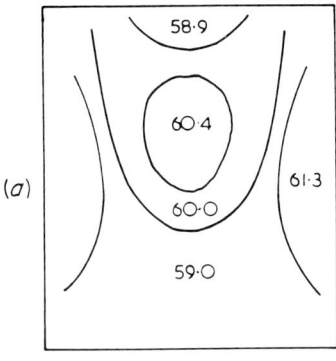
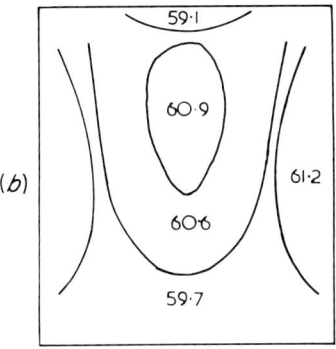

Figure 11.7.   Density distribution patterns for uranium dioxide compacts containing combined internal and die wall lubricant

Pattern	(*a*)	(*b*)
$P_A/MN\ m^{-2}$	401.82	398.33
H/D	1.128	1.126

region, each of high density, separated by an annulus of lower density. Changes in the axial region were closely related to changes in applied axial pressure whilst changes at the periphery were more dependent on lubrication conditions.

A common feature of density distributions deduced by Train[10] using magnesium carbonate, Unckel[13] using metal powders and MacLeod and Marshall[11] using uranium dioxide was the presence of a discontinuity in the peripheral density pattern. MacLeod observed that the discontinuity occurred at a distance from the stationary punch dependent on lubrication conditions and compact H/D ratio, but independent of applied pressure. For ceramic powders the distance varied from $0.2H$ in the presence of die wall lubrication to $0.3H$ in the unlubricated state. For magnesium carbonate the corresponding distances were $0.5H$ in the lubricated state and $0.67H$ for the unlubricated condition. Unckel found the discontinuity in metal compacts at a distance of $0.67H$ from the stationary punch. He deduced that the pressure front was uniformly distributed across the compact diameter at this point. Between this level and the moving punch the parabolae representing pressure isobars were concave upwards, gradually flattening as the level $0.67H$ was approached. Between $0.67H$ and the stationary punch the pressure isobars became concave downwards. MacLeod interpreted the discontinuity as the limit of material movement at the die wall and explained the different positions in terms of relative volume change. He showed that the discontinuity at the periphery coincided with the high density zone observed in the axial density pattern and proposed an alternative sequence for the development of density patterns in compacts.

11.2.3   Radial pressure effects

Radial force F_R is directly proportional to applied axial force F_A. Where η is the radial force coefficient:

$$F_R = \eta F_A$$

Since F_A decreases exponentially with distance from the moving punch one would expect F_R to decrease similarly at the die wall. Several workers demonstrated that this was the case[5,13–15]. Goetzel[16] considered that the proportion of applied force F_A transmitted as normal radial force F_R was a function of the Poisson ratio of the material being compacted but did not produce evidence to support his suggestion. Long[2] used a split die assembly in his studies of radial force and measured the force necessary to keep the two halves of the die body together during a complete compaction cycle. He extended Goetzel's suggestion that the Poisson ratio, v, defined the extent of radial force and proposed a simple theory to explain variations observed in radial pressure cycles. The theory predicted a characteristic cycle whose form was determined by the failure properties of the material (*Figure 11.8*).

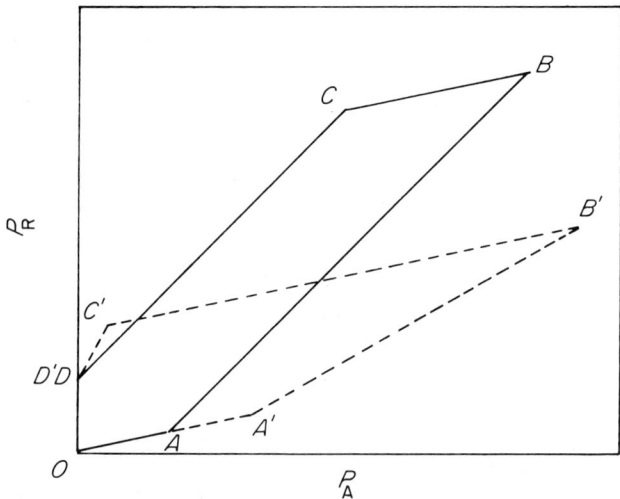

Figure 11.8.   Compaction profiles (after Long[2]); P_A and P_R are axial and radial pressures, respectively

Line $OA(OA')$ represents perfectly elastic behaviour of a solid isotropic plug of material such that if the axial stress were released the decay of radial pressure to zero would follow this line. The slope of the line is related to the Poisson ratio of the material by the relationship

$$P_R = \frac{v}{1 - v} P_A$$

When the elastic limit $A(A')$ is exceeded, the material yields and the slope changes owing to changing friction conditions in the plug. For a body with a constant yield strength in shear the difference between radial stress σ_R and axial stress σ_A becomes constant:

$$\sigma_A - \sigma_R = 2S$$

where S is the shear strength of the material. The slope of AB will therefore be unity. For a Mohr body the yield stress is a function of the normal stress on the plane of shear[17]. At A' the shear stress, τ, is defined as

$$\tau = S_0 + \mu_1 \sigma_n$$

where

σ_n is the normal stress on the plane of shear,
S_0 is the yield stress in pure shear,
μ_1 is the coefficient of internal friction.

In this case

$$\sigma_n = \frac{\sigma_A + \sigma_R}{2}$$

$$\tau = \frac{\sigma_A - \sigma_R}{2}$$

So we substitute to obtain

$$\sigma_A - \sigma_R = 2S_0 + \mu_1(\sigma_A + \sigma_R)$$

or

$$\sigma_R = \frac{\sigma_A(1 - \mu_1) - 2S_0}{1 + \mu_1}$$

and the slope of σ_R *versus* σ_A will be

$$\frac{1 - \mu_1}{1 + \mu_1}$$

Points B and B' represent the maximum applied axial pressure. When pressure is released the plug will recover elastically and the lines BC and $B'C'$ represent the extent of this recovery. The slopes will be parallel to BA and OA' respectively and equal to $v/(1 - v)$.

At points C and C' the induced radial stress, σ_R, exceeds the axial stress, σ_A, and the material again yields. In the case of a material with a constant yield stress in shear the slope CD will be parallel to AB and equal to unity. In the case of a Mohr body the slope $C'D'$ is the reciprocal of slope $A'B'$ since, for a stress reversal:

$$\sigma_R - \sigma_A = 2S_0 + \mu_1(\sigma_A + \sigma_R)$$

Hence

$$\sigma_R = \frac{\sigma_A(1 + \mu_1) + 2S_0}{1 - \mu_3}$$

The radial pressure does not return to zero on release of axial pressure when plastic deformation has occurred. For a body with a constant yield stress in shear the maximum radial pressure, *OD*, will be $2S$ but the observed value will depend on the applied axial pressure, P_A.

By putting σ_A equal to zero in the equation the maximum residual radial pressure in the case of a Mohr body is shown to be $2S_0/(1 - \mu_3)$ but the observed pressure will again depend on P_A, the applied axial pressure. Compaction of powders whose pressure cycles were typical of each of those model materials showed behaviour in good general agreement with Long's hypothesis.

Leigh and coworkers[18] used Long's technique to relate the compaction profiles of several pharmaceutical materials to their ability to form 'good' or 'bad' compacts in terms of capping and lamination faults. They found that powders which were prone to compaction faults exhibited a compaction profile typical of a Mohr body, while materials with good compaction properties showed a compaction profile typical of that of a material having a constant yield stress in shear. Addition of a binder or internal lubricant, in this case PVP, altered the behaviour of a Mohr body towards that of material with a constant yield stress in shear, at the same time minimizing capping and lamination faults. Leigh[19] proposed that the addition of the binder provided the necessary amount of plasticity to accommodate residual stresses. Marshall[20] has discussed the practical application of radial pressure cycles, which he termed 'compaction profiles', in terms of material and process variables.

The relationship between compaction behaviour and material failure properties was examined for metal powders by Suh[21] and by Schwartz and Holland[22]. Suh pointed out that the Mohr–Coulomb equation implied dilation of a material under shear stress and could therefore apply only under conditions of very low normal stress. He proposed a new failure criterion which provided for either dilation or densification, according to the state of compaction, and which took into account changes in material cohesion, C, owing to material work hardening and changes in angle of internal friction, ϕ, as densification proceeded. The yield criterion was of the form:

$$[(\bar{\sigma}_h + k)^2 + \tfrac{2}{3}\tau_{max}]^{1/2} - q\left[\cos\frac{\pi}{2} \cdot \frac{\theta}{\phi}\right]^n = 0$$

where

$\bar{\sigma}_h$ = mean hydrostatic stress
τ_{max} = maximum shear stress
$\tan\theta$ = $\tfrac{2}{3}\tau_{max}(\bar{\sigma}_h + k)$

ϕ = angle of internal friction
k, q, n = positive constants related to material properties

Schwartz and Holland[22] evaluated the proposed yield equation for the compaction of iron powder and showed good agreement with experimental observations. They pointed out that two limiting conditions of compaction could be defined. For an unconsolidated powder the Mohr–Coulomb equation applied and the shear stress necessary to cause failure was given by

$$\tau = C + \sigma_n \tan \phi$$

At complete densification the conditions for failure depended on the material itself. The shear strength τ equalled the cohesive strength, C, and the angle of internal friction, ϕ, became zero. The condition was defined by the Tresca yield criterion:

$$\tau = C$$

Between these limiting conditions there was a continuous range of values of C and ϕ.

Other authors[23,24] have critically examined failure equations and have proposed modifications. Mathematical analysis of stress and strain distributions in compressed powders in the limiting equilibrium states specified by the Mohr criterion have been reported[25,26] and the analyses extended to determine the resultant density distributions[27]. A fuller treatment of powder densification related to failure properties is presented elsewhere in this publication.

11.3 Pressure–volume relationships

Several authors have examined the relationship between applied pressure P_A and powder volume and have identified sequential stages in the densification process[8,28–31]. These may be summarized as follows:

1. Particle rearrangement leading to closer packing.
2. Formation of arches and vaults, protecting voids and capable of supporting the applied pressure.
3. A multifarious process of deformation, cold welding, fragmentation and percolation resulting in higher density.
4. Bulk compression of the material itself.

A schematic model for the consolidation process is shown in *Figure 11.9*. The slope and relative extent of each stage will depend on material properties and process conditions.

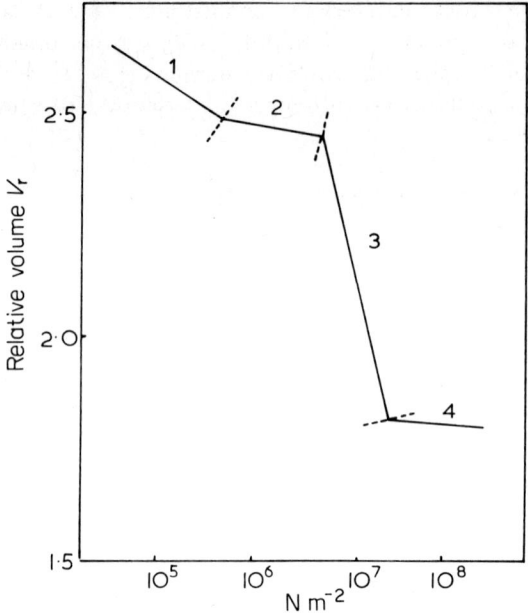

Figure 11.9.   Stages in the compaction process (for 1–4 see text)

11.3.1   Stage I: particle rearrangement

When a powder is poured into a container the bulk density it adopts depends on several variables. In terms of powder or granule properties the most important variables are particle size and size distribution, particle shape and surface properties. Process variables governing the poured density of a powder are rate of deposition and the relationship between container diameter, D, and particle diameter, d_1. When a poured bed is subjected to vibration or pressure, the particles move relative to each other to improve their packing arrangements. Ultimately a condition is reached when no further densification is possible without particle deformation. The bulk density at this point is termed the tapped density and is a fundamental property of the powder.

Several methods of elucidating gravitational packing of powders have been reported. Graton and Fraser[32] used a monosized spherical model to examine ordered geometrical packing and interpreted their results in terms of porosity and coordination numbers using classical packing arrangements. Mathematical treatments of this kind have been critically reviewed by Gray[33] and by James[3]. In general such treatments assume an 'infinite bed' condition where wall effects are absent. In practice the ability of a monosize spherical model to adopt an ordered packing arrangement is limited by the container wall. Recent investigations have emphasized the importance of wall effects. McGeary[34] examined a range of D/d_1 ratios using a 'whole number' geometrical model and demonstrated wide variations in packing fraction at D/d_1 less than 10 (*Figure 11.10*). Further differentiation of the packing curve under 'intermediate number' conditions[35] shows further variations over extremely small ranges of D/d_1 values (*Figure 11.11*).

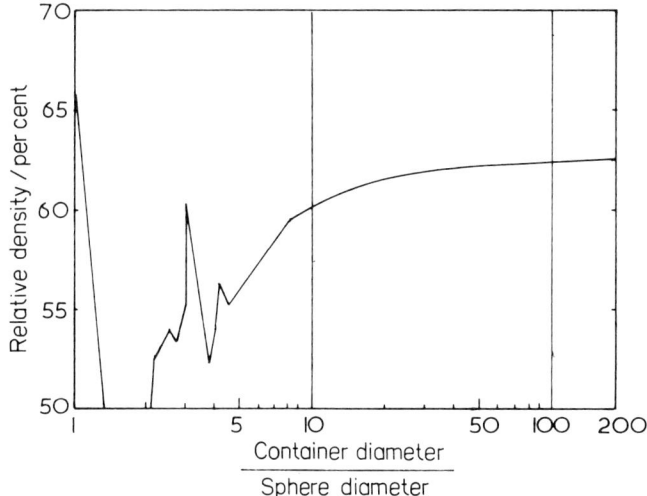

Figure 11.10. Packing of monosize spheres (after McGeary[34])

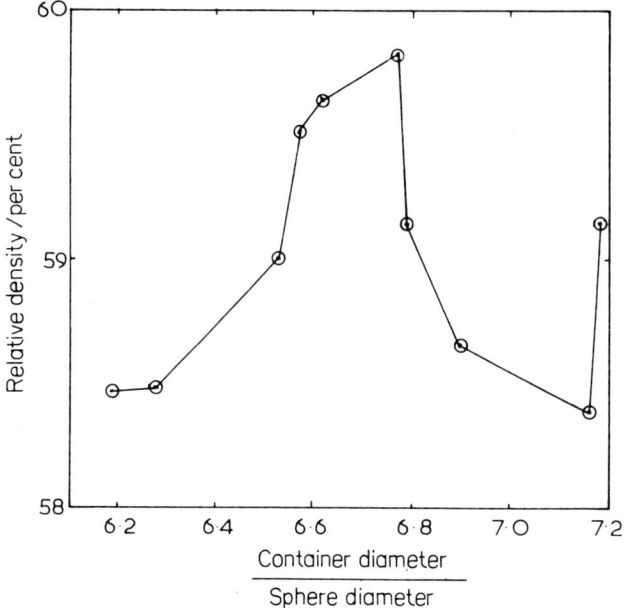

Figure 11.11. Packing of monosize spheres (after MacLeod[35])

The packing fraction of a monosized spherical powder at tapped density is approximately 0.60, i.e., the bed voidage is 0.40. If a secondary spherical powder component with diameter d_2 equal to or less than $(1/7)d_1$ is introduced into the void spaces between the primary particles it will fill a further 0.60 of the voidage to give an overall packing fraction of $0.60 + 0.24 = 0.84$. The process may be repeated with ternary and quaternary components to give further density increases to packing fractions of 0.936 and 0.974 respectively. For real powders with a continuous size distribution such packing efficiency is not possible but clearly some form of secondary or ternary packing can operate providing a diameter range of $50:1$ exists in the powder (*see also* Section 2.1.3).

Bernal[36] pointed out that the packing of real powders is a random process and used the technique of 'statistical geometry' to describe his investigations into the geometrical restraints of random powder packing. Similar statistical methods have been used by Blum and Wilhelm[37], Beresford[38], and Eastwood and coworkers[39] to derive equations predicting the porosity of packed beds.

Hausner[40] investigated the effects of interparticle friction on packing behaviour and used the ratio of tapped density ρ_t over poured density ρ_p, which he termed the *friction ratio*, to define material packing. The friction ratio was later used to classify powders by Gray and Beddow[41] who showed good correlation between it and the angle of repose of poured heaps of powder. In terms of particle surface properties, however, the reciprocal value is more informative, expressing initial packing behaviour as a function of tapped density. For smooth spheres the ratio then has a value close to unity and any depature from surface smoothness or sphericity

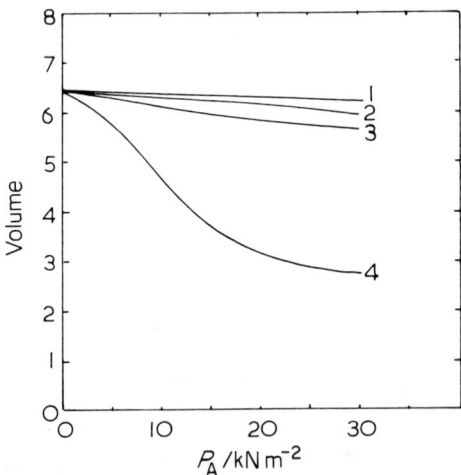

Figure 11.12.   Effect of particle properties on the packing of copper powder (after Kostelnik, Kludt and Beddow[43]): 1, Atomized copper; 2, reduced copper; 3, electrolyte copper; 4, copper flake

results in a decrease in its value. The ratio of poured density ρ_p over tapped density ρ_t has been used by MacLeod[42] to characterize the surface properties and frictional behaviour of monosize ceramic granules.

The relationship between applied pressure and volume change during densification of a powder bed from poured density ρ_p to tapped density ρ_t was investigated by Kostelnik and coworkers[43]. They measured punch displacement with pressure for different copper powders and demonstrated the effect of particle shape on particle rearrangement, finding it a more important parameter than particle size in defining the densification process (*Figure 11.12*). They used the friction ratio to define powder frictional behaviour and showed a smooth curve of punch displacement against ρ_t/ρ_p for a range of copper and iron powders (*Figure 11.13*). Their observations are in agreement with those of Duffield and Grootenhuis[44] who observed a 2—4 per cent volume reduction in copper powders when a load was applied, with evidence of particle rearrangement up to 6.9 MN m^{-2} giving rise to a further volume reduction of 1.5 per cent. Bockstiegel[45] considered particle rearrangement impossible at pressures above 24.8 MN m^{-2} for iron powders.

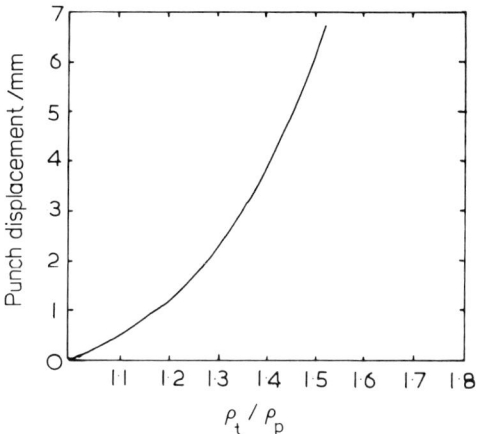

Figure 11.13. Effect of density ratio of copper powders on punch displacement at $P_A = 34.5$ kN m^{-2} (after Kostelnik, Kludt and Beddow[43])

Ceramic powders are typically hard and brittle and therefore more subject to fragmentation during particle rearrangement. Stuijts and Oudemans[46], however, showed that for uniform ceramic powders within the size range 0.3 to 3.0 μm, a punch travelled one third to one half of its compression stroke in the die at pressures less than 0.3 MN m^{-2} compared with a final compaction pressure of 50 MN m^{-2}. They found that particle breakdown at these pressures was slight, an observation supported by results from contemporary studies of ceramic compaction[47].

Some doubt was placed on these conclusions by the results of Hardman and Lilley[48], who found no evidence of particle rearrangement during the compaction of sodium chloride, sucrose and coal powders. It is difficult to accept that rearrangement of particles by slippage does not contribute to the densification process, particularly in the case of ceramic powders where there is ample evidence of the retention of feed particle identity at pressures up to 400 MN m^{-2}.

11.3.2   Stage II: formation of supporting structures

This stage of the consolidation process is characterized by a decrease in the rate of volume reduction with pressure, as particles lock together and resist the applied load. The duration of this stage at a uniform rate of pressure application is dependent on the failure properties of the material. Marshall's[31] results obtained under quasidynamic conditions also demonstrated the dependence of this stage on feed particle size.

11.3.3   Stage III: deformation and recombination of particles

This is the dominant densification stage of the consolidation process and its nature and extent are governed by the failure properties of the material. Metal powders, because of their high ductility, deform plastically when their yield strength is exceeded to occupy void spaces. Brittle ceramic powders are likely to fragment suddenly when their breaking strength is exceeded and occupy voids by rearrangement and percolation of the finer breakdown products. Other materials will exhibit deformation behaviour of a nature bounded by these two extremes.

Lawrence[49] examined the compaction behaviour of four mineral powders of widely differing mechanical behaviour and concluded that a combination of elastic, plastic and fracture deformation mechanisms occurred simultaneously in each. Hardman and Lilley[48] examined the change in specific surface, pore volume and bulk volume with increase in applied pressure, of sodium chloride, sucrose and coal powders. They found that for each material a different deformation mechanism predominated:

1. Sodium chloride densified by plastic deformation of the particles, the process of densification becoming progressively more difficult as work hardening of the particles increased.
2. For sucrose a fragmentation mechanism predominated with pore filling by percolation of fines an important secondary process.
3. Coal densified by a fragmentation mechanism, like sucrose, but the fragments produced were too coarse to permit percolation so that 'bridges' were formed which maintained an open pore structure.

Workers are generally in agreement that 'hard-brittle' materials are more difficult to densify than 'soft-yielding' material[50,51]. A possible explanation is that the filling of voids by fragmentation–percolation is a much less efficient process than filling by plastic deformation.

In view of the possible combinations of densification mechanisms it is not surprising that a large number of equations have been proposed to describe mathematically the consolidation process. Many of these are empirical in nature and relate to a particular material or range of compaction conditions while others attempt to define the complete process of densification. To facilitate evaluation and comparison of various equations, Kawakita and Lüdde[52] and Williams[53] independently expressed them in terms of relative volume and applied pressure. A similar treatment is presented in

Table 11.1. Summary of powder compression equations (*see* refs 54—56)*

No.	Equation	Authors
1	$\ln \dfrac{\rho_t - \rho_i}{\rho_t - \rho_c} = KP_A$	Athy, Shapiro, Heckel, Konopicky, Seelig
2	$\ln \dfrac{\rho_c}{\rho_i}\left(\dfrac{\rho_t - \rho_i}{\rho_t - \rho_c}\right) = KP_A$	Ballhausen
3	$\ln \dfrac{\rho_i}{\rho_t}\left(\dfrac{\rho_t - \rho_i}{\rho_t - \rho_c}\right) = KP_A$	Spencer
4	$\ln \dfrac{\rho_c}{\rho_i} = KP_A^a$	Nishihara, Nutting
5	$\ln \dfrac{\rho_t - \rho_c}{\rho_t} + K\left(\dfrac{\rho_c}{\rho_t - \rho_c}\right)^{1/3} = aP_A$	Murray
6	$\ln \dfrac{\rho_t}{\rho_c}\left(\dfrac{\rho_c - \rho_i}{\rho_t - \rho_i}\right) = \ln Ka - (b + c)P_A$	Cooper and Eaton
7	$\dfrac{\rho_i}{\rho_c} = 1 - KP_A^a$	Umeya
8	$\rho_c = KP_A^a$	Jaky
9	$\rho_c = K(1 - P_A)^a$	Jenike
10	$\rho_c - \rho_i = KP_A^{1/3}$	Smith
11	$\rho_c - \rho_i = KP_A^2$	Shaler
12	$\dfrac{\rho_c - \rho_i}{\rho_c} = \dfrac{K \times aP_A}{1 + aP_A}$	Kawakita
13	$\dfrac{\rho_t}{\rho_c}\left(\dfrac{\rho_c - \rho_i}{\rho_t - \rho_i}\right) = \dfrac{KP_A}{1 + KP_A}$	Aketa
14	$\dfrac{1}{\rho_c} = K - a \ln P_A$	Walker, Bal'shin, Williams, Higuchi, Terzaghi
15	$\rho_c = K + a \ln P_A$	Gurnham
16	$\dfrac{1}{\rho_c} = K - a \ln P_A$	Jones
17	$\dfrac{1}{\rho_c} = K - a \ln (P_A - b)$	Mogami
18	$\dfrac{\rho_c - \rho_i}{\rho_c} = KP_A\rho_i + a\left(\dfrac{P_A}{P_A + b}\right)$	Tanimoto
19	$\dfrac{\rho_c - \rho_i}{\rho_c} = \ln (KP_A + b)$	Rieschel

*ρ_t, net density of powder; ρ_i, initial apparent density of powder; ρ_c, density of powder under applied pressure P_A; K, a, b and c are constants.

Table 11.1. Obvious contradictions arise from comparison of these equations and the theoretical bases of some of them can be questioned. Several critical examinations of compaction equations can be found in the literature[54-56].

Cooper and Eaton[51] found that several sets of published data from the compaction of ceramic powders[57] did not fit proposed compression equations. They extended the theory of a diffusive pore filling mechanism proposed by Dallavalle[58] by developing an equation describing two independent diffusive processes. In the first stage of compaction, large voids of dimensions corresponding to the powder particle size were filled by particle rearrangement. In the second stage, occurring at higher applied pressures, small pores were filled by fragmentation–percolation or by plastic deformation of the powder particles. The probability of void filling was related to applied pressure by an equation of the form

$$\frac{V_i - V_p}{V_i - V_t} = a \exp(-kP_A) + b \exp(-mP_A)$$

where

V_i	= poured volume of powder,
V_p	= powder volume under pressure P_A,
V_t	= powder volume at theoretical density,
a, b, k, m	= constants for the material.

Very close agreement with this equation was demonstrated for the compact behaviour of calcia, magnesia, silica and alumina. Results published by Bruch[50] for a wide variety of powders showed a change of slope in the plot of compact density against applied pressure indicative of a two-stage process of this nature. No attempt was made in the paper, however, to explain the effect.

11.3.4   Stage IV: bulk compression

Stage III finishes when no further porosity reduction is possible by fragmentation–percolation or plastic deformation and further densification involves elastic compression of the material itself. The induced stresses will be isostatic in character since pressure transmission relies on material elasticity. Torre[59] considered isostatic densification in terms of pore closure, taking as his model a hollow sphere subjected to a uniform isostatic pressure and using the compression equation proposed by Konopicky[60]:

$$\ln \frac{\rho_t - \rho_i}{\rho_t - \rho_c} = kP_A$$

Bockstiegel[45] critically examined Torre's argument and found that a threshold pressure, P_{min}, existed below which material could not deform plastically:

$$P_{\text{min}} = \tfrac{2}{3}\sigma_0 \left(\frac{R_o^3 - R_i^3}{R_i^3} \right)$$

where σ_0 was the upper yield stress in tension and R_o and R_i were the outer and inner radii respectively of a hollow sphere with negligible internal pressure. As R_i decreases, P_{min} increases, i.e., large pores are likely to close before small pores as pressure is increased. James[3] pointed out that the equation is derived by assuming that Hooke's law applies whereas when deformation takes place the law becomes invalid.

In practical terms, elimination of porosities beyond a minimum of 10 per cent by either isostatic or die compression would be rendered impossible by the increase in frictional forces at the pressures required. Cooper and Eaton[5] reported porosities of 35 per cent in alumina compacts after compression at 520 MN m^{-2}. The densification process at pressures beyond 700 MN m^{-2} is not well understood but results suggest that pressure–volume relationships differ basically from those observed at lower pressures[61-63].

11.4   Friction and lubrication

Frictional forces arise during die-compaction of powders from two main sources:

1. Between material and die wall.
2. Between particles of material

Interparticle friction is complex in nature and is influenced by the surface characteristics of the feed material, its particle size, shape and forces acting on the particles. Its effect is generally regarded as being restricted to the first stage of the consolidation process since relative movement between particles is restricted as porosity is reduced. For material densifying by a fracture recombination mechanism, however, interparticle movement will continue well into Stage III of the process and interparticle frictional forces will continue to be important. Leigh and coworkers[18] demonstrated the change in compaction characteristics obtained by introducing an internal binder–lubricant into a powder before compaction. Pharmacists add glidant powders to promote a similar reduction in interparticle friction.

The major source of friction in the compaction process is that between material and the die wall. Its level depends on the surface condition of the die, the surface characteristics of the feed material and the area of contact between material and die wall. On a microscopic scale the interface between material and die wall is not smooth but consists of a series of interacting asperities so that the effective area of contact, A_{eff}, is only a small fraction of the total surface area A_t. When a normal load is applied across the interface, stresses induced at asperities cause them to yield, increasing the effective area of contact and thereby supporting the applied load. As radial normal stress increases there will be a corresponding increase in effective area of contact, depending on the yield strength of the material.

Neither effective area of contact, A_{eff}, nor total surface area A_t are conveniently measured in practice and much of the published work relating to die wall friction refers to a third area. This is the geometric surface area of the die bore, A_g, defined by πDH where D and H are compact diameter and length respectively. The use of A_g in deriving frictional relationships poses several problems since the ratio of A_{eff}/A_g varies with material yield properties as well as with radial normal stress.

When a lubricant is introduced, either by premixing with the feed material or by direct application to the die wall, its effect is to reduce die wall friction. At the interface, lubricant will fill depressions in the microscopic surface but there will inevitably be material contact with the die wall at asperities. Part but not all of the radial normal stress will be supported by the lubricant film. The principal components of die wall frictional force will be:

1. Friction between material and die wall.
2. Friction between lubricant and die wall.
3. Friction between material and lubricant.

During compaction the nature of the sliding interface will change as consolidation proceeds and, because of pressure transmission effects, will vary along the length of the compact. Frictional forces may also be studied during the process of ejecting the compact from the die. Since the state of consolidation is unchanged during ejection, measurements of friction made under ejection conditions are effectively isolated from consolidation effects.

Studies of die wall friction have involved two principal techniques

1. Measurement of transmitted force F_L or die wall reaction F_D to estimate frictional losses.
2. Measurement of force required to move the die wall relative to the compact at varying density levels.

The use by Duwez and Zwell[4] of the former technique in investigating frictional effects has already been mentioned. Beddow[64] used the dimensionless ratio P_L/P_A in work relating frictional behaviour to particle shape. Nelson and coworkers[6] termed the ratio F_L/F_A the 'R-value' and adopted this as a universal test of lubricant efficiency. Burr and Donachie[65] showed that, for metal powders, internal lubricant offered no advantage over die wall lubricant in improving pressure–density relationships. Hausner and Sheinharz[66] found that internal lubricants reduced the green density and green strength of metal compacts, observations reinforced by results from other workers[67,68].

Train and Carrington[69] investigated the compaction behaviour of regular arrangements of spherical metal powders. They found the die wall reaction, F_D, approximately proportional to the contact area between material and die wall but modified by other parameters such as particle size and work hardening of the material. By addition of lubricant they deduced that both interparticle friction and die wall friction affected die wall reaction. Leopold and Nelson[70] found an optimum level of internal lubricant to give enhanced compaction behaviour but for die wall lubricant no optimum level was reached. In either case very low levels of lubricant addition had a measurable effect on the pressure–density relationship. For low compaction pressures internal lubrication was the more effective whilst at high compaction pressures die wall lubrication gave improved compact densities. The greater the compact mass the more marked was the improvement in transmitted pressure as a result of lubricant addition.

Leopold and Nelson extended their work to examine the effect of lubricants on compact ejection from the die. In the absence of lubricant they found a linear relationship between ejection pressure, P_E, and compact mass, M, indicating that ejection pressure was directly proportional to the area of contact between material and die wall (*Figure 11.14*). Die wall lubrication was much more effective in reducing ejection pressure than internal lubricant, concentrations of stearic acid as low as 0.12 μg mm^{-2} at the die wall giving significant reductions.

Since the coefficient of sliding friction is less than the coefficient of static friction[71] one might expect a reduction in frictional losses and hence more efficient consolidation of material if shear were introduced at the die wall. Gregory and coworkers[72], studying the compaction of coal, found that if the compact, whilst under axial load, was subjected to shear by rotation of the die body, there was further densification without further load increase. This work was extended by Schwartz and coworkers[73] to examine the effects on die wall friction of direct lubrication by porous tooling, die rota-

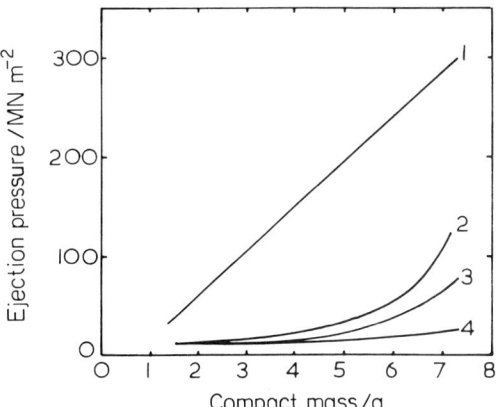

Figure 11.14. Variation of ejection pressure with compact mass and lubrication conditions (after Leopold and Nelson[70]): 1, No lubrication; 2, 0.25 per cent admix lubrication; 3, 0.5 per cent admix lubrication; 4, 2.0 per cent admix lubrication

tion during compaction and mechanical vibration of the die under load. Significant improvements in compact density relative to normal die compaction were reported.

Lewis and Train[74] evaluated powdered lubricants by forming compacts of them and examining their pressure–volume relationships. In common with other workers[6,73] they found that metallic stearates were superior to all lubricants examined including a wide range of waxes, talc, boric acid and graphite.

A complementary technique to the *R*-value for characterizing lubricant efficiency has been described by de Blaey and Polderman[75]. Information on compaction behaviour was obtained from the relationship between the axial applied force F_A and the punch displacement ΔH resulting from it. By plotting F_A against H for a complete compaction cycle a force–displacement figure or FD curve is obtained (*Figure 11.15*). The curve permits distinction between elastic and plastic work of compression. The total compressional work, W_t, is the sum of plastic work, W_p, and elastic work, W_e:

$$W_t = W_p + W_e$$

In *Figure 11.15(a)* the area XYO represents W_t, area XYZ represents W_p and area ZYO represents W_e. Area ZYO is the proportion of total work transferred back to the moving punch as pressure is released but it is time-dependent and difficult to determine exactly. If a second compaction cycle is carried out on the compact, however, an FD curve similar to that shown in *Figure 11.15(b)* is obtained. If relaxation has continued after the punch has lost contact with the compact in the first compaction the total work for the second compaction, area $X'Y'O$, will exceed the work of elastic recovery area ZYO in the first FD curve. Subtraction of area $X'Y'O$ from

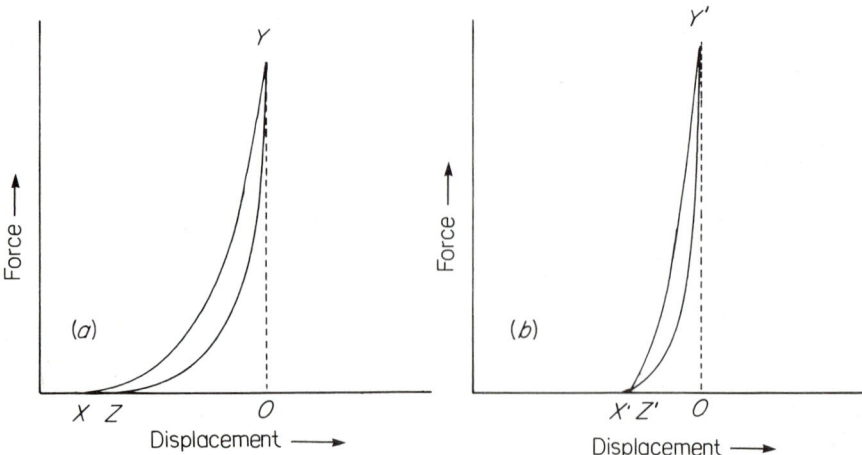

Figure 11.15.   Typical force–displacement curves for powder compaction for (a) *first and* (b) *second compression*

XYO gives the net work of compression for the material. In practice, the applied force, F_A, transmitted force, F_L, residual force on lower punch ΔF_L, and upper punch displacement relative to the lower punch, ΔH, are recorded during two compaction cycles. The net compressional work for both upper and lower punches is calculated and frictional losses during compaction found from the difference between them, i.e.

Frictional work = upper punch work − lower punch work

de Blaey and Polderman compared results from FD curve determination with *R*-value determinations and showed the FD curve technique to be more sensitive to changes in lubrication conditions.

In the compaction of ceramic powders the effect of lubricant addition on compact strength is negligible compared with the benefits of reduced die wall friction. In the case of pharmaceutical tabletting this is not so and the effect of lubricant content on compact strength has been extensively examined by Shotton and coworkers[76,77]. They found a measurable decrease in compact strength with lubricant additions of less than 0.5 per cent by weight. Increase in lubricant levels above this value had little additional effect.

11.5 Process variables

A typical ceramic fabrication route is shown in *Figure 11.16*.

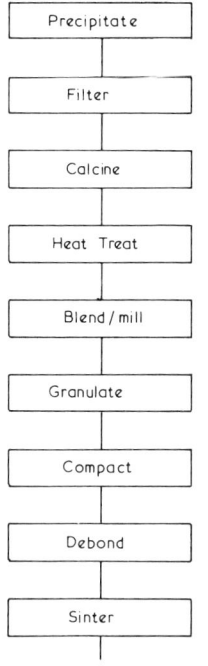

Precipitate

Filter

Calcine

Heat Treat

Blend / mill

Granulate

Compact

Debond

Sinter

etc.

Figure 11.16. Typical ceramic fabrication process

11.5.1   Precipitation

Precipitation conditions define the primary particle size of the powder, the crystallite size and the particle shape, all of which affect later stages in the fabrication process. Precipitation variables are:

1.   Solution concentrations.
2.   pH.
3.   Nature of anion.
4.   Method of mixing.
5.   Temperature.

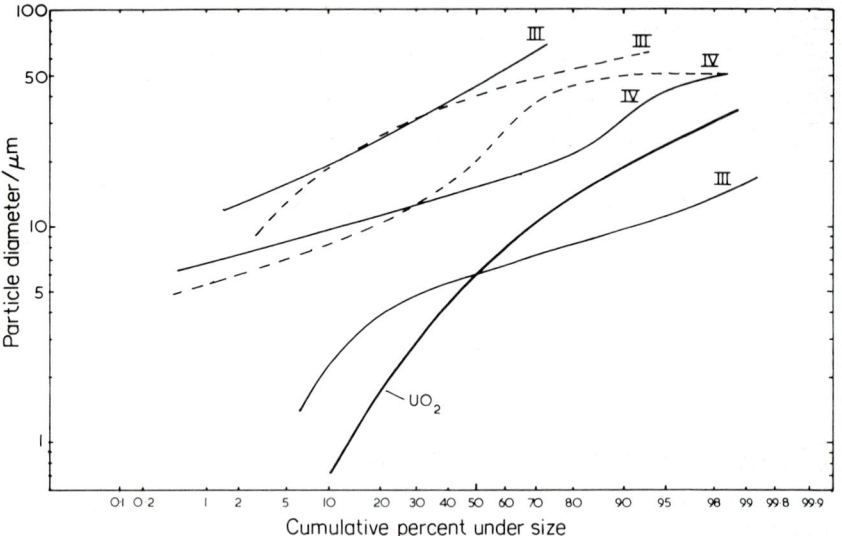

Figure 11.17.   Particle size distribution, as determined by a Coulter counter, for PuO$_2$ (valence states III *and* IV) *precipitated under various conditions as oxalate. The full curves are for reverse strike and the broken curves for direct strike precipitation*

Figure 11.17 illustrates the range of particle size distributions obtained for a plutonium oxalate precipitate under different conditions. Superimposed on the plots is a typical size distribution obtained for a uranium dioxide precipitate. The size of precipitate particle is usually chosen to permit ease of filtration and precipitation conditions are fixed accordingly. The effect of both valency state and method of mixing is shown in *Figure 11.17*. In direct strike precipitation, the acid is added to the bulk metal-salt solution and metal-salt solution to bulk acid in reverse strike precipitation. Particle and crystallite size additionally depend on solution temperature, aging and agitation after precipitation. The most common precipitate anions are hydroxide, peroxide, or oxalate while other organic acid radicals may be chosen for special circumstances. The criteria applied in choosing the anion is for ease of decomposition to form a pure oxide.

Figure 11.18.   Scanning electron micrograph of PuO_2 *prepared from an oxalate precursor*

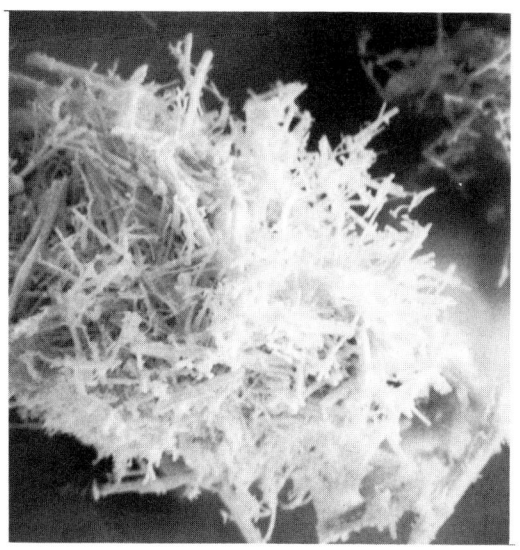

Figure 11.19.   Scanning electron micrograph of PuO_2 *prepared from an oxalate precursor in conditions different from those of* Figure 11.18

11.5.2   Calcination

Calcination involves decomposition of the precipitate under an oxidizing atmosphere to form the pure oxide. Air is the natural choice of atmosphere for most applications but pure oxygen, carbon dioxide and gas mixtures are also employed. During calcination the precipitate is heated to 500 °C or above and powder properties can be adjusted at this stage by varying the temperature and atmosphere. The basic particle shape is conferred during precipitation and the relict structure is retained through calcination. *Figure 11.18* shows a scanning electron micrograph typical of PuO_2 prepared from an oxalate precursor. The particles are crystalline and platelike. *Figure 11.19* shows a sample of PuO_2 from the same precipitation batch which experienced different conditions of solution strength, pH and aging treatment. The structures and hence the material properties are completely dissimilar. *Figure 11.20* shows PuO_2 prepared from a hydroxide precipitate, much less crystalline and of a smaller particle size. *Figure 11.21* shows a Linde 'A' alumina powder similarly prepared from a hydroxide precipitate.

During calcination the primary particles sinter together to form strong aggregates, resistant to breakdown during sieving operations or ultrasonic dispersion for particle size analysis, to an extent determined by the calcination temperature. *Figure 11.22 and 11.23* show the effect of increasing treatment temperature on a PuO_2 powder. Treatment temperature is chosen according to the specific surface area or tapped density desired in the product. *Figure 11.24* shows the relationship between calcination temperatures, specific surface area (SSA) and tapped density. If the SSA is too high the quantity of binder required is high, the green compact density consequently low and the subsequent shrinkage on sintering high and often non-uniform. Conversely, if the SSA is too low then the surface activity available for sintering is insufficient. Higher treatment temperatures also promote an increase in tapped density and aggregate strength so that higher compressive forces are required during die compaction to cause densification. An optimum SSA exists for each powder and in the case of nuclear ceramics this is typically in the range $3\text{---}7\ m^2\ g^{-1}$.

The process described assumes formation of a filter cake from the precipitation stage followed by calcination under a controlled atmosphere in a furnace. This technique offers the advantage that a wide variety of particle sizes and material properties can be obtained. Its application demands a subsequent milling stage to produce a homogeneous product and it is not suitable for high-volume production of sub-micrometre particles. However, the precipitate may be treated in other ways. Spray drying of the wet slurry under fluidized bed or batch reactor conditions is widely used for commercial production of ceramic powders, with or without an intermediate wet milling stage for adjustment of particle size. Thermal conversion of the precipitate to the oxide may be effected during spraying or as a subsequent stage. The granular product is then suitable for press feeding without further size adjustment. For specialized applications, controlled precipitation

Figure 11.20. Scanning electron micrograph of PuO_2 *prepared from a hydroxide precursor*

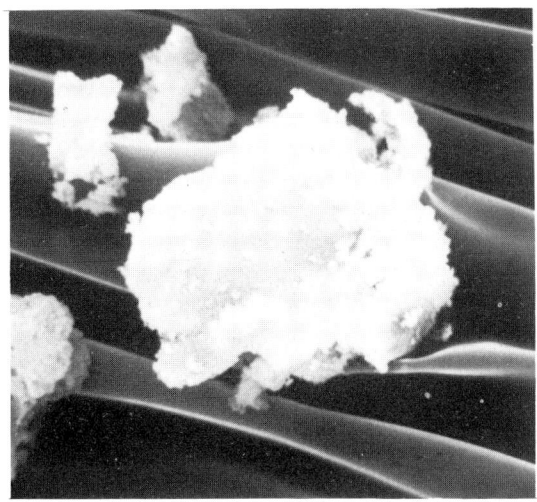

Figure 11.21. Scanning electron micrograph of Linde A *alumina prepared from a hydroxide precursor*

Figure 11.22.   Scanning electron micrograph of
PuO_2 *prepared from an oxalate precursor and cal-*
cined at 900°C

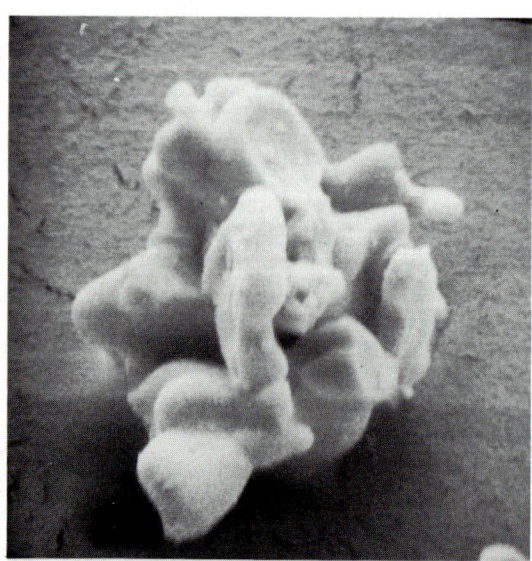

Figure 11.23.   Scanning electron micrograph of
PuO_2 *prepared from an oxalate precursor and cal-*
cined at 1050°C

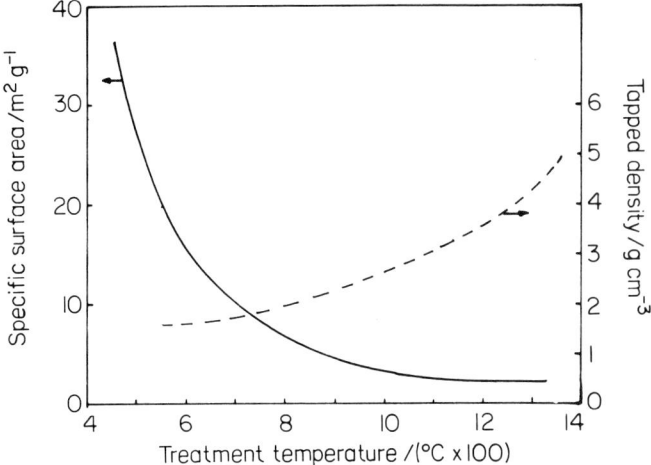

Figure 11.24. Effect of treatment temperature on surface area (full line) and tapped density (broken line) of PuO$_2$ *powders*

of sols to form gelled spheres of the required size in the range 5—100 μm has been reported[78]. The spheres are subsequently dried and calcined to form a powder of the required properties.

Alternatives to precipitation reactions for oxide formation are freeze drying, thermal decomposition of salts and oxidation of the metal. In freeze drying the salt solutions are sprayed into a freezing chamber and the resultant frozen droplets converted into the anhydrous salts by sublimation of the ice under controlled conditions of vacuum and temperature. The porous salt particles are then calcined to form the oxide. Thermal decomposition of the salts may also be applied directly to the liquid droplets or by pyrolysis of salts absorbed on a combustible medium such as cellulose or activated carbon. Control of powder properties is more difficult than with conventional techniques. PuO$_2$ powder produced by thermal denitration of the salt solution is shown in *Figure 11.25*. The particle structure is generally amorphous with non-uniform porosity.

Finally, for high purity oxides or for powders with ultrafine primary particle size, direct oxidation of the metal is employed, often in the vapour phase.

11.5.3 Blending and milling

In most cases ceramic powders are non-segregating and the two stages may be carried out simultaneously in a ball mill. Milling is necessary to break down coarse aggregates which may have high levels of included porosity. Typically a SSA increase of 10 to 20 per cent is observed when milling raw ceramic powders. For extremely hard ceramics such as nitrides it may be necessary to employ high-energy vibromilling in mills lined with tungsten carbide.

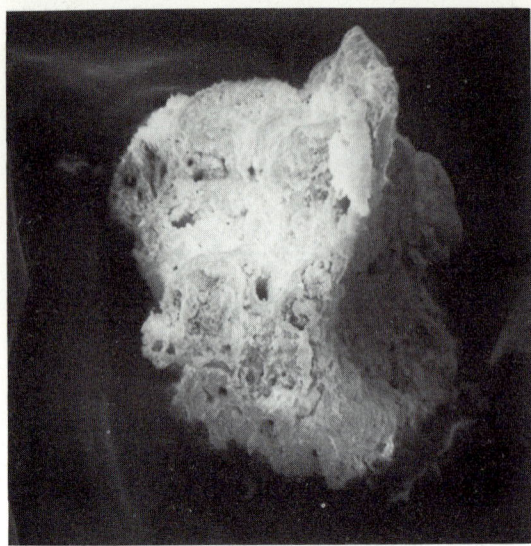

Figure 11.25.   Scanning electron micrograph of
PuO_2 *produced by thermal denitration*

11.5.4   Granulation

In order to provide a free flowing, dust-free feed to the compaction process to ensure uniform die filling and uniform compaction behaviour, the milled powder must be granulated. Two different approaches are possible. The usual choice is granulation by the addition of a binder when, as already indicated, the material compression characteristics may be altered to minimize pressing faults such as capping and lamination. In addition the binder acts as a lubricant both internally and at the die wall, improving pressure transmission and reducing die wall frictional effects. The level of binder addition is kept to a minimum since subsequent debonding leaves voidage which must be removed on sintering. The choice of binder is based on good wetting behaviour, compatibility with the powder, ease of removal and debonding behaviour with respect to undesirable breakdown products. Granules are typically formed by mixing with binder to a paste, drying to a crumbly 'dough' and passing through a screen granulator of appropriate mesh size.

If the powder already possesses good compression characteristics or if very high purity product is required, a binder need not be used. Granulation is effected by pre-pressing the raw powder into a convenient compact shape and crushing the compacts through screens to form granules of the required size. The technique is known as preslugging and is used extensively in ceramic fabrication. The preslugging pressure must be less than the proposed applied pressure for the final die compaction by about an order of magnitude. *Figure 11.26* shows the effect of preslugging alumina powder obtained by Bruch[50]. Significant improvement in density is demonstrated

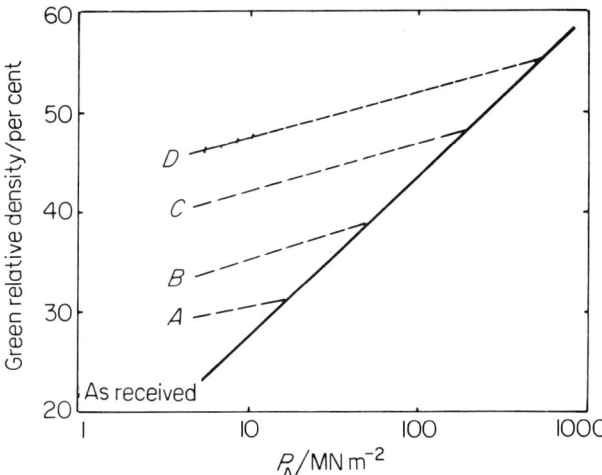

Figure 11.26. Increase in green density of alumina obtained by preslugging. The full line represents the alumina as received, and the broken lines refer to preslugging pressures A, 6.9; B, 27.6; C, 110.3; D, 441.3 MN m^{-2} (after Bruch[50])

by this technique. Further discussion of the handling and forming of powdered materials is presented by Niesz *et al*[79].

11.5.5 Compaction

Most studies of die compaction have been carried out under conditions of single-ended pressure application using a stationary die body and lower punch and a moving upper punch. Because of the pressure gradient inevitably generated in such a system it is rarely employed in practice. Usually the die body is fixed and both upper and lower punches move simultaneously equal and opposite distances into the die, compressing the powder column uniformly about its transverse midpoint. The density gradients are much less severe since the distance necessary for pressure transmission is effectively halved. The density distribution pattern is symmetrical about the transverse midpoint, the upper half being the mirror image of the lower half, and the lowest density is observed at the outer periphery of the compact at the midpoint where the powder has remained stationary relative to the die wall. Even with 'difficult' ceramic powders uniform compacts with $H/D = 2$ can be produced in this way. Some variants of the true double ended pressing technique are used in order to simplify press design. In the floating die press, the lower punch is fixed, the die body is spring loaded and is driven down on to the lower punch by die wall friction during the upper punch compression stroke. Correct spring adjustment permits a very close approximation to double ended pressing provided that the properties of the press feed remain constant.

11.5.6   Modified die compaction techniques

For compacts which are non-equiaxed, alternative die compression techniques have been devised. Isostatic compaction is a widely used technique for forming compacts with a large L/D ratio and will be discussed elsewhere in this publication. Koerner at Drexel University has reported improved compaction results using triaxial compression[80], a hybrid technique employing punch compression of powder in a die with a flexible, isostatically loaded wall.

So far only cold compaction techniques have been considered. The use of heated dies has been reported both for uniaxial compaction and isostatic compaction of ceramic powders. At elevated temperatures powder plasticity is greatly increased and pressure-enhanced sintering of the compact occurs during compaction. Using this technique components of extraordinary size may be formed. AB Atomenergie, Studsvik, recently reported[81] the use of a hot isostatic press, operating at 1350 °C under a pressure of 100 MPa, to fabricate alumina tubes closed at one end, 2.5 m long, of outside diameter 0.5 m, with a 10 cm thick wall. Each tube weighed 1600 kg and was designed to hold 150 nuclear fuel rods during their initial post-irradiation cooling.

References

1. BOUSSINESQ, J., *Mémoires couronnes et mémoires des savants étrangers*, Brussels (1876)
2. LONG, W. M., *Powder Metall.* **6**, 73 (1960)
3. JAMES, P. J., *Powder Metall. Int.* **4**, 2 (1972)
4. DOWEZ, P. and ZWELL, L., *J. Metals* **1**, 2, 137 (1949)
5. SHEINHARZ, I., McCULLOGH, H. M. and ZAMBROW, J. L., *J. Metals* **6**, 515 (1954)
6. NELSON, E., NAQVI, S. M., BUSSE, L. W. and HIGUCHI, T., *J. Am. pharm. Ass. Sci. Ed.* **43**, 596 (1954)
7. BAL'SHIN, M. YU., *Dokl. Akad. Nauk SSSR.* **67**, 831 (1949)
8. SEELIG, R. P. and WULFF, J., *Trans. Am. Inst. Min. metall. Engrs* **171**, 516 (1964)
9. COOPER, A. R. and GOODNOW, W. H., *Bull. Am. Ceram. Soc.* **41**, 11, 760 (1962)
10. TRAIN, D., *Trans. Instn chem. Engrs, Lond.* **35**, 258 (1957)
11. MacLEOD, H. M. and MARSHALL, K., *Powder Technol.* **16**, 1, 107 (1977)
12. van GROENOU, A. B. and KNAAPEN, A. C., personal communication
13. UNCKEL, H., *Arch. Eisenhütt. Wes.* **18**, 161 (1945)
14. SEELIG, R. P., *Metals Technol.* **14**, 5 (1947)
15. SHANK, M. E. and WULFF, J., *Trans. Am. Inst. Min. metall. Engrs.* **185**, 561 (1947)
16. GOETZEL, C. G., *'Treatise on Powder Metallurgy'*, vol. 1, Interscience, New York (1949)
17. HIRSCHORN, J. S., *'Introduction to Powder Metallurgy'*, American Powder Metallurgists Institution, New York (1969)
18. LEIGH, S., CARLESS, J. E. and BURT, B. W., *J. pharm. Sci.* **56**, 7, 888 (1967)
19. LEIGH, S., Ph.D. Thesis, University of London (1970)
20. MARSHALL, K., *'Proceedings of First International Conference on Compaction and Consolidation, Brighton'*, Powder Advisory Centre, London (1972)
21. SUH, N. P., *Int. J. Powder Met.* **5**, 1 (1969)
22. SCHWARTZ, E. G. and HOLLAND, A. R., *Int. J. Powder Met.* **5**, 1 (1969)
23. BOCKSTIEGEL, G. and HEWING, J., *Arch. Eisenhütt. Wes.* **36**, 751 (1965)
24. WILLIAMS, J. C. and BIRKS, A. H., *Powder Techol.* **1**, 199 (1967)
25. FUKUMORI, Y. and OKADA, J., *Chem. pharm. Bull.* **25**, 7, 1610 (1977)

26. van GROENOU, A. B., *Powder Metall. Int.* **10**, 4, 206 (1978)
27. STRIJBOS, S. and VERMEER, P. A., *'Proceedings of Crystals and Ceramics Conference, Rayleigh'* (1977)
28. HUFFINE, C. L., Ph.D. Thesis, Columbia University, S. Carolina (1953)
29. TRAIN, D., *J. Pharm. Pharmac.* **8**, 745 (1956)
30. SHOTTON, E. and GANDERTON, D., *J. Pharm. Pharmac.* **12**, 87T (1960)
31. MARSHALL, K., *J. Pharm. Pharmac.* **15**, 413 (1963)
32. GRATON, L. C. and FRASER, J. H., *J. Geol.* **43**, 785 (1935)
33. GRAY, W. A., *'The Packing of Solid Particles'*, Chapman and Hall, London (1968)
34. McGEARY, R. K., *J. Am. Ceram. Soc.* **44**, 513 (1961)
35. MacLEOD, H. M., *UKAEA Internal Report* (1977)
36. BERNAL, J. D., *Proc. R. Soc. A* **280**, 299 (1964)
37. BLUM, E. H. and WILHELM, R. H., *A.I.Ch.E. Symp. Ser.* No. 4, 21 (1965)
38. BERESFORD, R. H., *Nature* **224**, 550 (1969)
39. EASTWOOD, J., MATZEN, E. J. P., YOUNG, M. J. and EPSTEIN, N., *Br. chem. Engng.* **14**, 11, 1542 (1960)
40. HAUSNER, H. H., *Int. J. Powder Met.* **3**, 4 (1968)
41. GRAY, R. O. and BEDDOW, J. K., *Powder Techol.* **2**, 323 (1969)
42. MacLEOD, H. M., *UKAEA Internal Report* (1973)
43. KOSTELNIK, M. C., KLUDT, F. H. and BEDDOW, J. K., *Int. J. Powder Met.* **4**, 4, 19 (1968)
44. DUFFIELD, A. and GROOTENHUIS, P., *'Symposium on Powder Metallurgy'*, No. 58, Iron and Steel Institute, London (1954)
45. BOCKSTIEGEL, G., in *'Modern Developments in Powder Metallurgy'*, Ed. Hausner, H. H., vol. 1, p. 15, Plenum Press, New York (1966)
46. STUIJTS, A. L. and OUDEMANS, G. J., *Proc. Br. Ceram. Soc.* **3**, 81 (1965)
47. TURBA, E., *Proc. Br. Ceram. Soc.* **3**, 101 (1965)
48. HARDMAN, J. S. and LILLEY, B. A., *Proc. R. Soc. A* **333**, 183
49. LAWRENCE, P., *J. mater. Sci.* **5**, 663 (1970)
50. BRUCH, C. A., *Ceramic Age*, **83**, 10, 44 (1967)
51. COOPER, A. G. and EATON, L. E., *J. Am. Ceram. Soc.* **45**, 3, 97 (1962)
52. KAWAKITA, K. and LÜDDE, K. H., *Powder Technol.* **4**, 2, 61 (1971)
53. WILLIAMS, J. C., personal communication (1973)
54. SHEINBERG, H., *Powder Metall.* **12**, 269 (1969)
55. SHAPIRO, I., Ph.D. Thesis, University of Minnesota (1944)
56. SPENCER, E. S., GILMORE, G. D. and WILEY, R. M., *J. appl. Phys.* **21**, 527 (1965)
57. BERRY, T. F., ALLEN, W. C. and HASSETT, W. A., *Bull. Am. Ceram. Soc.* **38**, 8, 393 (1959)
58. DALLAVALLE, J. M., *'Micromeritics'*, 2nd edn., p. 161, Pitman, New York (1948)
59. TORRE, C., *Berg-U. hüttenm. Mh.* **93**, 62 (1948)
60. KONOPICKY, K., *Radex Rdsch.* 141 (1948)
61. BURR, M. F. and DONACHIE, M. J., *J. Metals*, **15**, 11, 849 (1963)
62. GREENSPAN, J., *'Metals for the Space Age'*, ed Benesovsky, Springer, Vienna (1964)
63. MORGAN, W. R. and SANDS, R. L., *Metals Mater.* **3**, 85 (1969)
64. BEDDOW, J. K., *Int. J. Powder Met.* **4**, 1, 27 (1968)
65. BURR, M. F. and DONACHIE, M. J., *Trans. Q. Am. Soc. Metals* **56**, 4, 863 (1963)
66. HAUSNER, H. H. and SHEINHARZ, I., *Proc. Met. Powder Ass.* **6** (1954)
67. LJUNBERG, I. and ARBSTEDT, P. G., *Proc. Met. Powder Ass.* **78** (1956)
68. YARNTON, D. and DAVIES, T. C., *Metallurgia*, 153 (1962)
69. TRAIN, D. and CARRINGTON, J. M., *J. Pharm. Pharmac.* **11**, 261T (1959)
70. LEOPOLD, P. M. and NELSON, R. C., *Int. J. Powder Met.* **1**, 4, 37 (1965)
71. BOWDEN, F. P. and TABOR, D. *'The Friction and Lubrication of Solids'*, Part I, Oxford University Press, London (1954)
72. GREGORY, H. R., JONES, D. C. and PHILLIPS, J. W., *Nature*, **184**, 120 (1959)
73. SCHWARTZ, E. G. *et al.*, *Quarterly Reports*, Contract SRO-475, College of Engineering, Columbia University, S. Carolina (1968)
74. LEWIS, C. J. and TRAIN, D., *J. Pharm. Pharmac.* **17**, 33 (1965)
75. de BLAEY, C. J. and POLDERMAN, J., *Pharm. Weekbl. Ned.* **105**, 241 (1970)
76. SHOTTON, E. and GANDERTON, D., *J. Pharm. Pharmac.* **13**, 144T (1961)
77. SHOTTON, E. and LEWIS, C., *J. Pharm. Pharmac.* **16**, 111T (1961)
78. FLETCHNER, J. M. and MASDY, C. J., *Chemy Ind.*, 48 (1968)

79. NEISZ, D. E., McCOY, L. G. and WILLS, R. R., *'Proceedings of Crystals and Ceramics Conference, Rayleigh'* (1977)
80. KOERNER, R. M., *Ceramic Bull.* **52**, 7, 566 (1973)
81. In *Process Engineering*, September 1978, p. 9

CHAPTER 12

Isostatic pressing and compacting techniques

D. E. Lloyd, I. K. Bloor and R. D. Brett
The British Ceramic Research Association, Queens Road, Penkhull, Stoke-on-Trent

The advantages of isostatic pressing relative to other shaping methods are discussed. Depending on size, geometry and production requirements of the articles the choice will be between a batch process using 'wet' bags or a (semi)automatic operation using 'dry' bags. The design and production of the required tool set has often to be carried out by the ceramic engineer, who will therefore require a good understanding of the compaction process and some familiarity with elastomeric materials. The former is illustrated on the compaction of a sphere and by the consequences of deviating from simplifying assumptions.

12.1 Introduction

Isostatic pressing is the process by which a powder contained in a flexible mould is compacted into a desired shape by a pressurized liquid or gas[1]. The present discussion will, however, be limited to a discussion on the process using liquids, e.g., water or an emulsion of oil in water at room temperature or temperatures very close to this. Most current isostatic pressing processes are carried out at room temperature but for certain specialist applications hot isostatic pressing uses gases as the pressure-transmitting medium and the tooling problems are quite different. Cold isostatic pressing is becoming a popular fabrication technique in ceramic technology since in many cases it provides a cheaper or better product. This is usually due to the following reasons:

1. A more uniform density distribution is achieved in the compacted article since there is less interparticle friction and no die wall friction. The greater uniformity of density achieved compared with other techniques reduces the likelihood of subsequent shape distortion during the firing cycle.
2. The cost of equipment for the compaction of the large articles is cheaper, i.e., techniques such as die-pressing, injection or transfer moulding.
3. Components with large length-to-diameter ratio can be formed because

the pressure can be applied in the desired direction, e.g., perpendicular to the wall of the tube. This is not possible by conventional die pressing.

4. The dimensional accuracy of a complex profile can be maintained, e.g., the bore of a spark plug.
5. In some cases the amount of final machining required can be reduced or avoided, e.g., producing spherical grinding media.
6. Compared with other shaping methods, e.g., slip-casting, extrusion, transfer and injection moulding, isostatic pressing requires comparatively small amounts of binder and no plasticizer. This means that intermediate stages in a fabrication process, such as drying-off water and debonding to remove organic binders etc. with attendant production losses due to rejects, are no longer required.

The development of cold isostatic pressing up to the present time has been along two distinct lines.

12.1.1   Batch process using 'wet-bag' tools

Large components (blocks, crucibles, high tension insulators, etc.) are formed by uniformly filling flexible tools or moulds with powder and isostatically compacting them by immersion in a fluid which is then subjected to pressure in the range of 1.38 MN m^{-2} to 105 MN m^{-2} (200 to 15 000 lbf in^{-2}). Sometimes pressures of up to 207 MN m^{-2} (30 000 lbf in^{-2}) are used, but unlike metals, ceramics usually reach their maximum 'as-pressed' density around 105 MN m^{-2} (15 000 lbf in^{-2}), and hence there is little to be gained by using very high pressures. With metal powders, however, it is beneficial to use pressures of up to 414 MN m^{-2} (60 000 lbf in^{-2}), since very high densities can be attained, e.g., up to 90 per cent of the theoretical value in some cases. This means that subsequent overall movement as the result of shrinkage during firing is reduced and distortion in the shape of the article is thus minimized.

12.1.2   Automatic process using 'dry-bag' tools

In this process several identical components, e.g., spark plug insulators, are produced per minute by automatically filling the tool with an accurately metered amount of powder, compressing it and ejecting the compact. This method differs from the above in that the flexible tool is an integral part of the pressure vessel and the operator never sees the wet side of the flexible tool.

In order to exploit to the full the versatility of the isostatic pressing technique the user must be able to design and manufacture his own tools. It is essential to do this since in many cases the optimum tool design will only be achieved by a reiterative approach, i.e., improving the tool by noting the defects, such as flaws, and dimensional inaccuracies of the article produced. This trial and error approach can in many cases be reduced and the procedure placed on a more fundamental basis by attempting to understand the processes occurring during a full compaction cycle.

12.2 Component shapes

Some shapes are more suitable than others for isostatic pressing, in particular those with a high aspect-ratio, e.g., tubes and crucibles. Other configurations which have been successfully formed by isostatic pressing include nozzle shapes, burner-heads for turbines, spark-plug insulators, and abrasive parts, e.g., cup wheels, and rod-shaped electrodes to name just a few. Articles whose shapes are particularly unsuitable for isostatic pressing include bricks and discs, since in both cases it is difficult to produce the flat surfaces and sharp corners required. These shapes have, however, been produced by using complicated tool designs, but ordinary steel die-pressing is generally a more suitable technique. Some shapes such as long to narrow walled tubes of high density and closed at one end can only be made economically by isostatic pressing.

12.3 Tooling

Tooling in isostatic pressing corresponds to designing the die set in a conventional press, and it determines to a great extent the accuracy of the as-pressed shapes. Depending on the shape of compact required the tool may be a simple toy balloon or a complex assembly of rigid and flexible parts. There is, however, a big difference between die-pressing and iso-pressing. Engineers are fully conversant with the metals available for the construction of dies, their properties and the machining methods. This is not the case with the elastomeric materials used in an isopressing tool. A very wide choice of elastomers is available commercially, and it is difficult for an engineer to decide on a suitable material. Work has been undertaken at the British Ceramic Research Association to select and assess suitable elastomers for use in isostatic pressing. The most important properties are hardness, tensile strength and resistance to compression set. The occurrence of compression set would be particularly unsuitable since the dimensions of the flexible tool would change with continued use. Adequate tensile strength is essential to withstand the stresses involved in use, whereas hardness determines the surface finish of a compacted powder. If, however, the hardness is too high, damage to the compact will occur during the decompression stage.

In order fully to appreciate the various problems that arise in tooling for isostatic pressing it will be best to consider first what happens when forming a sphere from a loose powder-fill and then consider what effects arise when deviating from this, which is theoretically the simplest case.

12.4   Isostatic pressing of a sphere

It is assumed that we have a spherical shell of an elastomeric material uniformly and loosely filled with a powder subjected to hydrostatic pressure. At low pressures the powder will not greatly resist compaction and the contraction of the sphere may also be influenced by the elastic behaviour of the spherical shell. The influence of the shell on the shape of the powder fill will increase with the thickness and hardness of the shell. The powder will at first compact very readily but will exert a steadily increasing resistance to compaction, i.e., the density–pressure curve will rise rapidly at low pressures but flatten out at high pressures. The densification of the powder will be the result of rearrangement, fragmentation and/or deformation of the particles, depending on their plasticity. There will also be some elastic deformation of the powder assembly, but this will be reversible, and when pressure is being released the compact will undergo the phenomenon of springback.

Assuming that the volume of the powder compact after having been subjected to the maximum pressure is half that of the original cavity of the shell, then during this procedure the density of the compact will have been doubled or the radius of the compact will be $R/(3\sqrt{2})$, where R is the radius of the cavity. In other words the volume compaction ratio is 2 and the linear compaction ratio $3\sqrt{2}$. Thus to a first approximation the problem of tool design resolves itself to the construction of a tool cavity equal to the desired compact dimensions multiplied by the linear compaction ratio. If we assume further that initially the pore volume amounted to two-thirds that of the cavity volume, i.e., bulk density was one-third of the true or theoretical density (for Al_2O_3 the theoretical density is $4.0 \times 10^3 \, kg \, m^{-3}$), then after compaction the pore volume will have been reduced to one sixth of the cavity volume, and the pore volume of the compact is reduced by a factor of four. Consequently during decompression the pressure in the pore system will be 4 atmospheres until the shell (tooling) releases itself from the compact. The exact stage when release of the tool will occur and the manner in which this will occur will depend on the elastic properties of the shell, the degree of compaction of the powder, the adhesion between tool and compact and on the rate of diffusion of the compressed gas through the compact.

The interaction of these factors and the strength of the compact produced determines whether compact damage occurs on release.

Release will generally take place at relatively low pressures and the following methods suggest themselves as procedures likely to obtain release without compact damage:

1.   A slow decompression rate within the critical range.
2.   Evaluation of the tool before compression.
3.   Addition of binder to strengthen compact.
4.   Provision of an intermediate layer which decreases the probability of tool-to-compact adhesion.

Several factors can exist which will lead to deviation from the above considerations, i.e.

1. The cavity has to be filled uniformly and completely. Once a uniform fill has been achieved care must be taken to prevent the fill becoming non-uniform owing to segregation, etc., i.e., the filled tool should not be left standing around but compacted immediately after the filling stage. Techniques such as vibrocompaction at the wrong amplitude and frequency can also lead to segregation.
2. The tool must be sealed effectively to prevent the ingress of the pressing fluid. The design of the closure is important from two aspects: apart from providing an effective seal it must not be sufficiently bulky or cumbersome so as to have adverse effects on the uniformity of tool contraction during pressing or affect the uniformity of pressure being transmitted to the powder. Non-uniformity of pressure transmission occurs when the hardness of the closure is significantly higher than that of the tool wall.
3. In many cases rubbers selected for use as tool materials do not behave elastically and may exhibit slow recovery or, worse, 'permanent-set' may occur which will lead to changes in cavity dimensions and the tool is thus not reusable.
4. The compact must be removed from the tool after isopressing; hence the opening must be of adequate size.

On many occasions therefore a tool may have to be made in several sections, and the assumption that the more complicated tooling will maintain its shape like the simple shell described above cannot be made. The magnitude of the deviations from the assumed geometry will depend to some considerable extent on the elastic properties of the tool material and its rigidity. (Rigidity is determined by the hardness of the tool material and the thickness of the tool wall.)

12.5   Tooling for rods and discs

12.5.1   The merits of using thick- and thin-walled tooling

In the fabrication of these shapes it has been shown that shape distortion, i.e., a change in the L/D ratio, occurs during compaction and that during decompression a situation may arise in which the tool transmits forces to the compact which are sufficient to cause a series of more or less regular cracks in the compact. Generally it was observed that the thicker the bag the greater is the likelihood of crack formation[2].

From the cracking point of view 'thin bag' tooling is preferred. However, only 'thick bags' are rigid enough to maintain the shape of the cavity when filled with powder. Thin bags have to be supported externally by a perforated can, the perforation allowing access of the pressurizing fluid. Another method of maintaining the shape of the cavity during loading is to apply a vacuum outside the bag and thus hold the thin bag by suction against a perforated or porous support[3].

12.6   Tooling for complex shapes

In isopressing as compared with die pressing accurate dimensions and good surface finish are generally to be sacrificed in favour or uniform density. (A dimensionally accurate pressed component of uneven density will become dimensionally inaccurate after sintering!) Frequently one resorts to a compromise by using flexible and rigid components in a tool assembly. With 'hollow-ware' (crucibles, tubes, etc.) one has to decide which surface requires the most accurate dimension and surface finish and this is then formed by pressing against a rigid part of the tool. Thus a crucible with a good internal surface will be produced by pressing against a steel mandrel forming the hollow part (*Figure 12.1*) using a 'contracting' flexible tool, whilst an electrical insulator with accurate external surface can be pressed by a 'dilating bag' against a steel outside shell[4]. Depending on dimensional require-

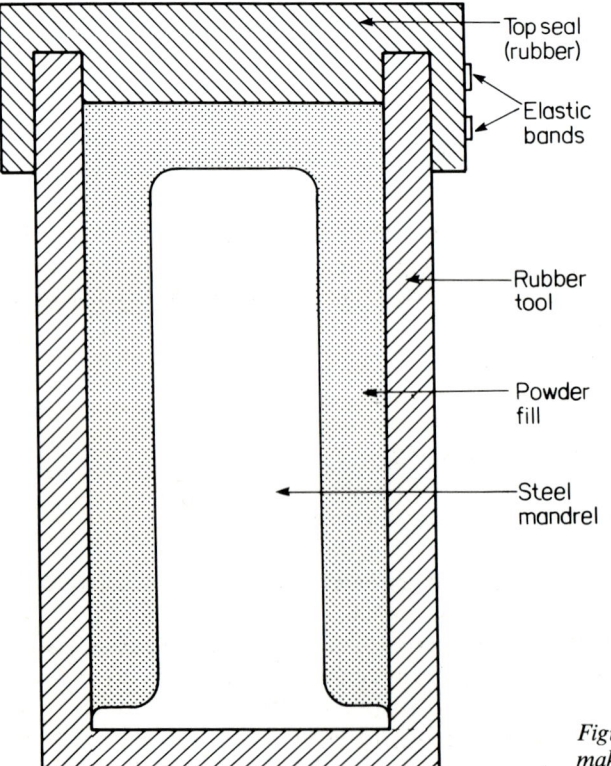

Figure 12.1   Tool set for making a crucible

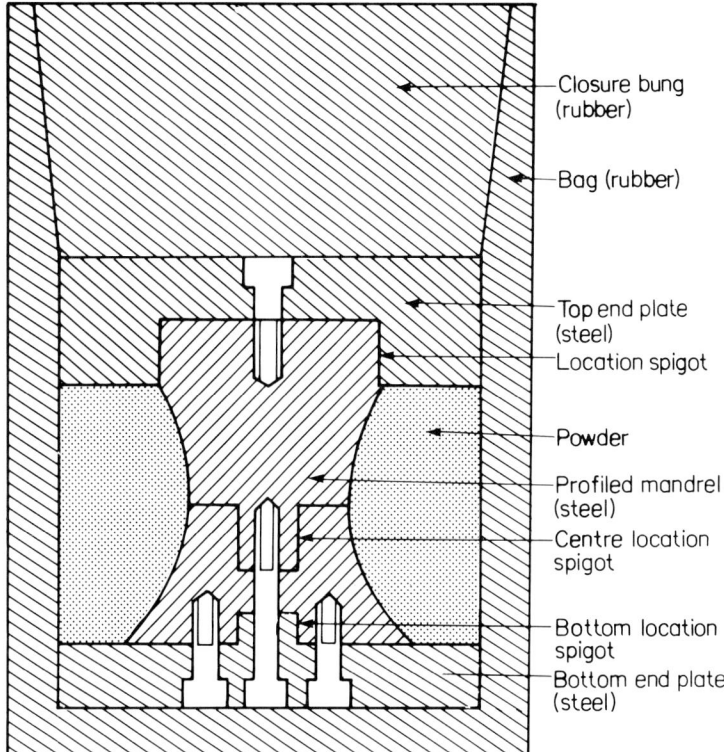

Closure bung
(rubber)

Bag (rubber)

Top end plate
(steel)

Location spigot

Powder

Profiled mandrel
(steel)

Centre location
spigot

Bottom location
spigot

Bottom end plate
(steel)

Figure 12.2   Tool set for making a rocket nozzle

ments the surface formed by the flexible part of the tool which exhibits random undulations may be left or improved by a machining finish.

A complex shape such as a rocket nozzle (*Figure 12.2*) requires an assembly of flexible and rigid parts. The latter must be disassembled before the pressed nozzle can be extracted.

12.7   General aspects of tool design

A satisfactory tool must fulfil the following requirements:

1. It must achieve the required dimensional tolerances and surface finish on the critical surfaces.
2. It must produce a component sufficiently oversize on those surfaces that are to be machined but must at the same time minimize the amount of machining.
3. It must produce a component free from flaws.

As in die-pressing, tool design may be simplified by appropriate article design, e.g., avoidance of sharp corners, etc. The dimensions of the tool cavity must allow for:

1. Shape distortion.
2. Compaction characteristics of powder.
3. Elastic expansion (springback) on decompression.
4. Shrinkage during subsequent firing.
5. Machining after pressing and/or subsequent sintering.

As a rough guide one may assume that:

1. For solid bodies compaction is uniform in all directions.
2. For hollow ware compaction of the walls is radial.

The tool wall should be made as thin as possible, bearing in mind that the tool will have to be be self-supporting or will be externally supported. The materials used for the rigid parts of the tool (mandrel) should have a surface which

1. Resists abrasion by the powder, and
2. Prevents adhesion.

It is not necessary that the materials used provide by themselves these properties; they might undergo a surface treatment to make them suitable.

Apart from the 'springback' already mentioned, which depends on the elastic properties of the compacted powder, the geometry of the rigid parts will be faithfully reproduced by the compact. Sharp corners should be avoided by supplying an adequate radius of curvature between adjoining faces. Recesses and reentrant cavities may be obtained by using disposable, e.g., soluble, decomposable or fusible parts, attached to the basic mandrel[5] (*Figure 12.3*). They may however represent a filling problem.

It has been suggested[6] that telescopic, spring-loaded mandrels may aid the accurate shaping of parts (*Figure 12.4*). This construction should lead to compaction which is more nearly isostatic than is the case when solid mandrels are used. A solid mandrel will limit axial compaction by providing a support for axial loads. Axial loads may also lead to bending of thin mandrels which can then not be extracted without destruction of the compact. Several eccentric axial holes may be provided in a compact by pegging mandrels against end plates while filling, but thereafter leaving the mandrel free to move laterally[7].

When pressing large compacts the weight of the rigid components of the tool may be considerable. Mandrels may be made hollow, allowing access of the pressing fluid. The rigid parts may to a first approximation be considered as incompressible.

Two factors must be considered in designing the closure of a tool:

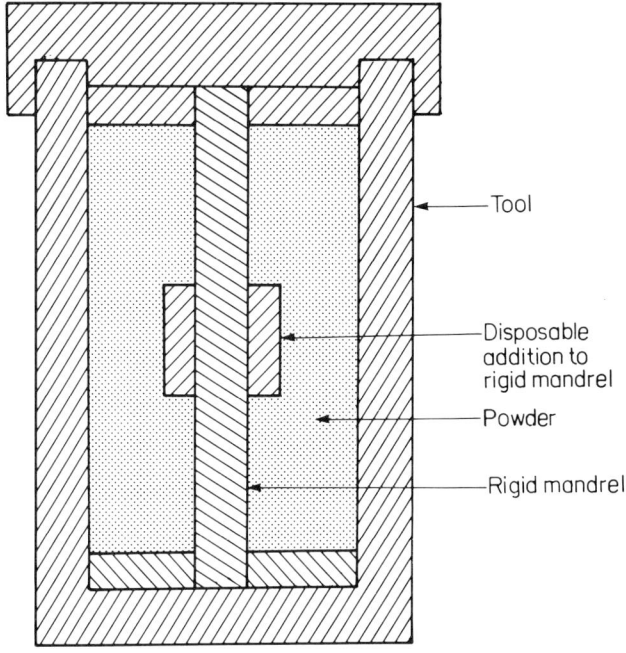

Figure 12.3. Tool set for production of a complex bore

Tool

Disposable addition to rigid mandrel

Powder

Rigid mandrel

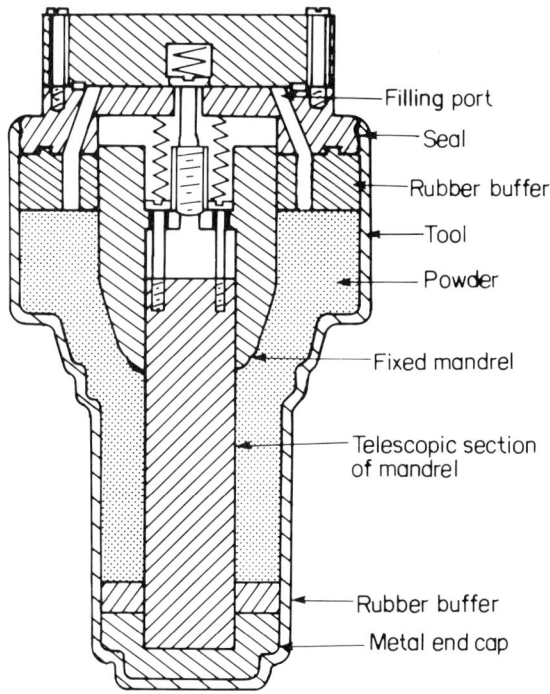

Figure 12.4. The use of a telescopic mandrel

Filling port

Seal

Rubber buffer

Tool

Powder

Fixed mandrel

Telescopic section of mandrel

Rubber buffer

Metal end cap

1.   The tool must not leak during compression or decompression.
2.   The closure should not introduce gross non-uniformity into an otherwise uniform tool since this may result in shape distortion.

A simple and most effective method of sealing a tool is to place it inside a heat-sealable polythene bag[8] or in a thin-walled latex tube which may be sealed by knotting. (This method is also most convenient when applying isostatic pressing to already shaped components for the purpose of equalizing and increasing the density.)

Other methods of sealing are:

1.   Sealing with adhesive tape.
2.   Plugs held in place by elastic tapes or metal clamps tightened up against a rigid tool component.
3.   Bag overlap held against a rigid part by elastic bands.
4.   C clamps.
5.   'O' ring seals in end plug.
6.   Cup-shaped seals supported by an internal metal ring[9] or a horizontal flange with vertical and inclined wall[10].

When it is desired to evacuate (de-air) a tool before pressing, a vacuum port and seal must be provided in the closure. The system may be simple or more elaborate (*Figure 12.5*).

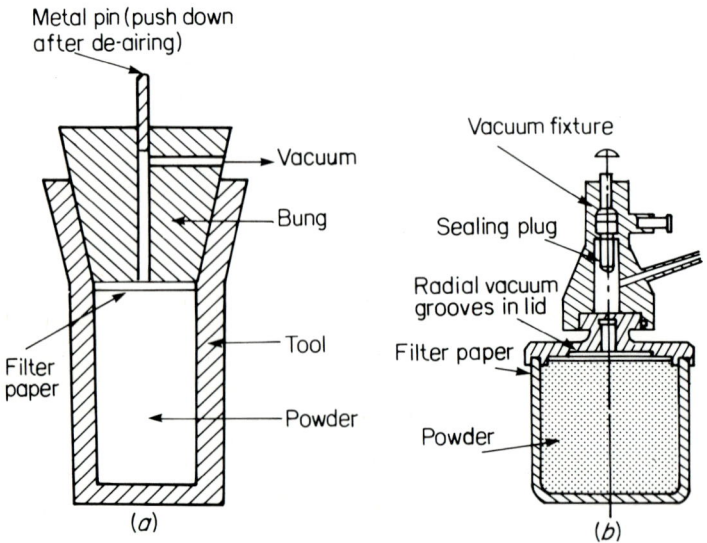

Figure 12.5.   A simple (a) and a more complicated (b) de-airing device

Acknowledgements

The author gratefully acknowledges the Director of Research, Dr D. W. F. Jones, for permission to present this Chapter.

References

1. POPPER, P., '*Isostatic Pressing*', Heyden, London (1976)
2. BLOOR, I. K., BRETT, R. D. and POPPER, P., '*Proceedings of First International Conference on Compaction and Consolidation of Particulate Matter*', p. 251, Powder Tech. Pub. Series, No. 4 (1972)
3. General Electric Company, *Br. Pat.* 828 066 (1960)
4. Ifoverken AB, *Br. Pat.* 1 227 572 (1971)
5. Coors Porcelain Co., *Br. Pat.* 1 093 205 (1067)
6. G.K.N. Group Services Ltd., *Br. Pat.* 1 260 583 (1972)
7. Coors Porcelain Co., *Br. Pat.* 1 200 522 (1970)
8. MINY, J.-C. and MARATHEU, A. *Industrie céram.* 417 (1969)
9. PAPEN, E. L. J., *Interceram* **4**, 296 (1967)
10. A.S.E.A., *Br. Pat.* 1 255 835 (1971)

Index

Acetaminophen, 237
Adsorption, 52, 67
  coefficient, 61
  equation, 55
  multilayer, 67, 81
  on non-porous surfaces, 69
  on porous surfaces, 81
Aerosil, 111
Agglomeration, 185
Alumina, 100
Aluminium, 68
  oxide, 68
  hydroxide, 68
Alyavdin equation, 15
Ammonia, 161
Anatase, 68
Angle of wall friction, 143
Angularity, 20
Appraisal of computer pore size models, 81
Argon, 63
Arithmetic mean, 11
Assessment of mixing, 126
Avicel, 114

Ball growth, 132
BDDT classification, 51, 52, 59
Bentonite, 104
Benzene, 63
BET,
  coefficient, 56
  equation, 56
Beta–autoradiography, 244
Bingham plastic, 163
Blending and milling, 271
Bonding, 150
  mechanisms, 129
Bonds,
  free liquid, 128
  solid bridges, 128, 130
Bridge formation, 124
British Standard 12, 5
British Standard 410, 5
British Standard 512, 18
British Standard 1902, 45
British Standard 2955, 2, 18

British Standard 3406, 2
British Standard 4359, 53
Brittle fracture, 152
Bulk,
  compression, 260
  density, 45
  modulus, 195
Bulkiness, 28

Caking, 181
Calcination, 268
Calcined alumina, 213
Calcite, 170
Calcium sulphate dihydrate, 54
Cam clay, 203
Capping, 151, 155, 158, 175, 252
Carbon black, 58, 60, 91
Catalysts, 161
Centrifuges, 8
Ceramics, 172, 241, 257
Chalk, 222
Charcoal, 61
Chrysotile, 81
Clays, 204
Co-ordination number, 100, 101
Coal, 106, 168
Cohan equation, 65, 73
Coherent packing, 43
Cohesive mixture, 121
Commercial instrumentation, 239
Compact defects, 174
Compaction, 150
  ceramics, 241
  cycles, 236
  effect of lubrication, 150
  effect of moisture, 149, 163
  effect of shape, 163
  effect of size, 149, 163
  equations, 259
  mechanisms, 161
  pharmaceutical, 158
  powders, 213
  process, 235
  profiles, 236
  uniaxial, 161

Compression,
  cycle, 151, 157
  ratio, 234
Compressor,
  multistation, 148
  radial, 148
  simulator, 155
  single punch), 148
Computer models, 75
Concrete, 151
Consolidation stress, 140
Contact,
  angle, 97
  area, 101, 106, 108, 150
Control,
  systems, 239
  of tablet machines, 239
Copper, 256
Coulomb equation, 199, 201
Critical state, 208
  line, 209
  void ratio, 209
Crushing,
  strength, 110
  test, 171
Cusum, 158

Decompression, 155
  effects, 151
Deformation,
  elastic, 215
  fracture, 169
  of particles, 258
  plastic, 169, 215
Degassing, 53
Density,
  apparent, 45
  bulk, 45
  distribution in compacts, 166, 245, 247
  particle, 44
  tap, 45
  variation with compaction, 107
Design of,
  fluidized bed granulators, 133
  mass flow hopper, 142
  mass flow plant, 145
  solid handling plant, 136
Deviatoric stress, 193
Diameter,
  drag, 2, 6, 7
  dynamic (centrifugal), 7
  dynamic (gravity), 6
  ESZ, 10
  free fall, 2
  image, 2
  mechanical, 3, 35
  microscope, 2, 35
  projected area, 2, 4, 35
  sedimentation, 6
  sieve, 2, 5
  Stokes, 2, 6, 7

Diameter (cont.)
  surface volume, 2, 9
Dicalcium phosphate, 109
Die,
  fouling, 176
  geometry, 164
  wall forces, 228
  wall friction, 243
Disintegrants, 154
Distribution,
  BJH, 75
  computer, 77
  cumulative, 15
  density, 166, 245, 247
  internal stress, 243
  log-normal, 13
  normal, 13
  pore size, 68, 73, 155
    mercury, 96, 97
    nitrogen, 75
DKR equation, 92
Dry bag, 278
Dry soils, 172
Dubinin–Radushkevich–Kaganer equation,
  92
Dynamic diameter,
  centrifugal, 7
  gravity, 6
  shape factors, 38

Ejection, 153
  force, 173
Elastic,
  deformation 215
  material, 163
  strain, 189
Elasticity, 195
Electrical resistance diameter, 10
  sensing zone (ESZ) diameter, 10
Elongation factor, 28
  ratio (n), 20
Energy,
  interfacial, 150
  surface, 150
Equation,
  adsorption, 55
  Alyavdin, 15
  BET, 56
  compaction, 259
  Cohan, 65, 73
  Coulomb, 201
  Dubinin–Radushkevich–Kaganer, 92
  Halsey, 69
  Huttig, 58
  Kamack, 8
  Kapteyn, 15
  Kelvin, 64
  Langmuir, 55
  Mohr–Coulomb, 201, 252
  Rosin–Rammler, 15

Equation (*cont.*)
Washburn, 81
Weibull, 15
Young–Laplace, 95
ESZ diameter, 10

Failure,
function, 140
locus, 140
properties of particulate solids, 138
Flashing, 176
Flatness ratio (*m*), 20
Flow factor, 143
Flow,
into dies, 149
of solids, 136
Flowers of sulphur, 214
Fluidized bed,
dryer, 180
granulation, 179
Fluorite, 170
Force,
balance, 243
measurement, 231, 241
systems, 231
Force–displacement curves, 235, 264
Formulation, 158
Fourier analysis, 29
Fractal, 29
Fracture, 152, 169
Free flowing mixture, 121
Friction, 261
angle of wall, 143
die wall, 243
interparticle, 243
Fumed silica, 149

Gasil, I and II, 54
Gelatin, 150
Geometric mean, 11
Glass,
capillaries, 58
spheres, 69
Glidants, 149, 154
Granta gravel, 203
Granulation, 129
dry, 128
extrusion, 135
fluidized bed, 128, 179
pan, 128, 133
pharmaceutical, 148
process, 131
wet, 128, 154
Graphite, 172
Graticules, 4
Gretag TPG 400, 239
Gypsum, 170

H/*D* ratio, 243
Halloysite, 100

Halsey equation, 69
Hardness, Vickers, 170
Harmonic mean, 12
Heat balance, 182
Height–diameter ratio, 243
Helium, 45
Histogram, 12
Huttig equation, 58
Hvorslev surface, 206
Hydrodynamic focusing, 11
Hydrostatic stress, 193
Hysteris, 88

Illite, 101
Image forming instrument, 3, 35
Inert nuclei, 187
Instrument for,
attenuation diameter, 2, 8, 38
dynamic diameter, 2, 6, 7
electrical diameter, 2
image forming, 2, 3, 4, 35
mechanical diameter, 2, 5
scattering, 2, 9
surface diameter, 2
Instrumentation,
machine, 230
punch, 230
rotary tablet, 230
single acting, 227
Interfacial energy, 150
Internal stress distribution, 243
Interparticle friction, 243
voidage, 104, 115
Intraparticle porosity, 104, 115
Iron,
oxide, 65, 166
oxide–chromium oxide, 99
powder, 172, 257
Isostatic pressing, 277
Isotherm,
Type I, 51, 88, 91
Type II, 51, 52, 59, 61
Type IV, 64
variation with compaction, 102

Jenike failure locus, 140
Jenike shear cell, 138

Kamack equation, 8
Kaolin, 100
Kapteyn equation, 15
Kelvin equation, 64, 85
Krypton, 63

Lactose, 157
Lady Windsor coal, 106
Lambert–Beer law, 8
Lamination, 151, 152, 174, 252
Langmuir equation, 55
Linde silica powder, 90
Lineal analysis, 113

Lippens–de Boer micropore method, 91
Liquid bonding,
  capillary, 129
  droplet, 129
  funicular, 129
  pendular, 129
Localized cracks, 175
Lubricants, 149
Lubrication, 261

Machine instrumentation, 231
Macropores, 53
Magnesium,
  carbonate, 166, 247, 249
  oxide, 102
  stearate, 157
  trisilicate, 54, 104, 114
Magnetite, 113
Mass balance, 182
Mass flow hopper, 137, 142
Material properties, 163
  elastic, 163
  hardness, 163
  plastic, 163
  toughness, 163
  viscous, 163
Mean,
  arithmetic, 11
  geometric, 11
  harmonic, 12, 36
Mechanical diameter, 5, 36
Median, 13
Mercury, 97
  intrusion, 95
  penetration, 95
Mesopores, 53
Mica, 99
Microcrystalline cellulose, 153
Micropore filling, 93
Micropores, 53, 91
Minimum fluidization velocity, 179
Mixers,
  ribbon, 122
  scooping blade, 122
  tumbler, 122
  V, 122
Mixing, 120, 124
  quality, 120, 126
Mode, 13
Modified die compaction, 274
Modulus,
  rigidity, 195
  shear, 195
  Young's, 188
Mohr,
  body, 237, 251
  circle, 219
  diagram, 201
Mohr–Coulomb,
  equation, 201, 252
  failure, 201

Moisture, 149
Molecular area,
  adsorbates, 63
  nitrogen, 63
Molybdenum disulphide, 215
Monolayer, 56
Multilayer adsorption, 81

Nickel, 166
Nitrogen, 63
  adsorption isotherm, 52, 99
  molecular area, 62
Non-coherent packing, 43
Non-free flowing mixture, 121
Normal distribution, 13
Nuclei,
  creation, 132
  growth, 132

Oedeometer, 205
Onion ring growth, 185
Optimization,
  computer, 158
  formulation, 158
  process, 158
Outgassing, 53

Packing, 47, 255
  coherent, 43
  cubic, 48
  effect of shape, 253
  effect of size, 257
  non-coherent, 43
  orthorhombic, 48
  rhombohedral, 48
  spherical particles, 49
  tetragonal, 48
Paracetamol, 157
Particle,
  density, 43
  growth, 183
  properties,
    porosity, 163, 185
    shape, 163
    size, 163
    size distribution, 163
  rearrangement, 167
  size, 2
Particle size distributions, 2–11
Pelleting, 161
Penicillin, 211
Pharmaceutical granulation, 148
Photoelasticity, 235
Photosedimentation, 9
Physical adsorption equation, 53
Piezoelectric force transducer, 232, 235
Pilot plant, 187
Plaster of Paris, 151
Plastic,
  deformation, 169, 215
  material, 163

Plastic (*cont.*)
  solid, 164
  strain, 190
Plutonium dioxide, 266
Plutonium oxalate, 266
Point B, 56
Poisson ratio, 108, 151, 188, 236, 242, 250
Polystyrene, 223
Pore shape, 88
Pore size,
  computer models, 81
  distribution, 68, 155
    mercury, 96, 97
    nitrogen, 75
Pores,
  macro, 53
  meso, 53
  micro, 53
Porosity, 43, 46, 113, 185
  lineal analysis, 113
Porous particles, 185
Portland cement, 214
Potassium chloride, 68, 157, 219
Powder,
  compaction,
    equations, 150, 174, 259
    high stress, 221
    low stress, 213
    medium stress, 217
  mixing, 120
  preconditioning, 153
  preparation, 53
Precipitation, 265
Pressure transmission, 241
Pressure–volume relationships, 150, 174, 253, 259
Prilling, 128, 135
Principle stress space, 195
Probability, 13, 15
Process variables, 265
Proportionality constants, 20
Pseudo plastic, 163
Punch instrumentation, 230
Punch position, 229

Quartz, 68
Quenching, 181

R-values, 264
Radial pressure effects, 250
Ratio, *H/D*, 243
Relaxation time, 157, 158, 165
Rendulic failure surface, 220
Resolution, 16
Rheological properties, 168
Ribbon mixer, 122
Rigidity modulus, 195
Rock salt, 172
Rosin–Rammler equation, 15
Roughness, 20, 26
Roundness, 20, 23

Sand, 39, 51, 68, 225
Saran charcoal, 90
Scale of scrutiny, 120, 126
Scooping blade mixer, 122
Sedimentation,
  centrifugal, 7
  cumulative, 7
  diameter,.6
  gravity, 6
  incremental, 7
Segregation, intensity of, 120
Shape, 18–34, 88
  factors, 18, 21, 38
    applications, 35
  pores, 88
Shear,
  cell, 138
  modulus, 195
  strain, 139
  stress, 139
Silica gels, 66
Silver, 68
Simulator, 155
Single acting tablet machines, 227
Size enlargement, 128
Sodium chloride, 60, 157, 168, 225, 237, 258
Soil plasticity, 210
Soils, 199
Solid bridges, 128
Spherical,
  alumina, 80
  silica, 51, 94, 99
Sphericity, 23, 25
Spray drying, 128, 135
Statistical thickness, 69
Steel balls, 221
Stick–slip motion, 121
Storage hopper, 136
Strain, 163, 193
  elastic, 189
  gauges, 231
  plastic, 190
  rate, 163
  shear, 139
Strength producing mechanisms, 171
Stress, 163, 190
  consolidation, 140
  deviatoric, 193
  distribution, 166
  hydrostatic, 193
  relaxation, 172, 173
  shear, 139
Stress–strain curve, 139, 163
Sucrose, 168, 237, 258
Supercentrifuges, 8
Surface area,
  from image diameter, 35
  from shape factors, 35
  from sieve diameter, 35
  from Stokes diameter, 37, 38
  variation with compaction, 102

Surface energy, 150
Surface tension, 97

Tablet defects, 174
   capping, 152, 155, 175
   cracks, 175
   lamination, 175
Tablet machine, 149
   control, 238
Talc, 149
Tap density, 45
Tensile strength, 151
   agglomerates, 124
   variation with compaction, 107
Tesa yield criterion, 196, 253
Thermal efficiency, 183
Thickness of adsorbed layers, 67
Thomas tablet control, 239
Titanium dioxide, 53, 211, 212, 213
Tooling, 282
Topography, 52
Transmission electron microscope, 4
Triaxial cell, 205
Tumbler mixer, 122
Tungsten, 68
Type I isotherm, 51, 88, 91
Type II isotherm, 51, 52, 59, 61
Type IV isotherm, 64

Unidimensional consolidation, 205
Unmixing, 121
Uranium dioxide, 201, 217, 245, 249

V-mixer, 122

$V_a - t$ method, 91
Van der Waals,
   bonding, 171
   forces, 125, 126
   mixing, 125, 126
Vickers hardness, 170
Viscoelastic, 168
Viscoplastic, 164
Viscous flow, 168
Voidage, 43, 46, 104, 115
Volume filling of micropores, 93
Volume reduction, 205
Von Mises, 196

Wall friction, 143, 218
Wall slope, 143
Washburn equation, 81
Watchdog tablet press, 239
Water, 63
Waveforms, 29
Wax powder, 172
Weibull equation, 15
Wet bag, 278
Wetting agent, 154

X-radiographs, 245

Yield point, 163
Yield stress, 109
Young–Laplace equation, 77, 95
Young's modulus, 151, 188, 195

Zirconia, 51